THE
HORSE-BREEDING
FARM

THE

South Brunswick and New York: A. S. Barnes and Company
London: Thomas Yoseloff Ltd

Larryann C. Willis

HORSE-BREEDING FARM

© 1973 by A. S. Barnes and Co., Inc.
New material © 1976 by A. S. Barnes and Co., Inc.

A. S. Barnes and Co., Inc.
Cranbury, New Jersey 08512

Thomas Yoseloff Ltd
Magdalen House
136-148 Tooley Street
London SE1 2TT, England

Library of Congress Cataloging in Publication Data

Willis, Larryann C
    The horse-breeding farm.

    1. Horse breeding.   I. Title.
SF291.W55   636.1'08'2   72-5179
ISBN 0-498-01164-X
ISBN 0-498-01977-2 (paper)

reprinted 1979

Cover photograph of stallion:

**Moolah Bardell #488063
Long's Training Stables
Martinez, California**
    Owner: Charles Schreiner

Printed in the United States of America

To my parents,
whose help and encouragement
made this book possible

CONTENTS

For quick reference, the text of each chapter is preceded by a very general synopsis of the material contained in each principal section of that chapter.

ACKNOWLEDGMENTS

I would like to express my appreciation to all who directly or indirectly contributed to this book. Special acknowledgment goes to Jess Long, who provided legal information, researched and edited a large portion of the work; and to Marion Long for her diligent research and clerical assistance. I wish also to express my gratitude to William Nissen, DVM, Paul Van Eeckout, DVM, William Gibford, head of the Horse Department at California Polytechnic University, Mary Vero, Kenneth and Ada Brown, Kent Long, and my husband, Porter.

To others, whose names are not stated, thank you for your helpful comments, suggestions and assistance.

L. C. W.

INTRODUCTION

For centuries man has been closely associated with the horse. At first the animal was a source of food and later a beast of burden. The equine became a living vehicle of rapid travel for long distances and could draw heavy cargoes on sleds or wheels. Soon the horse was indispensable to man's endeavors, ranging from the planting of crops to the waging of wars. Less than a century ago men throughout the world were helplessly dependent upon the horse for agriculture, commerce, trade and travel.

With the advent of steam and gasoline engines man turned away from his old friend to the newer, swifter and more powerful machines. Man no longer seemed to need the horse. The equine population began to decline rapidly. Many soothsayers predicted the extinction of the horse within a hundred years. Indeed, had the number of horses continued to fall as it did about 60 years ago, extinction might well have been the result. However, the machine could not entirely replace the horse. We have not yet seen the engine that can do the work of the cow pony, jump fences and follow the fox, maneuver on the polo field or traverse rough mountain trails without disturbing the balance of nature. No piece of steel can fill the place in people's hearts reserved for their friend the horse.

As people's lives become easier and they have more leisure time, many turn their attention back to the horse. The animal is no longer strictly a beast of burden; it is a companion and a source of great enjoyment to an ever-increasing number of people.

In the last several years interest in horses has grown dramatically. The heaviest human density centers also have the most horses. Los Angeles county in California, which has the largest number of registered automobiles, also has the greatest number of horses per square mile. The horse population of the United States is now estimated at approximately 7,000,000, which is projected to double in the next ten years.

With the horse population increasing at such a fast pace, it is obvious that many people are raising horses. Horse breeders in the United States fall into one of three main categories:

A. Professional breeders who are devoted solely to raising high-quality horses for showing, racing and selling at a worthwhile profit. These breeders usually operate on a large scale and anticipate a long-term operation.

B. Breeders who raise horses to meet their own business needs. The most notable in this group are the larger cattle companies, which

11

produce the working horses for use by the ranches. Another example is Anheuser-Busch, Inc., which raises Clydesdales for their famous beer wagon.

C. Casual breeders who raise one or two foals a year purely for personal enjoyment. The ever-increasing number of these casual breeders has contributed greatly to the growth of the horse industry.

This book, although primarily designed as a guide for the establishment and efficient operation of the large, quality breeding farm, furnishes the essential information required by the breeder in every category.

THE
HORSE-BREEDING
FARM

1
ESTABLISHMENT

(SYNOPSIS)

This chapter considers the advisability of raising registered horses, farm location, qualifications of manager, other personnel and specialists; type of breeding stock, genetic inheritance; the selection of bloodlines, mares and stallions and promotion of the ranch reputation. The following synopsis by section number is provided as an aid for quick reference to the specific subjects.

1.1 General Statement. All business operations are the result of ideas
conceived, developed, refined and carefully structured through advance
planning. Horse breeding for profit is not an exception. The basic commercial elements—investment level, supply, demand, probable cost, tax
consequences and anticipated profit—are as fundamental to the business
of horse breeding as to any other business for profit. Here attention is
directed mainly toward the breeding and raising of fine horses.

1.2 Registered vs. Unregistered Horses. The venture of breeding horses
that will not qualify for registration under the rules of the breed associations is almost doomed to financial failure at the start. The purposes
of breed associations are: the improvement of the breeds, the protection
of the pedigrees of worthy horses and the maintenance of authentic
records for the certified proof of identity, age, ancestral line and registered ownership. The great divergence in price between a registered
and an unregistered animal of the same quality is not the result of
chance or accident. The price difference began because of the premium
people were willing to pay for documentary assurance that an outstanding animal was not a fluke and that it was likely to pass its traits on to
its offspring. Recorded registration as known today has replaced the
old, unreliable pedigree-by-memory passed on orally and often falsified
by unscrupulous "horse traders." Long ago farsighted horsemen acted
to establish registration systems designed to protect the purchasing
public as well as the ethical breeders and dealers. This was purely a
nongovernmental function, but most if not all states now recognize the
need for authentic, dependable registration and have enacted statutes in
support of the system. As an example, the California Agriculture Code
Section 16501 is quoted:

> It is unlawful for any person by any false or fraudulent pretense to do any
> of the following:
>   a) Obtain from any association organized for the purpose of improving the
>   breed of domestic animals a certificate of registration of any animal in
>   the herd register, or other register of such association or a transfer of
>   any such registration.
>   b) Give, for a valuable consideration, a false pedigree of any animal, with
>   intent to mislead.

The general acceptance of horse registration is borne out in numbers. As of January 1972 a total of 772,641 horses were registered by
the American Quarter Horse Association (A.Q.H.A.) alone (a gain of
approximately 65,000 over the previous year). Registration logically
stimulates pride of ownership and influences the purchasing public to

such an extent that the price of unregistered horses can often be less than the cost of production.

The prospective horse breeder has the choice of many breeds, but whatever type he chooses, he must be prepared to support the quality of the horses he raises with registered documentary proof of pedigree or suffer the economic consequences.

1.3 Feasibility Survey. There is no point in investing a great deal of time and money in a product that will not sell at a profit. The decision to establish a horse-breeding farm requires a careful survey of all aspects of the industry coupled with a detailed analysis of investment, supply and demand. It is not the purpose of this book to advise on detailed matters of investment, location, supply and demand, but to point out a few of the things that must always be considered when establishing a new horsebreeding farm.

1.4 Farm Location. The location of the farm often has a direct bearing upon the breed to be raised. A cattle-ranching community probably demands working horses such as the Quarter Horse. Metropolitan areas are more inclined toward pleasure horses and may prefer Arabians or Morgans. Location is less important for Thoroughbred and Standardbred breeders, provided that the ranch is reasonably accessible. A farm with a reputation for producing fine animals will probably not lack for customers no matter where it is located. However, new establishments will find it much easier to sell the type of animals preferred by local inhabitants.

1.5 Availability of Breeding Stock. Speed and ease of transportation today makes it possible to obtain breeding stock from almost any part of the world. It is true that cost is increased by seeking stock at distant points; however, the opportunity to obtain superior breeding animals should not be lost because of transportation costs, for high-quality animals are not easy to locate and are even more difficult to purchase. Occasionally a very good farm will disperse, making several outstanding animals available. The breeder should study the feasibility of purchasing some of these animals even if they are of a different breed from what is being raised. For example, the Quarter Horse breeder who has an opportunity to purchase an outstanding Thoroughbred, Appaloosa or Pinto stallion should look into the possibility of establishing an additional branch for his operation.

1.6 Capital for Livestock. The most expensive breed is the Thoroughbred, closely followed by the Standardbred. Both Thoroughbred and Standardbred racing are supported by tremendous amounts of money. Purses are high and winning animals are extremely valuable.

Quarter Horse: Doc Bar. Sire of ten AQHA Champions and numerous outstanding cutting, trail, reining, hackamore, working cow horse, western pleasure and halter horses. Owned by Double J Ranch, Pacinies, California.

Thoroughbred: *Siempre. A Classic Stakes Winner of 12 races; eight consecutive wins; 3/4 in 1:09 flat; a mile in 1:34 3/5 (under top weight). Stallion. Rancho Paraiso, Walnut Creek, California.

Arabian: Serafix (shown at age 22 years). Champion, sire of Champions. Leased by Lacey's Training Center, Alamo, California (formerly Rogers Arabian Ranch).

Thoroughbred or Standardbred breeding can produce exceptional profit but the cost of the breeding stock is correspondingly higher. Some Thoroughbred stallions may bring prices in excess of a million dollars and mares often sell for upward of a hundred thousand. This business requires a very strong financial position. If the venture is not capitalized with strength needed for the Thoroughbred or Standardbred breeds, it should deal in a less expensive breed rather than sacrifice quality.

1.7 American Quarter Horse. The largest and fastest-growing breed in the United States is the American Quarter Horse. Because of its large number, people are more familiar with Quarter Horse production and breed activities than with those of any other breed. For this reason the Quarter Horse is frequently used in this book to illustrate particular points involving bloodlines and registration.

A Quarter Horse breeding farm is less difficult to establish than most other farms for the following reasons: 1) excellent breeding stock is quite readily available at comparatively reasonable prices, 2) the

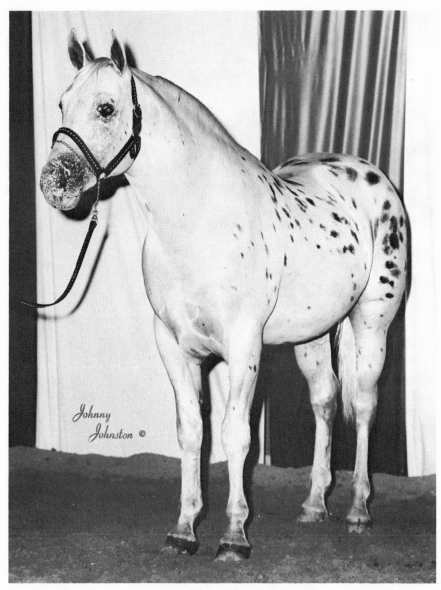

Appaloosa: Real Kuhl Spots. Champion in Halter, English, Western, Competitive Trail Riding. Outstanding sire owned by Rolling Oaks Farms, Concord, California.

Quarter Horse is exceptionally popular, thus creating a large market, 3) there are many races, sales and shows held especially for Quarter Horses, providing unlimited opportunities to advertise the product. Along with the benefits there are two main drawbacks to raising the breed: 1) there are many farms producing outstanding animals; competition is keen; acquiring a favorable reputation is correspondingly more difficult for the new establishment; 2) because of the large numbers of high-quality animals being marketed, top Quarter Horse prices are somewhat lower on today's market than prices of animals of the same quality of several other breeds. These major drawbacks are the reasons why many new breeders are raising Appaloosas, Pintos, Peruvian Pasos or other light horse breeds that are gaining in popularity.

Whatever the breed, it is fair to say that truly outstanding horses are the exceptions and will always be in demand at top prices, even during times of slack economy.

1.8 Ranch Manager Qualifications. The capability and skill of the ranch manager are perhaps the determining factor in the success of the breeding farm. Although preferable, he need not have had his past experience with the same breed or type of horse that will be raised. However, he must have had many years of experience in all phases of the industry before being entrusted with the total operating responsibility of breeding, maintaining and marketing valuable animals. While a good educa-

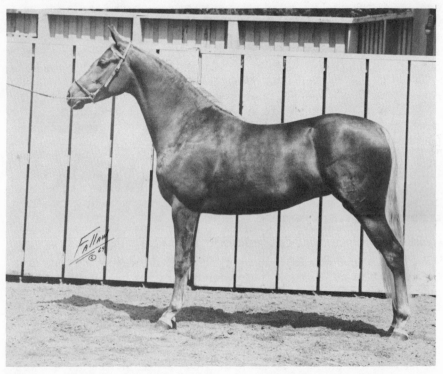

Morgan, "Truly a Versatile Breed": Oakhills Missy (shown as a two-year-old). Driving, English Pleasure, Trail and Combination Classes winner. Owned by Summerhill Farm, Diablo, California.

tion can help produce a good manager, a college diploma does not necessarily qualify a person for this important job. There is no substitute for firsthand experience. No one can be an effective ranch manager unless he is well versed in every aspect of horse production, including genetics, conformation, pedigrees, health, feeding, training, showing, racing, marketing and veterinarian procedures.

The manager will supervise all ranch personnel. He will be required to deal and contract with many suppliers and he must become well known and respected in those circles having equine interests. He must be a coordinator and a salesman. His honesty, integrity and trustworthiness must be apparent to the ranch employees as well as to those he will deal with during business transactions. He must be able to maintain his position of authority without losing the respect of his employees. The ranch manager must be available night and day and be willing to work long hours. He must be energetic, enthusiastic and knowledgeable, as well as receptive to new and progressive ideas and methods.

Where is such a qualified man to be found? Advertising in the news media usually proves to be fruitless. Such a person if not committed will make his freedom to deal known in equine circles. He may advertise in the breed journals or other livestock publications. Veterinarians, trainers, farriers, feed suppliers, show judges, auctioneers, farm managers and others who circulate to many ranches or who are in frequent contact with the business side of the horse industry may know of an available manager. The announcement of a complete dispersal sale or an estate sale may lead to a manager who is just winding up his former obligation. The investor should not run the risk of launching the business without a well-qualified manager. When he is found, he will not come cheap. These men are in great demand. If there is no doubt of the prospect's managerial ability or standard of social conduct, it may be worthwhile to consider granting him an incentive interest such as a small percent of net profits.

Even before there is a ranch to operate, the manager should be hired to assist in establishing the ranch location and facilities and purchasing the breeding stock. The investor is paying for this expert advice and cannot afford to make these important decisions without considering the manager's suggestions.

1.9 Ranch Personnel. At the commencement of ranch operations or if the venture is quite small, the owner and manager will probably be able to cope with many of the daily chores without much additional help; but as the number of ranch animals increases, barn help is necessary to free the manager for more important tasks.

Barn help is usually not a great problem, but must be reliable, willing and able to follow instructions and above all else have genuine concern for the animals. If the help is concerned for the horses' well-being, there is much less likelihood of negligence or disregard of instructions.

Even with a competent manager all farms will need specialized advice and assistance from time to time. Larger farms may need to employ a full-time studman and hayman. No farm can do without a competent farrier or a veterinarian who specializes in horses.

a) *Studman.* A qualified studman is more than the term implies. He is a specialist in breeding procedures, genetic prepotency, estrus cycles and stallion management. Of course he cares for the stallion, but his knowledge includes the mare's readiness and physical capacity to breed as well as fertility problems and difficulties that can arise.

Most ranches are totally dependent upon foal production and stud fees for income. During breeding season in particular the studman is as responsible for the success of the ranch as is the manager.

b) *Hayman.* Quality horses cannot be produced on low-grade feed. A hayman who is especially knowledgeable about all feeds, including hay, grain and pasture, is a great asset. He will be responsible for the quality of feeds purchased and grown by the ranch as well as proper and safe storage. The planting and maintenance of good pasture and control of weeds and poisonous plants will be his responsibility. The cracking, crimping and rolling of whole grains purchased in bulk or raised, as well as their protection from rodents and the elements, are part of the hayman's assignment.

c) *Veterinarian.* Few horse farms are large enough to warrant the expense of a full-time veterinarian, but no horse farm, in fact no horse, can do without the needed services and specialized knowledge of the professional equine practitioner. Unless the veterinarian is a reputable equine specialist, he is probably not equipped with the instruments and knowledge vital to the horse-breeding farm. The farm must arrange for a qualified equine veterinarian to make routine periodic visits to the ranch for the purpose of treating minor injuries and ailments. The veterinarian is usually bulging with valuable information of particular interest to the breeder, is well advised on disease trends in the district, can constructively criticize the condition of the animals and advise on the value of feeds as well as economy in their purchase.

Of course, the farm cannot commit its valuable herd to the care of an inexperienced or incompetent practitioner. A careful survey of other horse farms, trainers, horseshoers and feed supply houses will usually point out those in the area most qualified.

d) *Farrier.* The ranch will need a farrier, horseshoer, who is adept in corrective trimming and shoeing. Good farriers are sometimes very difficult to locate; nevertheless, the ranch should settle for nothing but the best, for a poor shoer can permanently damage the feet and legs of young horses. A farrier will travel a considerable distance if there is enough work to make his trip worthwhile. An experienced farrier can shoe from 15 to 18 horses a day if no corrective work is involved, but the exacting technique of corrective shoeing or trimming is time-consuming and often requires frequent return visits. The large farm may employ a full-time farrier or contract for regular visits. The local veterinarian, other farms in the area or trainers may be of assistance in locating a competent farrier.

e) *Trainer.* An accomplished and reputable trainer will be essential for the handling, showing and marketing of ranch horses. The trainer may be a permanent employee of the ranch with training facilities on the property, or the horses may be delivered to an outside trainer's stables. The training of the ranch horses is a most important phase of the operation and should not be delegated to anyone not thoroughly qualified and recognized for the results of his work. The cost of training on a monthly basis ranges from $150 for colts to $300 or more for racetrack training. Cattle work and other specialized training averages approximately $200, adjusting, of course, to the economy. This may be a considerable expense, but an expert trainer is indispensable if top-performance horses are to be produced. The rancher will be the loser if he tries to lower costs by hiring a second-rate trainer. Poor training will reduce sales and income and will degrade the ranch horses in the eyes of the viewing public. An otherwise fine animal will be downgraded if poorly trained. Trainers can be measured in terms of winning results; those whose entries consistently place high in shows or races are the desirable ones. A good trainer should not be interfered with; he spends his life schooling horses and is in the best position to determine what should be incorporated into the training program. Most trainers will not accept horses unless they are to have complete control of the schooling.

1.10 Breeding Stock. Qualified, dependable and experienced personnel will make the ranch an efficiently run operation, but the horses produced will determine the ranch reputation. Skillful handling can bring out the full potential of each horse, but it cannot put something into an animal that was not there to begin with. The demand for horses goes up and down with the economy cycle. In times of plenty any "crow bait" will sell, but only the best horses will find a market during times of recession.

Although the initial investment is many times more for outstanding stock, the return is far greater. The cost of raising a horse is essentially the same whether it is worth a few hundred dollars or many thousands. Ten inexpensive animals cost about ten times as much to raise as one valuable horse. The valuable horse will always sell at a handsome profit, for quality is scarce and people are willing to pay for it. Mediocre horses are ordinarily nonprofit items.

A specific breeding plan is essential if the expenditure of large sums for the purchase of unqualified mares is to be avoided. Two points must be decided upon before establishing a breeding plan: the type of horse to be raised, and the methods of crossing to be used in producing the desired type of foal. The decision to raise working horses, halter horses or race horses should depend on the demand for that type of horse, the prevalence of breeding stock, the amount of capital available for the investment, the expected return on investment and the investor's personal preference. There are several ways to produce champions, but no matter what the approach, a full understanding and application of genetics enhances the probability of producing winners on a consistent basis.

1.11 Genetic Inheritance. All animal characteristics are governed by genes. A thorough understanding of the principles of genetics is needed before the manager can hope to develop a good breeding program. The subject will be dealt with only briefly here. A much more comprehensive study is required before proficiency in determining desirable crosses can be reached.

Upon fertilization, genes unite in pairs and work together to control the development and characteristics of the unborn foal. One member of each pair is contributed by the stallion and the other by the mare.

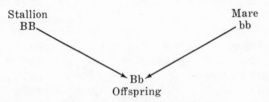

The following hypothetical example selects color to demonstrate how genes control traits. Solely for the purpose of this brief discussion it is assumed that genes are either dominant or recessive. If one of the two genes is dominant and the other recessive, the dominant one will be expressed, masking the presence of the other. For example, if B (black) is dominant and b (white) is recessive, the colt produced with the genotype Bb will be black. This dominance and recessiveness cause characteristics to skip generations. The offspring of a BB (black) stallion and a bb (white) mare would all have the genotype Bb but would appear black like the sire. However, by crossing two black animals having the genotype Bb, three different genotypes and two different colors or phenotypes may be produced.

♂ / ♀	Stallion	
	B	b
Mare B	BB offspring	Bb offspring
Mare b	Bb offspring	bb offspring

The color of the BB animal is black, the color of the two Bb animals is black, and the color of the bb animal is white, which is unlike either parent and skips back to the grandparents.

Obviously it cannot be foretold from appearance whether the black animal has a BB or a Bb genotype. If the animal has a pedigree of all black animals for several generations, or if the animal has never produced a white offspring, chances are that the genotype is BB and the horse will produce only black animals. Of course, the more generations traceable or the more foals by that animal observable, the better the chance of a correct assumption.

Unfortunately genes are not just dominant or recessive. There are a great many complications. Some genes are dominant in the presence of one gene and recessive in the presence of another. Other genes exhibit incomplete dominance, resulting in a compromised characteristic. If B (black) exhibited incomplete dominance over b (white) the animal Bb might be gray. (Note: The colors black, white and gray have been arbitrarily selected to simplify the discussion of inheritance. In practice the breeding of white to black animals will produce a variety of colors in different offspring.) A great many genes may control one characteristic. Eye color, for example is dependent upon the combination of many gene pairs all working together or against each other to create a shade. Genes are hooked together on strands called chromosomes. Each cell of a horse's body contains 32 pairs of chromosomes for a total of 64 individual strands. Each gene has an assigned position on a specific chromosome, which means that a certain characteristic occurs linked with the other characteristics on the strand. The male animal is lacking part of one chromosome strand. This determines the sex of the animal. Some genes are missing so the characteristics that those missing genes are coded for are determined by an unmatched gene. For example, if one gene for the body color was lost due to the male-determining chromosome, the color would be determined by a simple B instead of BB. Many characteristics are determined in this way and are referred to as sex-linked characteristics.

The importance of the animal's pedigree should not be underestimated. The chances are that an animal from a long line of top individuals on both sides has a genetic makeup of mostly good genes. The greater number of poor or unknown individuals in the animal's pedigree, the more undesirable genes the animal is likely to possess. The greater the number of poor individuals produced, the more bad genes the parent animal possesses. An ideal sire or dam would possess all desirable genes, but the chances against the ideal are astronomical. Despite the unattainability of perfection, selective breeding will increase the number of desirable genes.

1.12 Closebreeding, Linebreeding, Inbreeding and Outcrossing. Closebreeding is the quickest way to accomplish the increase of desirable genes. Closebreeding is generally understood to be the breeding of sires and dams back to their own offspring or the breeding of brothers to sisters. Closebreeding is very severe inbreeding that will concentrate outstanding genes into one animal and greatly increase the number of pairs of identical genes. Many recessive undesirable genes are also brought out strongly this way, and by vigorously culling individuals that are not up to quality the number of good genes in the herd can be greatly improved. Because more of the genes are alike, animals bred in this way consistently produce similar offspring. The ability of an animal to pass on its own characteristics is called prepotency. Prepotency is the most valuable asset an outstanding individual can have. However, the poor inbred or closebred individual is just as prepotent for bad characteristics and must be eliminated from the breeding herd immediately. Since a great many defects are brought out this way, a large number of animals must be eliminated. Unless one is an expert in this field, has a large herd of outstanding horses to work with, and can afford to take substantial losses at first, the method of closebreeding for herd improvement is not recommended.

A slower but more feasible method of improvement, followed by most progressive breeders, is that of linebreeding. Linebreeding, also a degree of inbreeding, is the breeding of animals that trace back to one common ancestor, with the idea of concentrating that one animal's genes into the herd. The relationship of linebred animals is more distant than closebred animals, but is in no sense a remote relationship.

Rigorous culling of linebred horses must still be employed so that fewer poor individuals are produced. However, getting rid of undesirable genes is a much slower process, because they tend to skip more generations than in the method of closebreeding.

While closebreeding and linebreeding produce genetically superior animals, the crossing of different inbred and linebred lines usually results in offspring that are far superior to either parent in appearance and ability. Outcrossing animals to conceal faults consistently produces animals close to ideal. Even though the genes for the faults are still there, they are often hidden by superior dominant genes.

The phenomenon of hybrid vigor also results from outcrossing. When two closebred or linebred but unrelated animals are crossed, their offspring tend to be stronger, healthier and generally more vigorous than the average animal with no linebreeding at all in his pedigree. An intelligent breeder can use outcrossing to much advantage. Theoretically the crossing of carefully chosen linebred and closebred breeding stock will probably produce more outstanding individuals than the crossing of champions on the basis of performance alone.

It bears repeating that the breeding farm manager must carefully study the problem of matching the sires and dams to best advantage. This is especially true when the foal crop does not measure up to the farm's expectations.

1.13 Selecting Bloodlines. Choosing the best lines to put into the herd is not easy. The lines chosen must first be known to produce outstanding horses, and second, they must be popular. Fortunately the two qualities usually go hand in hand.

The prospective breeder should look at the pedigrees of the individuals now winning and try to match the combination of lines in these pedigrees. Many nicks (crosses that consistently produce champions) can be determined in this way. Utilization of these nicks can help increase production of winners.

When considering the purchase of a mare or stud, it is wise to obtain a record of the animal's past performance and production. Most breed associations maintain complete records on all animals registered with them; the information may easily be obtained by telephoning or writing to the registration secretary. Leading horse magazines often publish Breed Association and Registry addresses.

Though pedigrees are very important, many breeders overemphasize the bloodlines and forget that, while ancestors play a significant part in an individual's makeup, the parents exert the most influence. The animal bred for paper qualities isn't worth much to the ranch if it cannot produce quality offspring. The most feasible way to insure that breeding stock will produce outstanding horses is to purchase proven producers. A starting place to determine the horses that are producing quality is the breed magazine. For example, the annual May issue of the *Quarter Horse Journal* lists the leading animals of all events, the leading sires and dams of winners and the leading breeders of winners. Other breed publications perform this same service. Some of these individual animals or animals closely related to them could be a sound foundation for a breeding herd. However, most of these horses are not for sale, or if they are, the prices are very high. It is not unusual for an owner to turn down more than fifty thousand dollars for a proven broodmare.

1.14 Selecting the Mares. It has been said that the mares are the determining factor in a successful breeding program. This may very well be true, for while the stallion contributes one-half of the genes to the foal, the mare supplies the other half and provides the environment for fetus and foal development. In terms of stock replacement, it is easier to replace a single failing stallion than it is to replace a poorly chosen band of broodmares. Thus, whether buying one or many mares, quality should be considered first and foremost.

When buying mares there are several points to consider: 1) pedigree, 2) progeny, 3) performance, 4) conformation, 5) age, and 6) mothering ability. A mare may have a championship pedigree, may have produced champions, may have been a champion herself, her conformation may be flawless and she may have years of production left, but if she doesn't fit into the breeding program for the type of horse to be produced, she will be useless as a broodmare prospect. If the stallion has been selected, the mares chosen should have strong characteristics that will compensate for any weaknesses of the stallion. Stallions and mares that are deficient in some of the same points, have the same undesirable characteristics in their background or have any of the same serious faults should not be crossed. The offspring of such a cross is very likely to be strong in the inherited faults.

A breeding evaluation chart such as the one following Section 1.15 should be of assistance when determining the advisability of purchasing a particular animal or of crossing certain individuals.

Probably it will not be possible to acquire a large band of outstanding producing mares immediately. The farm may have to resort to purchasing some performance horses. Only outstanding champions should be selected, with preference to those that possess some linebreeding. There is a good chance that an outcrossed champion mare will not produce foals comparable to herself for she may be exhibiting hybrid vigor that will not be so strong in her offspring. This is not to say that such an

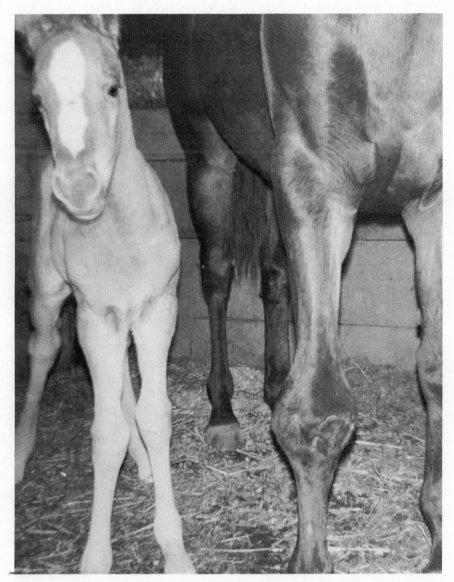

An unsoundness such as this mare's calcified knee does not interfere with services as a broodmare.

animal cannot be an outstanding producer too, but she would be a safer choice if she were more inbred.

Also of importance in selecting breeding stock is conformation. A poorly built mare cannot be expected to produce outstanding offspring even if she is bred to a top stallion. When purchasing breeding stock, excellent conformation is most essential, but nonhereditary unsoundness is important only to the extent that it hampers fertility. Certain unsoundnesses are hereditary and should not be introduced into the herd (see Breeding Evaluation chart following Section 1.15). Other unsoundnesses due to injury or past disease are not too important unless they impair the animal's fertility or health, or greatly restrict its mobility, all of which hamper the breeding procedures or interfere with carrying the foal.

Age is perhaps the least critical factor in buying breeding stock. An old mare, if well cared for, may produce many more fine foals. Because of age and poor appearance, a good mare can often be obtained for a very reasonable price. One foal will more than pay for the mare and the cost of raising the offspring. Mares over eight or nine years of age that have never been bred should be avoided; they may be difficult to settle. Many breeders prefer young mares because they give more years of service, but unproven young mares are a gamble, and proven young mares are expensive.

An often-overlooked but very important consideration when buying a broodmare is her mothering ability. Here the term is used very broadly to cover all phases of motherhood. The most desirable broodmare is very healthy and vigorous. She is free from breeding problems and is easily settled. She is a very roomy mare that foals quickly and easily and recovers rapidly. She accepts her foal immediately, cleans it well and produces a great deal of milk. She has a calm disposition but watches her foal carefully, protects it from danger and shows it how to get along in the world. Unfortunately, man in his eagerness to produce beautifully conformed performance animals has often disregarded this important quality in mares. Outstanding mares that are difficult to settle are given hormone therapy and are kept for years in hopes that they will produce just one prize-winning foal. While the therapy may induce physical readiness for conception that has not been provided by nature, it may not bring about complete psychological preparedness for motherhood. When the foal is born it enters the world among attendants, veterinarians, respirators and protective inoculations. If its mother doesn't look after it well or provide it with milk, man takes mother's place.

Because so many mares and foals are pampered in this way, problems of infertility and weak foals are occurring in greater numbers, especially in Thoroughbreds. The Thoroughbred has an added problem in that the very best racing mares are usually more masculine. Masculinity indicates a tendency toward low fertility and poor mothering ability. Because of the value of the animals involved, the problem is very difficult to solve. The breeder can alleviate the problem to some extent by selecting mares with emphasis on femininity and mothering ability.

1.15 Selecting the Stallion. Selecting a stallion is quite different from selecting a mare. A breeder can sometimes afford to gamble on a mare because he has several, but the stallion is a much more significant part of the entire breeding program. It is far too risky, especially when getting into the business, to gamble on an unproven stud. Buying or leasing one of the top studs in the nation is the best policy. He will be the farm's biggest drawing card. He must be a popular horse that will draw a number of outside mares for a respectable fee. No young stud can do this and a little age on a stallion, unless it affects his fertility, is not a deterring factor. A horse does not acquire a good reputation until he has proven himself over a period of years. A breeder who has a proven stud in service and feels he can afford to gamble on a young prospective stallion should nevertheless select the very best. The potential stallion must have a flawless pedigree, be sired by a top stud out of a fine producing mare that is sired by an outstanding stud. He must be unbeatable in performance and conformation and be free from serious hereditary defects, in himself or in his background. This combination is difficult to find and if such a stud cannot be located, the rancher may have to resort to attempting to raise one himself; if he is successful the animal will be in demand as a stud and people will be willing to accept him and his offspring.

In the case where the stallion being acquired is for servicing an existing band of broodmares, the selection becomes somewhat more complex. Each mare will have strong and weak characteristics, which should be compensated for by the stallion. This may be virtually impossible because of the number of differences in the mares. Of course, other things being equal, the stallion that will offset the faults of the most mares or the best mares should be the choice. The breeding evaluation chart following this section may be of considerable assistance when deciding upon the stallion and its ability to cross favorably with certain mares.

If the breeding farm has only one stallion, it may find itself in a situation where, in order to avoid closebreeding (discussed in Section 1.12), it will be required to engage the services of outside stallions. Sometimes different ranches will exchange stud services until the herds develop a generation of horses that are not more closely related than grandparents and grandchildren. The ranch wishing to have complete control over its breeding program should have more than one stud.

Dry Doc (pictured as a two-year-old). This horse is an excellent example of what to look for in an unproven stud prospect. He is by Doc Bar, famous Quarter Horse sire of halter, cutting and reining horses. His dam was the immortal Poco Lena, great cutting mare shown by B. A. Skipper, Jr. Poco Lena placed in 395 N.C.H.A. (National Cutting Horse Association) approved cuttings, winning a total of $99,782.13, the most money ever won by a cutting horse. This mare was Reserve World Champion Cutting Horse five times and N.C.H.A. World Champion Cutting Horse Mare three times.

This top-stud prospect demonstrates outstanding Quarter Horse conformation. Dry Doc was purchased, untrained, from the Double J Ranch at Pacinies, California, for $25,000, a sum that attests to his new owner's confidence. He won the 1971 six-day Texas Cutting Horse Futurity (competing against 245 entries) with winnings of $17,246. His brother, Doc O'Lena, won the 1970 Texas Cutting Horse Futurity.

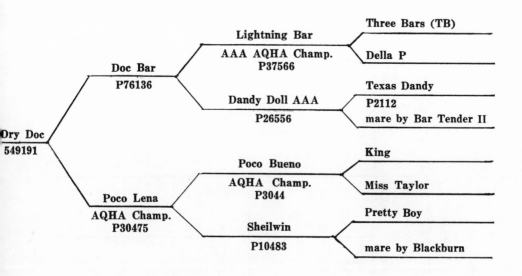

		Three Bars (TB)
Doc Bar P76136	Lightning Bar AAA AQHA Champ. P37566	Della P
	Dandy Doll AAA P26556	Texas Dandy P2112 mare by Bar Tender II

Dry Doc 549191

		King
Poco Lena AQHA Champ. P30475	Poco Bueno AQHA Champ. P3044	Miss Taylor
	Sheilwin P10483	Pretty Boy mare by Blackburn

POINTS OF THE HORSE

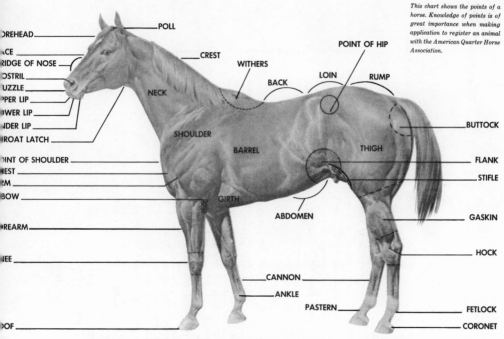

This chart shows the points of a horse. Knowledge of points is of great importance when making application to register an animal with the American Quarter Horse Association.

Courtesy of the American Quarter Horse Association.

This type of chart is often helpful in determining which individuals to breed in order to produce the best possible offspring.

BREEDING EVALUATION CHART

Name of Horse _____ Reg. number _____ Sex Mare ☐ Stallion ☐ Age ___ Date ___

Conformation point	Individual				Parents		Description of problem
	Ideal	Good	Poor	Serious Fault	Ideal	Serious Fault	
Head—well proportioned, refined							
Eyes—large and set well apart							
Ears—small and well shaped							
Teeth—meet evenly							
Throatlatch—refined and roomy							
Neck—long, pleasing curve, head set on well							
Shoulder—long, sloping, well muscled							
Withers—well defined							
Back—not too long, straight and strong							
Loin—wide and well muscled							
Underline—long, gently sloping up to flanks							
Hindquarters—powerful, pear shaped, plenty of width							
Length of hip—long							

Breeding Evaluation Chart (Continued)

Conformation point	Individual				Parents		Description of problem
	Ideal	Good	Poor	Serious Fault	Ideal	Serious Fault	
Length of hip to hock—long							
Muscling—powerful, wide at stifle							
Tail set—not too low							
Heart girth—deep, great circumference							
Rib spring—wide, roomy body							
Chest—deep and wide							
Front legs—true, straight, squarely under body, refined but strong							
Forearm—long, well muscled							
Knee—clean and strong							
Cannon—short, flat, good bone							
Fetlock—refined, strong							
Pastern—relatively long and sloping							
Foot—well rounded, open at heel, adequate size							

Breeding Evaluation Chart (Continued)

Conformation point	Individual				Parents		Description of problem
	Ideal	Good	Poor	Serious Fault	Ideal	Serious Fault	
Rear legs—true, straight, squarely under body, refined but strong							
Stifle—clean							
Gaskin—well muscled inside and out							
Hocks—clean and strong, no fleshiness							
Cannon—short, flat, good bone							
Pastern—relatively long and sloping							
Foot—well rounded, open at heel, adequate size							
Flank—well let down							
Breed type—							
Balance and symmetry—equally well muscled in fore and hindquarters							
Action—lively							
Front legs—track straight							
Rear legs—track straight							

Breeding Evaluation Chart (Continued)

	Individual				Parents		Description of problem
	Ideal	Good	Poor	Serious Fault	Ideal	Serious Fault	
TEMPERAMENT							
PREDISPOSING FAULTS							
Straight shoulder							
Straight pastern							
Easy pastern (too sloping)							
Calf knees (back at the knee)							
Tied in below the knee							
Light bone							
Weak joints							
Interfering							
Cow hocks							
Extremely sickled hocks							
Small vulva							
Crooked legs							

Breeding Evaluation Chart (Continued)

HEREDITARY UNSOUNDNESSES	Individual				Parents		Description of problem
	Ideal	Good	Poor	Serious Fault	Ideal	Serious Fault	
Hernia							
Cateract							
Heaves (allergies)							
Roaring							
Navicular Disease							
Bone spavin							
Stringhalt							
Ring bone							
Side bone							
Susceptibility to disease							
Susceptibility to indigestion							
Low fertility							
Low milking ability							
Poor mothering ability							

1.16 The Search for Breeding Stock. There are many places to search for quality stock. Horse shows, horse races, horse dealers, reputable breeding farms, trainers and livestock sales are some of the noteworthy places to look. The search may be far and wide but should be extended until the best stock available is acquired. A venture started with second-rate animals will have a long and costly struggle before true quality production can be attained.

The old-fashioned local horse auction is a very poor place for a prospective breeder to look for breeding stock. All too many horses are found there because they cannot be sold elsewhere. Few real bargains are ever found at a local auction. Today great efforts are being made by the various breed associations to improve the image of auction sales of registered horses. Many breed associations hold their own sales. Such sales are designed to offer outstanding animals of the breed for sale to the general public. Large breeders are encouraged to consign only the best animals so that the sale averages will be high, thus publicizing the value of the breed as a whole and also bringing together many excellent representatives of the breed for the public to view and

Many high-quality Quarter Horses are sold through the All Western Quarter Horse Sale held every summer in Sacramento, California. Breed sales such as these are an excellent outlet for young stock and provide the breeder with an opportunity to advertise his product.

purchase. Many of these sales will not accept even average quality animals for sale and maintain strict inspection of all animals for quality and soundness before allowing such horses to be consigned. The Keeneland Thoroughbred sale is a very famous example of this type of auction. The 1971 Keeneland Summer Sale of Thoroughbred yearlings had a price average of $31,775 on 333 head, which indicates the quality of animal that is sold at this Kentucky sale. Purchases made at such sales are usually quite safe. Many consignment sales will accept all animals; however, consignment fees are quite high, thus the practical effect is limitation to those worth enough to make sale through consignment profitable. Many ranches or groups of ranches offer yearly production sales as an outlet for the animals they have produced during the year. These sales are usually held on or near the ranch and offer the buyer a chance to view the sires and dams of the horse being sold. Few proven producers will ever be offered for sale at either the consignment or production sales; however, excellent potential breeding stock may be purchased there.

Exceptional quality is not usually found at the near-dispersal sale. Often such a sale is held by a ranch that is drastically upgrading its breeding program and wants to rid itself of low-quality animals. If such animals are not suitable for them, they will be unsuitable for another breeding farm.

The most desirable type of sale is the Complete or Estate Dispersal sale. All animals are sold and all bids are taken. Such sales provide the breeder with the rare opportunity to purchase outstanding proven producers.

When purchasing animals through sales, the buyer must be constantly on the defensive. Once an animal has been purchased, it is difficult if not impossible to be reimbursed if the animal proves unsatisfactory. Animals sold by reputable ranches are usually presented honestly; however, it is recommended that any animal the buyer intends to bid on be thoroughly inspected prior to the sale by both the buyer and a qualified veterinarian. Some states have statutes making it illegal for an animal to receive prior to the sale any medication that is intended to deceive the buyer. If the buyer suspects his new horse has received illegal medication, he should have the veterinarian examine a blood, saliva or urine sample immediately after purchase. Where a large amount of money is involved, such a test is always recommended.

1.17 Ranch Recognition and Reputation. Fine horses will not completely guarantee a successful operation. People must know of them or no one will come to buy them. The new breeder should enter the business with as much recognition as possible. He should make public the quality of his herd by advertising the outstanding individuals he owns. He should be hospitable and encourage visits to his ranch. He should become active in horse clubs and breed associations so that other members will become aware of his horses. He should show or race as many of his animals as he can. It is very important to place horses with people who will show and advertise them, even if it means a few losses at first. His horses must be in the minds of the purchasing public. He should send as many horses as he can to qualified trainers, especially his prospective breeding stock. Most important is maintaining honesty and integrity. Honest representation of quality and complete disclosure of defects are the building blocks of the successful horse farm. Unethical practices in the horse industry are quick to be recognized and word is passed along. Horse trading has been suspect by the public since the days of pedigree-by-memory. Although breed associations and supporting laws have greatly destroyed unethical practices, suspicions still linger to some extent. It takes very little to reignite the anxiety of the purchasing public. Excellent quality and good-faith dealings can assure stability of the investment and good long-range profit. The people who deal with the reputable ranch will become its best sales force.

2
BUILDINGS, EQUIPMENT AND PASTURES

(SYNOPSIS)

This chapter considers ranch location, building sites, structures, stalls, fencing, ranch roads, paddocks, pastures, grazing and plants poisonous to horses. The following synopsis by section number is provided as an aid for quick reference to the specific subjects.

2.1 Farm Location. Many important factors must be taken into consideration when choosing a location suitable for the breeding farm. Since the horses produced will probably be sold to the public, the animals must be where the public can see them. Therefore, the farm should be located near a well-traveled road and within reasonable hauling distance of noteworthy horse shows, races and sales. Water, electricity and telephone services are necessities that must be considered. If the location cannot produce sufficient water from springs, streams or wells it must be situated where public water can be obtained. Very remote areas often are not serviced with electricity, in which case a private generator must be installed and maintained. Locations where temperatures reach over 100° are not conducive to horse health or comfort; moreover, a little icy weather is desirable as a deterrent to many types of parasites and disease-causing organisms that do not survive cold winters. Some industries emit pollutants into the air or water, where they are ultimately consumed through grass or drinking water. Lead and aluminum emissions are particularly harmful to horses and should be avoided.

Attention should be paid to the adjacent area and to land prices with respect to the possibility of later expansion. There is no point in establishing a breeding set-up where the farm may be locked in or even forced out by foreseeable land developments or zoning ordinances. Of course, personal preference will be the final determining factor in location, for no matter how well the site fits the requirements, if it is not where the breeder wants to locate he should pass it by for the sake of his own contentment.

2.2 Soil and Terrain. Good soil is essential to the horse farm. A sandy loam does not pack or become deep with mud in wet weather, but many sandy soils are leachy, lose nutrients easily and result in high incidence of sand-colic. Clay or adobe is slippery when wet and cups badly; both characteristics increase chances of injury. Gravel soils are not advisable because they lack nutrients and tend to cause injuries, but even so, a successful farm can be established on gravely soil if proper planning is employed. Generally speaking, productive grassland is indicative of good soil and therefore is highly desirable. Sufficient drainage is important, since bogs or wet, marshy, poorly drained areas are unsanitary and are breeding places for mosquitoes and disease-causing organisms. Hilly terrain is preferred, since it is conducive to drainage and helps growing colts develop muscling.

2.3 Building Locations and Main Structures. Often when a ranch is purchased there are several buildings on the premises that can be utilized, but when building new structures, care shoud be taken to place them where they will take advantage of existing conditions and provide the most economic use of labor. General attractiveness of the farm should be included in the planning, because horses presented in attractive surroundings tend to bring higher prices. The farm buildings, fences and grounds must be well maintained, reflecting a prosperous and efficiently run operation.

The manager's or foreman's residence should be a comfortable up-to-date house or mobile home situated near the visitor-calling area and upwind of the barn area to avoid odors and flies.

The number and types of structures needed will be determined by the size of the operation, the climate, amount of capital to be invested and the owner's preference. Precut stalls or stables are available and often prove very satisfactory. If properly engineered they are strong, insulated against weather extremes, easy to assemble and are quite portable. Preconstructed pipe corral or paddock sections can also be obtained commercially and are convenient for temporary use.

All structures should be placed on high ground so that adequate drainage is afforded the year around. If the ground water table is high, damp floors can be overcome by sub-draining with a layer of drain rock (smooth, rounded river rock) laid down prior to building. If the problem is severe, concrete floors can be installed in the barn.

Pre-cut stalls or stables, if properly engineered, are strong, insulated against weather extremes, easy to assemble and quite portable. (Lacey's Training Center, Alamo, California)

In mild climates open shelters are practical and inexpensive.

Prevailing winds should also be taken into consideration when constructing barns so that the narrow end faces the wind, thus reducing exposure to cold weather but allowing cool air to circulate through in warm weather. Partially open shelters should have the solid side facing the wind, but it must be noted that in certain areas the storm winds blow from a different direction from the prevailing winds. Buildings, especially hay barns, should be separated so that in the event of a catastrophe, such as fire, only one building is likely to be lost. To assure labor efficiency, the buildings should be easily accessible for feeding, cleaning and maintenance.

Barns constructed of concrete blocks with metal roofs are practical and popular. Upkeep on such barns is inexpensive, they are sturdy and attractive, and will stand a great deal of wear and tear. Horses may test the concrete but will not persist in kicking or chewing it. Concrete barns are cool in the summer, warm in the winter and are nearly fireproof.

In order to assure fire protection, many farms install automatic fire-extinguishing equipment. Sprinkling systems are effective although somewhat expensive to install and must be protected from freezing in cold climates. Automatic fire-detection systems with sensors that activate the fire alarm if a 15° temperature change occurs within one minute are used in many areas. Fire-retardant lumber, pressure-impregnated with special mineral salts to reduce flammable elements in the wood, is considered by insurance companies to considerably lessen the danger of fire. "No smoking" signs should be posted and the rule strictly enforced. Should a fire occur, a blanket, jacket, feed sack or other available blinder can be placed over the horse's eyes if it refuses to leave its stall. Care should be taken to insure that it does not break loose and return to the burning structure; horses are apt to do this.

A concrete barn should have the foundation and walls reinforced with steel rods and the blocks making up the exterior walls should be filled with cement or exploded rock for heat-and-cold insulation. Adequate ventilation can be easily built in by providing for a gap of a foot or more between the top of the exterior walls and the roof. Such a gap in the windward wall should be equipped with a shutter for use in the winter. Plastic or fiberglass skylights built into the roof provide natural lighting, but skylights in a flat roof become catch areas for leaves, dust and other debris unless installed with one side elevated to allow for rain flushing.

After construction, the block walls should be coated with a commercial paint sealer, which makes the surface smooth, nonabrasive and easy to clean. Unless treated with a sealer the cement blocks tend to collect moisture. The coating of the interior walls should be a fireproof sealer. Intense heat will practically explode a combustible sealer and this, added to an otherwise manageable interior fire, will result in an uncontrollable inferno. Concrete blocks are available with a baked ceramic tile surface that is quite inexpensive and ideal for laboratories and foaling stalls. Adjustable ceiling ventilators are particularly helpful for controlling air circulation in stalls or other close areas.

2.4 Stall Construction. Box stalls within the barn should be no smaller than 12' x 12'. Foaling mares and stallions need at least 16' x 16' stalls. The minimum roof or ceiling height should be at least ten feet, for a rearing horse can severely injure his poll on a low ceiling. In addition, high ceilings provide better ventilation. The stall walls should be smooth and free from projections that might cause injury. Sills and ledges encourage cribbing and wood chewing and should be avoided or protected with metal strips. Partitions made of 2"-thick tongue and groove lumber with vertical planking are practical. Vertical planking is less likely than horizontal planking to cause injury if a board should be kicked out or loosened. Kickboards can be installed as facing on concrete block walls to lessen the incidence of fractured third phalanges (coffin bones) and cracked hoofs, although many horsemen maintain that horses will not kick concrete blocks. By sloping the lower four feet of the partition twelve to sixteen inches into the stall, or by putting cleats on the walls, problems with cast horses can be greatly reduced. Hallways and alleyways should be at least ten feet wide to allow for manure removal by motor-drawn equipment. When designing a barn it should be noted that wider hallways and alleyways can be utilized in winter for walking or exercising the horses.

This well-planned, attractive barn provides a wide alleyway for easy movement of equipment, large airy foaling stalls and quarters for attendants, as well as tack and feed storage.

Sliding stall doors mounted on small wheels or rollers are not in the way when open. Skylights built into the roof provide natural lighting.

Many horsemen prefer to discourage close attachment between neighboring horses by installing solid partitions between the stalls, but provide a view of the various daily activities and of other horses by allowing the animals to see out the front of the stalls. Partitions should be seven feet high, but if a view of the next stall is to be provided, the top two feet can be a heavy-mesh wire screen, vertical piping or cold-rolled steel rods spaced no wider than three and one half inches so that horses cannot put their noses in between them.

Internal swinging doors and door panels must, as a safety precaution, open outward from the stall. Doorways must provide sufficient room for easy passage. Doors less than four feet wide and eight feet high will hamper the passage of equipment and materials. Narrow doorways may result in hip injuries to horses. The most popular type of door is the two-panel (Dutch) door that allows the top and bottom panels to be opened independently of each other; the bottom panel should be five feet high and the top one three feet. Many breeders prefer the sliding door mounted on small wheels or rollers, which never sags and which is not in the way when open. Every type of door installed should be of heavy construction with no sharp edges on either the door or the latches. An outside door for each stall is desirable for providing fast evacuation of horses, facilitating cleaning and providing cross-ventilation in hot weather. Full-size screens or webbing can be useful when

Strong screens can be used in place of doors to increase air circulation and give the public a better view of the animal.

the door is left open for ventilation or for public viewing of the animal. Webbing is equally valuable at shows or races for showing animals to the public.

Wooden stall floors should be avoided; they are slippery, rot easily, harbor germs and are difficult to disinfect. Packed clay floors are preferred by most horsemen because of their resiliency and relative warmth. Red rock or concrete can be used but is much less desirable than clay. A drain rock or tile leach bed should be placed under the clay to provide adequate drainage. If a solid, waterproof floor is installed, there should be an even floor slope of about one inch for every five feet from the front of the stall to the back, or an approximate three-inch slope for a 15-foot stall. A drain line or gutter placed at the end of the slope to catch the drain-off of urine can carry the waste to a main drain at either or both ends of the stable. Breeding sheds, foaling stalls and recovery stalls require floor surfaces that provide traction for safe footing and can be easily cleaned and disinfected. Resilient, nonabsorbent mats that are especially suitable are produced under commercial brand names such as Tartan and Ensolite.

Paving of rough concrete or fine aggregate macadam will serve well for smaller aisleways used only for passage, while a large aisleway can be filled with sand to provide an indoor place where horses can be exercised.

Windows are best located at least eight feet above the floor out of the horse's reach. They should be a minimum of 2' x 4' in size, hinged at the bottom and mounted to open inward, deflecting air upward to avoid drafts.

Stall accessories should be kept to a minimum and all lights should be placed high, preferably recessed and protected by screen to prevent accidents. A fresh supply of water is absolutely necessary and is easily provided with the use of automatic water basins. Grain should be fed in removable rubber tubs securely fastened to the wall. Rounded tub bottoms make it easier for the horse to pick up all grain. Hay is best fed at ground level, but not on the bare ground, for that is the horse's natural eating position and stretching will help prevent development of heavy necks, especially in stallions. Hay racks located so the animal must reach up will cause a greater incidence of eye problems due to chaff's falling out of the hay into the animal's face. All troughs, bins and racks should have rounded edges to prevent injury. Salt blocks may be placed loose in the manger or in separate containers.

2.5 Fencing. Many types of fencing have been developed, but one of the most practical and inexpensive for a horse farm is 5' high, 12½-gauge, galvanized 2" x 4" V-mesh. When installing this type of fence, wooden posts should be set about twenty feet apart with capped metal posts between. The wire should be raised about six inches off the ground, and a white metal strip at the horse's eye level running through it will insure that the horses will see the fence when playing. When constructing alleyways it is recommended that wooden posts be set closer together and that three 2" x 6" boards be used between posts. This type of fencing is stout and will last for years, as well as being attractive to the eye. If wooden fences are used, they should be creosoted or treated with another commercial preparation to prevent chewing and wood eating. Animals should be kept away from freshly treated lumber to prevent skin and eye irritations. Any paint used around livestock should contain no lead. Horses that chew fences painted with lead paint will eventually suffer from lead poisoning. When fences are made of wood, four horizontal boards should be used for pastures and five or six boards for stud corrals. Most horsemen avoid wooden fences because they are expensive, require a great deal of maintenance and encourage wood eating. Chain-link fences with pipe posts set in cement are attractive when new, but when stretched or bagged cannot be repaired. Barbed-wire fences, if tightly strung, can be used in very large pastures but are not recommended where valuable horses are housed; they do cause injuries. Electric fences are popular, but some horsemen feel they make the horses cranky and nervous. Contact with an electric fence may startle the horse, causing him to bolt in another direction and charge through another fence. A few farms have successfully used army-surplus steel emergency landing mats for fencing. Pipe fencing for small paddocks

This paddock area demonstrates how practical and attractive the well-planned use of chain link fences, alleyways and inexpensive open shelters can be.

Five-wire, tightly strung barbed-wire fences may be used in large pastures. However, barbed wire is potentially dangerous and not acceptable for use with valuable animals.

Pipe fences are attractive, easily maintained and safe when properly constructed.

Commercially available aluminum gates are light weight, safe and practical. They require no upkeep and eliminate wood chewing.

attached to the barn has been used with great success in many areas. Injuries are quite uncommon from such a fence if well constructed.

Pipe or aluminum gates hold up best and are safest, and therefore are highly desirable. Gates should be large enough to permit the entrance of machinery for cleaning or pasture maintenance, and should swing both ways.

2.6 Internal Ranch Roads. Gravel roads provide easy access to corrals and pastures in any weather and gravel pads around gateways and feeding areas are beneficial as well as convenient. Number 2 stone aggregate or red rock has proven safe and practical although others may be used satisfactorily.

2.7 Loading Chute. Since many horses will be coming and going from the breeding ranch in large trucks and vans, a loading chute will be needed. It should be constructed of 3-inch pipe or heavy, treated lumber and should be about four feet wide and perhaps twelve feet long, with a ramp incline rising to a height where it will meet the vehicle

Well-built loading ramps of different heights will make loading a much easier and safer task.

bed. A sand floor is preferred for the ramp because it provides a more natural, secure footing. Ramps or chutes constructed side by side with different elevations will be convenient when loading vehicles of variable bed heights.

2.8 Stallion Barn and Paddock. The stallion barn should be located upwind of the other horses and not so close to the breeding operation that the stallions can see and hear what is happening. It should be impressive and easily accessible to visitors. Stalls should be large, at least 16′ x 16′ and spacious runs should be provided. Grass paddocks (one for each stallion) three or four acres in size are conducive to health and pleasant disposition. The studs should be allowed to see each other; they are usually easier to handle when they are allowed the visual company of other horses. Fences should be at least 6′ high with a 5′ to 8′ alleyway provided between stallion paddocks to prevent fighting.

2.9 Foaling Paddocks and Foaling Stalls. Many breeders feel that the best place for a mare to foal is in a clean grassy paddock that can be

Stallions are often housed in attractive barns separate from the other horses and readily available to the viewing public.

Alleyways between paddocks facilitate feeding, herding and sorting animals, and prevent fighting between animals in different fields. An alleyway should always be provided between stallion paddocks.

easily seen at all times and is equipped with lights for night observation. The fences of such foaling paddocks should be close enough to the ground to prevent the foal from rolling under it if born nearby. Occasionally weather or other circumstances will not allow outdoor foaling and a foaling stall must be available. It should be a minimum of 16' x 16', equipped with adequate lighting and heating. A heat lamp is suitable if the barn does not provide central heating. Infra-red chicken brooder bulbs can be used satisfactorily. If the stall is enclosed, a small window will allow the mare to be observed without disturbing her. The stall should be free from drafts. A composition-rubber mat on the floor greatly facilitates cleaning and disinfecting. Straw bedding for foaling is preferred over shavings because the straw particles do not adhere so easily to freshly exposed membranes.

2.10 Weanling and Yearling Barn. A weanling and yearling barn is beneficial to the horse ranch. In such a barn young animals can be kept available for prospective buyers to observe. Needless to say, the barn should be constructed and maintained as a show place, for this is where

Open, airy foaling stalls such as these are excellent in mild climates. Protective shutters may be rolled down during cold or rainy weather. Foaling stalls should be a minimum of 16′ x 16′.

most of the horses will be sold. The barn can also be used to advantage for conditioning horses for shows and sales. It should be equipped with a cross-tie area, wash rack and equipment storage space. Box stalls and grassy paddocks should be provided for the young stock. In warm climates three-sided structures with large adjoining paddocks are quite adequate if the youngsters are not being conditioned or maintained for halter classes.

2.11 Mare Paddocks. Barren mares should be kept separated from foaling mares. They require less feed and attention and are more frisky and likely to cause trouble. Some barren mares will try to steal a foal and a fight often ensues, resulting in injury to at least one if not all three animals. Foals have been known to starve to death because the self-appointed mother refused to let it nurse the natural mother.

Large paddocks of not less than 150 square feet per animal, complete with a teasing area are most practical for both barren and foaling mares. Aggressive or bossy mares should be kept separated from timid mares to prevent injuries and provide equal distribution of feed. Be-

In warm climates a three-sided stall such as this is very practical and is easily cleaned.

cause animals tend to pair off, it is unwise to allow an uneven number of horses to run together; the odd one will be tormented by the others. This rule, however, does not apply to mares with foals.

2.12 Quarantine and Isolation Barn. An isolation or quarantine barn is necessary. New arrivals should be kept away from the main farm for a minimum of three weeks to assure freedom from disease, and the farm's sick animals should be moved to the isolation area as soon as an illness is detected. However, in the event that a large number of the animals become ill, the usual procedure is to leave them all together where they are to prevent a new area from becoming infected or cause additional animals to be exposed. Facilities should be provided at the unit for the separation of sick animals from quarantined new arrivals. To decrease the chance of spreading disease, the quarantine barn should be accessible without passing through the main part of the ranch and should be located downwind of the farm to lessen the chance of airborne germs being transmitted to other parts of the ranch.

Large farms sometimes use a closed-circuit television monitoring system to keep the foaling stalls, stallion quarters, isolation and quarantine units under surveillance day and night.

Teasing to establish heat cycles for isolated mares is conducted at the quarantine unit. A different teaser should be used for isolated mares, but if the farm is not large enough to maintain an extra teaser for this purpose the teasing of quarantined mares should be restricted to apparently healthy animals. The unit should maintain a teasing area, examination chute, first-aid supplies, storage and tack room, a loading chute, warm stalls and outside paddocks.

2.13 Breeding Shed or Barn. The structure or area used for breeding should be covered to allow use in bad weather. Ceiling height should be a minimum of 15 to 20 feet. The structure may be enclosed or open, but a round shape is preferred so that handlers will not be trapped in corners. The floor should be covered with clean peat moss or sand to keep dust at a minimum. The facilities should be equipped with box stalls for waiting mares and an out-of-the-way stall for the stallion. Examination and teasing chutes should be 8' long and 4' wide, with solid front and sides. An 8' x 8' foal cage must be placed where the mare with a foal at her side can easily see the foal during the breeding procedure. If properly constructed such a barn can also be used for training colts.

A laboratory with office space is a necessary addition for the efficient breeding facilities. The laboratory should include a sink with hot and cold water, autoclave (sterilizer), refrigerator, semen storage facilities, microscope, artificial vagina, vaginal speculums, pipets, gelatin capsules, medication and first-aid kits. The space should be kept absolutely clean and dust free.

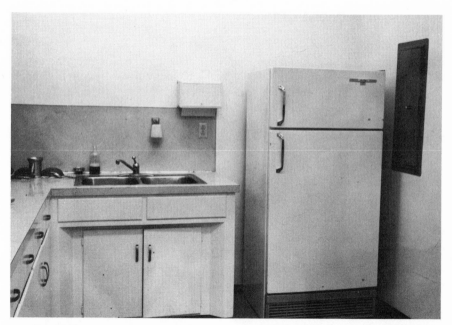

A refrigerator and running water are essential for a laboratory.

Every farm should have a well-equipped laboratory where instruments can be kept and sterilized and have an incubator for culturing mares. This lab is also equipped to handle rats for pregnancy testing.

2.14 Miscellaneous Facilities, Structures and Equipment.

a) *Barn Office.* Large farms or those having large horse barns regard a barn headquarters office as almost indispensable. Often the barn headquarters is quite elaborate. All pertinent records of the horses stabled in the barn, as well as current feeding schedules, tables and ration formulas, animal location charts and special individual treatment reminders, are maintained there. The headquarters should be equipped with toilet and washroom and some have permanent sleeping arrangements. The office should have a direct telephone line to the various barns or each barn should have a separate phone. Emergencies are common occurrences on the horse farm and the loss of a few moments for lack of a telephone may mean the loss of a valuable animal. The office should keep handy a first-aid kit and supplies and a portable fire extinguisher as recommended by the local fire department.

b) *Tackrooms and Equipment.* Although tackrooms should contain a flashlight, small hand tools and light hardware used in minor on-the-spot barn maintenance and repairs, their main purpose is to house the equipment needed for grooming and handling horses. The point of a tackroom in each horse barn is to keep the often-needed items in close proximity to the horses stabled there. The equipment used for showing horses must be kept in "show shape" at all times and should be stored separately from everyday equipment.

Items not used on the horses in that barn should not be kept there.

An attractive and prominently located office reflects a well-managed, prosperous enterprise. It is a headquarters for all ranch operations and is in direct contact with the various outbuildings.

Portion of stable with tackroom, electrical outlets, fire extinguisher, framed pedigree charts and convenient tie rings.

Many items are common to all horses and should be at hand in all tack-rooms. Grooming tools are among the common items, which usually consist of:

brushes (dandy and water)
curry combs (rubber and metal)
electric clippers (large and
    small)
extension cords (50' long)
hoof picks
scissors (blunt ends)

shedding blades
soft cloths
sponges
sweat scrapers
vacuum cleaner (for horses)

All tack rooms should have first-aid kits for both horses and people. Assorted items that are generally stored in the tackrooms, depending upon the type of horses stabled in the barn, are:

bell boots
blankets (fly sheet to heavy
    water proof)
breeding hobbles
bridles
cavassons
chain lead shanks
crops
feed bags
hackamores
half hobbles
halters
hobbles

lead ropes
leg wraps
longe lines
neck sweat hoods and bands
popper or scat bats
ropes (cotton 1" x 30')
saddles
saddle blankets and pads
stallion rings
tail wraps
twitches
whips

Material and tools for servicing and maintaining tack should also be kept handy in the tackroom. Often needed are:

awls
hole punch
knife
latigo leather
neat's foot oil
nylon string
rivets

saddle leather
saddle soap
sheepskin
waxed linen string

The nearby tackrooms save time and labor if they are large enough and maintained in an orderly manner so that the needed items can be found quickly. They are in constant daily use and should not be used for long-term general storage. All items in the room should be kept repaired. Equipment used on or with an animal having a contagious or infectious disease must be thoroughly disinfected before being put back in the tackroom, and contaminated material should never be allowed in with the tack. The area should have good ventilation and the equipment prone to absorb moisture and sweat should be frequently aired and exposed to the direct sun rays.

c) *Barn Feed and Bedding Supplies.* Grain processed and ready for use should be kept at each horse barn in an amount that can be conveniently stored but used before it deteriorates. From time to time some individual animals may need a more laxative feed or may require minerals or vitamins in larger quantities than contained in the feed mix; it is advisable for each barn to have an ample supply of supplements on hand to take care of these special cases.

Hay and bedding are fire hazards; the supply kept with the horses should be a minimum, usually not more than a week's, supply. Local barn storage is labor saving, but the risks to the horses is too great if much of the combustible material is housed with them. As a safety measure, a fire wall of cement block, fire-retardant lumber or metal can be constructed between the storage area and that portion of the barn where horses are stabled. Hydrated lime may be stored in the barn with the extra bedding for use when cleaning stalls. Section 5.8 should be consulted for a full discussion of the types of bedding and health considerations. Light-weight aluminum scoop shovels, wheelbarrows, manure forks and two-foot push brooms are needed in all barns and should be within easy reach to encourage prompt, continual clean-up.

d) *Hay Barns and Storage.* Hay and feed storage should be adequate, but large amounts of hay—a considerable fire hazard—should not be stored near other buildings, especially horse barns. In addition to fire due to carelessness, damp hay stacked too tightly to allow for adequate air circulation is an invitation to fire produced by spontaneous combustion. Stored hay should be well ventilated to permit the hay to dry thoroughly and as a means of reducing the heat generated; pole barns with six-foot roof overhangs on the sides exposed to the wind provide inexpensive, adequate hay storage fairly safe from fire, but even these should be separated enough so that a fire in one will not ignite another. Hay barns having two or three thousand bales capacity are the most practical.

Large supplies of bedding material are often stored in a portion of the hay barn. This material creates too much of a fire risk to be stored in large quantities in the horse barns because it also can be ignited by spontaneous combustion if stored when damp.

e) *Granary, Equipment and Storage.* Grain and supplements ready for use should be stored in the horse barns or very near by to reduce the labor at feeding time. Grain and supplements can be stored centrally but, whether in bulk or ready for use, all should be stored in large parasite- and vermin-proof bins set well off the ground or floor and free from dampness. Vitamin and mineral supplements, salt and other feed additives may also be stored in the granary.

The farm having many horses often finds it economical to purchase whole grain in bulk, which is processed, prepared and mixed as needed for feeding, but not many weeks in advance of use. Many ranches prefer to purchase pre-mixed grain by the truckload rather than by the sack. If this is the case, the storage bin should be situated so that it can be easily filled through a chute running directly from the truck.

f) *Blacksmith Equipment.* Regardless of arrangements with a farrier, the ranch must be equipped to care for horses whose hooves need prompt attention. Most farriers carry a fairly complete blacksmith shop and supplies on their trucks, but in the farrier's absence the farm will often need certain basic tools and supplies. A specific area such as a shed or barn section should be set aside and equipped for trimming and shoeing. Only qualified personnel should be allowed to work on an animal's feet. Incorrect hoof care can damage the animal's feet and legs, which may take months to repair.

Hot shoeing requires a forge (portable preferred), fire rake, farrier's tongs and fuel (coal or coke). Whether hot or cold shoeing is preferred, the following equipment will be necessary:

alligator clincher	nippers
clinch cutter	parer
fullering iron (creaser)	pincer
hammers (driving and rounding)	pritchel
hardy	rasp
hoof knife	shoeing anvil
leather apron	

Shoeing nails and blank shoes of assorted sizes and weights should be on hand.

g) *Repair Shop, Equipment and Storage.* A building or shed for machinery storage and repair is likely to be unsightly and should be located as inconspicuously as convenience will permit. A pole and corrugated metal structure complete with fuel supply and work shop is usually adequate; however, not being insulated, it is difficult to heat in cold weather. In this building there should be a sturdy workbench equipped with a large general-service vice, a pipe vice and assorted tools commonly needed in repair and maintenance work.

Particular items included should be:

Fencing Material	*Irrigation Equipment*
cement	pipe cutter
fence stretcher	pipe fittings
gates	pipe threader
hinges	pipe wrench
latches	*Tools*
metal post caps	anvil (multi-purpose)
post driver	electric drill
post-hole digger	electric saw
posts	forge
tamping bar	grinder
wire	hand tools
wire cutters	welding equipment

Near the repair shop should be housing and yard space for the ranch motor vehicles, trailers and other heavy equipment. Unless such a parking place is carefully planned and maintained, it can take on the appearance of a junk yard and will undercut public opinion of the breeding farm.

h) *Miscellaneous Equipment.* A motor-driven feed cart is often used for efficient distribution of feeds from the hay barns and granary to the horse barns and pastures. It may even be used to service the stalls in barns having sufficiently wide aisleways. Some farms use automatic grain feeders as a labor-saving device, but close observation of horses at feeding time should not be overlooked.

A skip loader with scraper and manure cart greatly facilitates stall and paddock cleaning and removal of soiled material. The skip loader is indispensable in handling manure, whether from the stalls and paddocks or from the compost pile to the manure spreader.

A manure spreader for distributing manure through the pastures will efficiently and easily accomplish an otherwise difficult chore.

The ranch intending to raise feed or cultivate pasture must either own or lease heavy equipment such as a tractor, baler, rake, disc, harrow, seed drill, mower and other specialized machinery.

2.15 Arena. The majority of large successful ranches consider an arena almost indispensable and some, having sufficient capital, develop large, covered showplace arenas complete with grandstands and restrooms. These showplace arenas are prestige items; they attract the horse-minded public and are assets for built-in advertising but are economically impractical for most small ranches. If staging shows or rodeos is in the planning, the arena should be ruggedly constructed, with particular attention given to good sandy footing and even drainage. It should be at least 100 feet wide and 225 feet long to allow for adequate working and roping room and should have entering and leaving facilities for horses and cattle at both ends. If covered, the ceiling must be very high; many horses will not work under low roofs. Adequate, convenient parking space is almost as necessary as the arena itself.

A skiploader with a scrapper attachment is a very versatile piece of equipment that saves labor and will more than pay for itself over a period of time.

An indoor arena provides a dry area for working or exercising horses during wet weather. Working a performance horse under a roof will familiarize the animal with conditions it will often encounter during campaigning.

2.16 Pasture Arrangement and Facilities.

a) *Arrangements.* Quality pastures indicate effective, economical management. Although the initial expense may seem high, in the long run the savings in feed and labor costs as well as the improved health of the animals prove its worth. A profitable farm provides its horses with the best possible pasture grazing, for, in addition to good health, horses on pasture develop and retain better dispositions. From a health standpoint growing pastures abound in minerals and vitamins, especially vitamins A and E, and are unexcelled as sources of high-quality nutrients. The sunshine provides a valuable source of vitamin D and the exercise contributes to the improvement of bone structure, strengthens the muscles, improves circulation and reduces incidence of digestive disturbance. Fertility of both mares and stallions is improved by the consumption of green grass, probably due to the estrogen present in the lush growth. Respiratory infections are less likely to affect pastured horses and they are generally much healthier than those maintained in the corrals and barns.

Several small pastures of 5 to 20 acres each have decided advantages over large pastures. Large pastures are seldom evenly grazed and tend to be over-grazed in some parts and under-grazed in others. The horses

Roomy pastures provide exercise and nutritious forage. These alfalfa fields are harvested and then used as pastures for mares, colts and young horses. Courtesy of Double J Ranch, Pacinies, California.

The contrast between a well-managed pasture and an overgrazed pasture is obvious here. Note the encroachment of brush and weeds as well as the lack of grass in the pasture in the foreground.

can be encouraged to utilize the pasture better if the salt and supplements are moved to under-grazed areas. Development and distribution of drinking water in several parts of the pasture also greatly encourage the horses to use all sections of the field. The grasses need occasional rest from grazing to allow for regrowth and the sun and soil microorganisms need some time to destroy parasites, worm eggs and germs deposited in fresh droppings. Pastures will deteriorate rapidly if over-grazed and in the interest of health must be kept reasonably sanitary. Small pastures are more evenly grazed than larger ones and several pastures make rotational grazing possible so that one field can be allowed to recover while grazing is rotated to others. This provides time for fertilizing, dragging, reseeding, weeding or complete replanting.

Separation of visiting mares from ranch mares and segregation of foaling mares from barren mares, colts from fillies and young horses of different ages is desirable and is simplified when several small pastures are available. The number of pasture accidents is related to the size of the pasture and the number of horses together; within reasonable limits the fewer horses in a pasture the fewer the injuries.

b) *Facilities.* The general discussion of fencing in Section 2.5 applies to pasture fencing. Corners create hazardous areas for the horses; these can become traps in the event of a fight and even a playful romp in a corner can result in injuries. Each field should be provided with a catch-pen, and teasing areas should be in the plans for all pastures.

A teasing partition should be built into every paddock. When this teasing section is used, the small gate at the top is opened so that the teaser can reach over the partition.

When herding horses to another field, a portable panel can be set up next to the gate through which the handler wishes the horses to pass. A panel such as this is also useful for teasing, closing off an alleyway as a temporary paddock or providing a wing to aid in loading unruly animals into a trailer.

It is more feasible to take the teaser to the mares than to move all the mares. A teasing area need not be elaborate; a section of solid fencing with a space for the teaser's head to reach through will be adequate for each pasture. Alleyways between pastures provide excellent avenues for feed distribution and for herding animals from one field to another. In addition, an alleyway separating colts from fillies will prevent injuries due to unsupervised teasing over the fence.

Wherever horses are kept there must be free access to fresh, clean water. Clear, clean running streams are ideal for pastures, but if a stream is not available, clean water troughs that have no snags or sharp corners and are accessible from all sides are needed. Perhaps one trough will be sufficient for a small pasture but large areas require several well-separated troughs to encourage even grazing. Unsanitary water dispensers will spread disease faster than any other facility. They must be inspected daily and thoroughly cleaned on a regular schedule, whether they appear to need scrubbing or not. Short sections of large-diameter concrete pipe stood on end, the lower end sealed with concrete

Water troughs should be accessible from all sides and should have no sharp corners.

and a drain outlet for cleaning provided at the bottom will serve well for watering. Large, round metal watering tanks are more costly but due to the smooth surfaces are easily maintained in a sanitary condition. Both arrangements provide use around the entire circumference. Metal watering tanks, relatively small and mounted on wheels, are commercially available for farms needing water in an area temporarily. Mosquito larvae and dead insects can be controlled by the introduction of mosquito fish (gambusia) into the water. Copper sulfate (bluestone) added to the water retards algae growth and, if used as specified in the package directions, will not harm the horses or the fish.

Pastures should not have ponds or other depressions that will collect run-off water, which the horses will drink. Rain water, melting snow run-off and irrigation water pick up the germs, worm eggs and intestinal parasites left by the horses' droppings; these will be ingested by horses drinking such water.

Pastures should have some field shelter, although in mild climates open pole structures with roofs may be adequate. These shelters provide weather protection and are necessary for supplemental feeding during rain. Iodized salt and minerals should be kept under cover to avoid deterioration from precipitation. The feed troughs, whether outside or

Hay feeders should be placed near the fence for ease of feeding. Enough room must be left to allow horses to move around the feeder to avoid the possibility of a timid animal being trapped by aggressive horses.

A creep feeder surrounded by a corral with an opening too small for the mare to pass through is an excellent way to supplement the growing foals.

under a roof, should allow for approach from all sides and be situated to avoid places where a horse may be cornered. The shelter should be built so that wind cannot blow rain into it. A single solid wall on the windward side may suffice but it should be noted that in many areas winds that accompany storms blow from a different direction.

Creep feeders are essential in the pastures where the mares and foals graze together, as is explained in Section 11.2. Foals will be more likely to use the creep feeders if they are located near where the mares congregate.

As mentioned in Section 2.6, gravel can be utilized beneficially in all traffic areas such as alleyways, gateways, shelters and feeding or watering spots; however, some farms do not gravel around watering troughs but prefer to allow the troughs to overflow in the summer to create a place where, in order to drink, the horses must stand in mud. This procedure can prevent brittle, cracked hooves during dry weather.

2.17 Establishing New Pastures. Pasture planning should be carefully considered even before the building layout is established. Ease of access to the pastures will be time- and labor-saving and a broad panoramic view of pastures from the main farm buildings will enable the manager and farm help to observe the grazing horses more easily.

Brush and undergrowth should be cleared from the pasture areas well in advance of planting. Such vegetation is not only low in feed value but also harbors ticks and other parasites detrimental to horses. Weeds should be removed, the low places should be filled and the high spots leveled. The flat areas should be graded to slope about three inches in 100 feet to provide enough drainage to avoid standing water and bogs. Proper drainage discourages the types of weeds that thrive on wet land, prevents the drowning of some desirable grass species and eliminates breeding areas for parasites and other harmful organisms. If time and circumstances will permit, it is excellent procedure to first plant an annual crop such as Sudangrass or grain hay. Such a crop is available for use the first year while the land fills are settling and weak growth areas are discovered. The extra cultivation also eliminates any weeds that survive the initial weed removal. At the end of the crop season special attention given to the problem areas disclosed will assure a much more satisfactory permanent pasture.

Expert assistance should be sought when planning pasture establishment. The cost of such assistance is easily repaid by increased pasture production and labor saved by more efficient cultivating and planting.

The basic factors of most concern in pasture development are the local climate, precipitation and the soil characteristics of the proposed pasture land. Nothing can be done to improve the weather conditions but often the soil can be treated to enhance plant productivity. A soil analysis should be made well in advance of planting. Commercial soil laboratories or the U.S. Department of Agriculture Extension Service can be contacted to conduct such tests and make recommendations. A soil expert will take soil samples from several places and will make a

Broodmares and foals do well on summer pastures of Sudangrass.

record of the samples identified as to the field location they represent. The subsequent analysis of each sample will indicate what type of soil is present, what type of renovation and fertilization is needed and what species of grasses or legumes will grow well in the soil.

The properties of the soil cannot be completely changed but the deficiencies can be greatly overcome by chemical treatment and fertilizing prior to planting. The manager should discuss the soil analysis with the soil expert and may also wish to consult the County Farm Advisor, the Commissioner of Agriculture and the local agricultural colleges or universities. These consultations will point out the necessary initial treatment and subsequent fertilization needed to establish and maintain a high level of forage production from the plants best suited for growth under the climatic conditions.

Sometimes soil analysis will indicate a problem of excess acidity, which can usually be corrected with an application of limestone; however, if time is short the use of "burnt" or hydrated lime will lower acidity much more rapidly. Occasionally the study may indicate that the soil contains an element in sufficient quantity but it may be chemically bound and unavailable to the plant. Sandy soil, as an example, may need more potash than the test indicates. Such a condition suggests that although the tests are extremely valuable in determining the type and ratio of fertilizer to use, actual crop experience is the final proof of soil treatment needed.

Growing crops consume nutrients and weather conditions leach them from the soil at unpredictable speeds. The soil of both new and old pastures should be retested each year or two if maximum production is to be maintained. While perhaps a bother, soil testing takes the guessing out of growing pasture and helps avoid the cost of unnecessary fertilizer.

Plant species best suited to the climate, precipitation and soil conditions are the most dependable. Knowledge of plants favorable for the area can be gained by visiting neighboring farms and by examination of the foliage growing in uncultivated areas such as ditches and along the roadsides. These areas give a good idea of the climax grasses in the area. Government agencies and agricultural colleges are excellent sources of information and should be consulted before the pasture species are selected. Except for the bearded grasses such as bearded barley and mouse barley, which are injurious to eyes and gums, plants that do well in the area should be used heavily.

Regardless of the advantage one species may have over others, complete reliance upon a single variety is a mistake. Since weather conditions do not remain constant, a mixture of varieties will assure pasture in the event of unusually dry or wet seasons. If only one species is planted and the weather is not favorable for it, the chances are that the ranch will have very poor grazing that year. It should also be noted that grasses do not all have the same growing season; consideration should be given to varieties that will provide pasture for the entire year. Summer grasses and winter grasses suited to the local climate and soil if seeded in the same field will ordinarily provide continuous year-around pasture. Some areas find that summer varieties such as needlegrass or rescue grass combine well with winter rye or winter clover to assure year-long grazing. Although some species of Sudangrass may present some danger of prussic acid poisoning if affected with drought or frostbite, and has been criticized as causing cystitis syndrome, Sudangrass is a popular summer grass. Many recently developed strains, such as Piper Sudangrass, are relatively free from prussic acid and are considered quite safe for grazing. The local Farm Advisor should be consulted regarding its use in the area.

Soil treatment must be done in advance of seeding, and fertilizer worked or watered into the soil to assure best results. Inoculation with a nitrogen-fixing bacterium is essential when legumes are to be planted. For most grasses the seeding should be very shallow and well concentrated for maximum forage production. Where the ground is easily worked, seeding by drill is the most economical and effective; however, broadcast seeding is not uncommon. Pasture land can be effectively seeded by helicopter, which may be the preferred method for reseeding if the terrain is too hilly or rocky for ground machinery, but this method will require more seed. Almost all crop-dusting firms furnish helicopter seeding service. These firms can be contacted through the local Farm Advisor.

Cover crops of grain hay are often planted with new pasture because the grain comes up sooner and protects the young pasture from sunburn. By the time the hay is harvested the young grass is well established, thus the farm gains not only new pasture but also an additional supply of hay.

Proper time for planting depends entirely upon the type of plants to be raised and the climate of the area. Summer pastures should be planted in the spring and winter pastures in the fall. Grazing of new pastures planted in the spring should be delayed until the latter part of summer when the seedlings will have a very strong start. The pasture crops reach their peak of production in the spring and can be heavily grazed unless they are first-year pastures. After the middle of summer, unless the pastures are irrigated, production will be declining and overgrazing becomes a greater problem unless carefully watched and avoided. By rotating pastures, idle fields will be able to recover before the desirable species are grazed out.

2.18 Pasture Maintenance.

a) *Irrigation.* Pasture plants must not be waterlogged, but must have an adequate supply of moisture. When they are not provided with natural rainfall, irrigation is needed. The soil should be tested occasionally for moisture content and depth since retention differs greatly between soils having different textures. Heavy irrigating at longer intervals

Sprinker irrigation systems are desirable, because they require less water and do not cause the erosion that most methods of flood irrigation do. They can be transferred from pasture to pasture. Photography by Mary Ann Czermak.

develops much better root systems and hardier pasture than more frequent light watering. A portable sprinkling system with rain-bird type of sprinkler heads is preferred by most ranches. The horses should be kept off recently irrigated fields until the ground has become firm enough to support their weight without indenting deeply, otherwise the hoof prints when dry create dangerous holes and much of the grass is trampled into the mud or ripped out by the roots as the animals graze.

b) *Fertilizing*. Existing pasture soil continuously undergoes loss of nutrients through use by the plants and leaching by water. The lost nutrients must be replaced if the quality of production is to be maintained at acceptable levels. Refertilization is more beneficial if done in the early spring, but most areas require one more application later in the year. Once a year will not be sufficient for strong plant growth. As stated before, intermittent soil analysis is a reliable means of spotting nutrient deficiencies, which must be built up to the proper level. Most fields composed primarily of grasses will need a complete fertilizer, but those having one third evenly distributed legumes can do well without added nitrogen. Well-rooted legumes fix nitrogen into the soil, making it available to all plants in the pasture; therefore, less expensive nonnitrogen fertilizers can be used for such crops. Phosphate should be applied yearly, preferably in the fall. Phosphate sweetens the grass so that the horses will eat foliage mildly tainted by urine and manure. Stall litter is high in nitrogen and organic matter beneficial to the soil and should not be overlooked as a fertilizer. The litter, however, when spread in the field, must be free from worm eggs, larvae and disease-causing organisms, and time should be allowed for the sun and soil microorganisms to destroy the harmful content before grazing is permitted.

c) *Clipping*. Horses tend to avoid the high, coarse grasses in preference for shorter, more tender grass. Better use will be made of the pastures if they are clipped or mowed once or twice annually before the stems become so coarse and stiff that a field of stubble will remain, which discourages grazing. Clipping should be done before the top foliage becomes so thick that it smothers the under plants. Some crops may be cut as short as three inches but most should be left at a height of eight or nine inches. Clipping is a means of weed control, will increase plant growth, provide better maintenance of legumes, assure more uniform grazing and add to the beautification of the fields. Most ranches prefer the rotary mower to the sickle mower because rotary mowers are easily maintained, are faster and leave about the right amount of mulch.

d) *Weed Control*. Weeds are a constant threat to good pasture and unless kept under control may drastically reduce productivity. Clipping or mowing if done at the right time prevents the weeds from reseeding and gives the crop an opportunity to choke out most of the remaining weeds, provided the crop is strong and free from over-grazing. Cutting for weed control should be conducted prior to budding of the perennial weeds and before the annual weeds have formed seeds. If either type of plant is cut near maturity the weed seed will continue

This permanent pasture is in process of being cultivated for replanting.

to mature in the pod and reseed the soil. If weeds gain a strong hold it may be necessary to treat the pasture with herbicides that are not harmful to the horses or the crop. A herbicide known as 2,4-D is commonly used to control many of the broadleaf weeds. An application of 2,4-D at the rate of a half pound per acre is usually sufficient and will not retard or damage most pasture crops. Alfalfa and clover are rather broadleafed and may be damaged by herbicides. Spraying in the early winter when these crops are dormant will control late weeds. Poison ivy, brush and woody undergrowth can generally be controlled by the use of 2,4-D and Silvex. Many such herbicides are marketed and should be investigated. The best source of information is the local Farm Advisor, who can make recommendations for the specific problem and will be able to furnish information regarding regulations governing the use of such chemicals.

e) *Dragging or Disking.* Pastures develop dangerous holes, rough ground and areas of concentrated manure which should receive attention at the end of the grazing period. By dragging or disking the area these conditions can be greatly improved. If a problem of holes or roughness is present, the most effective time for the chore is while the soil is damp following the first rain in the fall when grass production is at its lowest. At this time the ground offers less resistance and the holes will be filled more easily. Dragging spreads the droppings for better fertilization and breaks the dung balls so that the sun can destroy the parasites and bacteria. Chain link drags or tooth harrows are the most popular tools. When disks are used they should not be set for deep penetration to avoid killing the roots of permanent pasture plants or causing erosion.

2.19 Pasture Rodent and Snake Control. A good program of rodent control must be maintained. Field rodents that cause a great deal of damage are pocket gophers, ground squirrels, tree squirrels, mice, kangaroo rats, muskrats and rabbits. In large numbers any rodent is of concern to the ranch, but the ground squirrel causes by far the most trouble for the horse farm. The squirrel not only digs dangerous holes and consumes forage but also carries dangerous diseases such as bubonic plague, tularemia and Rocky Mountain spotted fever.

The principle means of control are: poison baits and poisonous gases, trapping, shooting, and encouragement of natural predators such as hawks and coyotes. The County Agricultural Commissioner will be able to provide information on county and state rodent control programs. In some counties the commissioner's office will take over the control program without charge. Poisonous baits are the most popular method of control and can usually be obtained through the agricultural commissioner's office. When using poisonous baits the utmost care must be taken to avoid harming livestock, handlers and pets.

Poisonous snakes feed on rodents so a good rodent-control program will also reduce the number of snakes. King snakes and road runners, natural predators of the rattlesnake, can be introduced and encouraged in the area. Grazing pigs in locations infested with pit vipers will effectively reduce their population, because pigs kill and eat poisonous snakes.

2.20 Poisonous Plants. Pastures should be observed carefully for poisonous plants. The County Farm Advisor can provide information on which types of poisonous plants are in the area and which methods of control are best. Some of the more common plants poisonous to horses are:

1. Arrowgrass (*Triglochin maritima* L.)
2. Black henbane (*Hyoscyamus niger* L.)
3. Black locust (*Robinia pseudo-acacia*)
4. Blue-green algae (*Cyanophyta*) (blooms of *Nostoc, Anabaena, Anacystis* and *Aphanizomenon*)
5. Boxtree (*Buxus sempervirens*)
6. Bracken fern (*Pteridium aquilina*)
7. Buckeye (*Esculus*)
8. Buttercup (*Ranunculus* spp.)
9. Castor bean (*Ricinus communis*)
10. Cheeseweed or Mallow (*Malva parviflora* L.)
11. Choke-cherry (*Prunus virginiana*)
12. Climbing bittersweet (*Celastrus scandens*)
13. Common rabbit brush (*Chrysothamus nauseosus*)
14. Coyote tobacco (*Nicotiana attenuata*)
15. Curly dock (*Rumex crispus*)

The castor-bean plant can prove deadly to livestock as well as to humans.

Horsetail is a common plant toxic to horses.

16. Death camas (*Zygadenus* spp.)
17. Desert tobacco (*Nicotiana trigonella*)
18. Ergot (*Claviceps purpurea*) (fungus of grains and meadow grasses)
19. Fiddleneck (*Amsinckia intermedia*) (Tarweed spp.)
20. Fitweed (*Corydalis caseana*)
21. Flax (*Linum usitatissimum* L.) (Linseed. Under certain conditions)
22. Golden Corydalis (*Corydalis aurea*)
23. Ground ivy or creeping ivy (*Nepeta hederacea*)
24. Horse nettle (*Solanum carolinense* L.)
25. Horsetail (*Equisetum*)
26. Indian hemp (*Apocynum cannabinum*)
27. Jessamine (*Gelsemium sempervirens*)
28. Jimson-weed (*Datura Stramonium*)
29. Johnswort (*Hypericum perforatum*) (St. Johnswort)
30. Knotweed (*Polygonum avicular* L.)
31. Larkspur (*Delphinium* spp.)
32. Locoweed or rattleweed (*Astragalus* spp.)
33. Lupines (*Lupinus* spp.)
34. Milkweed (*Asclepias galioides*)
35. Mushrooms (of the genus *Amanita* spp.)
36. Nightshade (*Solanum* spp.)

Loco weed or rattle weed is addictive and toxic to livestock.

The oleander is a popular but poisonous ornamental shrub often used to line driveways or fence lines. Other ornamental plants that cause poisoning when eaten by horses are aconite, castor-bean, daphne, English ivy, foxglove, larkspur, lily-of-the-valley, matrimony-vine, meadow saffron, and narcissus.

37. Oleander (*Nerium oleander*)
38. Penny-cress or Fanweed (*Thlaspi arvense* L.)
39. Pigweed (*Amaranthus retroflexus*)
40. Plume tree (*Cercocarpus montanus*)
41. Ragweed (*Ambrosia artemisiifolia*)
42. Ragwort (*Senecio* spp.)
43. Rough pea (*Lathyrus hirsutus*)
44. Russian knapweed (*Centaurea repens*)
45. Saint Johnswort or Tipton weed or Goatweed (*Hypericum perforatum*)
46. Sneezeweed (*Helenium hoopesii*)
47. Sour dock (*Rumex acetosella*)
48. Star of Behlehem (*Ornithogalum*)
49. Star thistle or knapweed (*Centaurea* spp.)
50. Stink grass (*Eragrostis cilianensis*)
51. Tansy (*Tanacetum vulgare*)

52. Tarweed (*Amsinckia intermedia*)
53. Tree tobacco (*Nicotiana glauca*)
54. Water hemlock (*Cicuta* spp.)
55. Yew (*Taxus*)

Most poisoning occurs in the spring when the grass is still short or in the late summer when the grass is dry or parched. Usually horses will not eat poisonous plants unless they are very hungry and no other feed is available. Overgrazed pastures become propagating areas for poisonous plants because the animals avoid them and they are left to seed and multiply.

It is important to note that the mature seeds of several species of weeds may cause poisoning if harvested with hay or grazed with the pasture crop. Small amounts of such seeds will not cause poisoning but if the pasture is allowed to become over-infested with them, the concentration of toxic substances is increased and poisoning is much more likely to occur. Some of the weeds that produce poisonous seeds are corn cockle, false flax, fan-weed, flax, field peppergrass, hare's-ear mustard, lupines, mustards, rattle-box, smartweeds, tarweeds, thornapple and tumble mustard.

In some areas the county authorities may insist that property owners control certain poisonous plants. If the owner is negligent the county will remove the plants at the owner's expense and possibly fine the offender.

3
RESTRAINTS

(SYNOPSIS)

This chapter contains practical methods of restraint and immobilization.

The following synopsis by section numbers is provided as an aid for quick reference to specific subjects.

Section 3.1   General Statement
Section 3.2   Tying (precautions; equipment; cross tying)
Section 3.3   Tying up legs (foreleg; hind leg)
Section 3.4   Hobbles (use; training; equipment; precautions)
Section 3.5   Twitching (twitch description; use)
Section 3.6   Squeezing (use; equipment)
Section 3.7   Throwing; Casting (total immobilization; tranquilizing; methods; equipment)
Section 3.8   Improvised restraint (without special equipment)

3.1 General Statement. Many instances arise in which the horse must be restrained or immobilized. The most common occasions requiring restraint are times when the horse undergoes routine or specialized care, medication or examination.

The degree to which the horse must be restrained and the temperament of the animal determine which method should be used. This chapter deals with frequently used methods of tying, hobbling, twitching, squeezing and throwing.

3.2 Tying. The most common method of tying a horse is through the use of a halter and lead rope securely tied with a slipknot to a very strong and immovable object. The rope should be short enough so it is taut when the horse's muzzle is lowered to the ground. A longer rope poses the danger of entanglement, which can lead to injuries. The more unruly the animal is, the shorter it must be tied. The tie should be above wither height to reduce the danger of dislocating the neck should the horse pull back. A turn or two of the rope around the anchoring object before tying the slipknot will help prevent the knot from being pulled tight and made difficult to loosen. The tie should always be to a strong, immovable object such as a post, tree or extremely heavy object. A weak anchor may break or come loose should the horse pull back. Such an accident is the forerunner of serious injury. The horse is liable to go over backward when suddenly released, but whether he falls or not the mishap is likely to frighten him, resulting in a wild runaway. A dragging post or a flying object at the end of the rope striking the running animal will add to his fright.

If the anchor is not firmly embedded in the ground, it should be much heavier than the horse, and if mounted upon wheels it should be adequately blocked with the brakes set. The ordinary horse trailer is not such an immovable object unless hitched to a truck. A horse has the weight and strength with which to move a horse trailer even though the wheels are blocked. Horses have been known to dislodge horse trailers resting on sloping ground; severe injury to the horse and great damage to the trailer have resulted from such incidents. Trailers are mentioned, not because they are the only improper tie anchors, but because many accidents involving them have occurred due to the convenience of tying to an unhitched trailer. If an adequate anchor is unavailable, the tie should be such that the rope will come loose under light pressure.

A horse expected to struggle against the tie should not be held by the halter alone unless it is specifically designed for that purpose. The safest and most practical tie is by the use of a one-inch cotton rope about 25 feet long, knotted with a bowline around the heart girth, run between the forelegs, through the halter ring and tied with a slipknot after a wrap or two around the anchor. Pullbacks can also be secured by a rope tied in a bowline knot around the throat latch and passed through the halter ring; however, if the horse is tied in this manner or with the halter alone, a violent struggle may damage its neck to the extent of dislocation or breaking. For this reason it is always a safer procedure to tie horses expected to fight by the heart-girth method.

Whatever tie is used for a pullback, the rope must be tied high enough on the anchor so that it will be above the top of the withers, otherwise there is danger that the vertebral column in front of the withers may be greatly damaged by a downward pull.

A more restrictive tie when needed can be made by using two ropes attached to a very strong halter or by putting two halters on the horse with a rope attached to each. In either case one rope should be anchored to the right and the other to the left of the animal and they should be pulled tight, allowing no slack. This method of restraint is known as "cross-tying."

Proper tying methods should be used under *all* circumstances. Even gentle horses are sometimes frightened and if tied improperly can severely injure themselves or others.

1. The manger tie or slip knot

2. Manger tie or slip knot

3. Manger tie or slip knot

4. Manger tie or slip knot

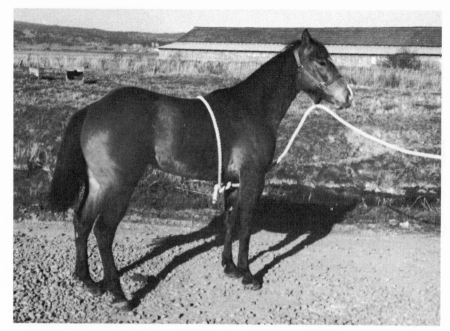

A horse expected to fight while tied should be secured with a strong rope tied around the heart girth, run through the forelegs and out through the halter ring.

1. The bowline knot

2. The bowline knot

3. The bowline knot

4. The bowline knot

Proper method of cross-tying.

3.3 Tying up Legs.

a) *Tying up Foreleg.* Foreleg treatment, if minor in nature, can usually be accomplished by holding up or tying up the opposite foreleg so the treatment can be administered. This procedure forces the body weight upon one foreleg so that the animal is less likely to move or strike out if the application is painful or irritating.

The foreleg to be tied up can be lifted, doubled tightly to the fore-arm and fastened to it by a strap or short rope. The strap should be placed far back from the knee and should be removed as soon as the restraint is no longer needed. A figure-eight leather hobble is excellent for tying up a foreleg. When hobbles are used the cross should be be-tween forearm and pastern. While a tied foreleg does reduce the chance of striking or rearing, either action is still quite possible; this capability of the horse must not be overlooked.

b) *Tying up Hind Leg* (sideline or Scotch hobble). A kicking horse or one that is likely to kick during medication or examination is very dangerous. Such a horse is unsafe to treat unless a chute is used or a hind leg is tied up in such a manner that it cannot kick easily. When one hind leg is tied high, there is less chance that the horse will kick with the one leg carrying the weight, but kicking with the free leg and rearing are nevertheless still possible. These possibilities must never be ignored. Tying up the hind leg of a horse that will permit the use of a half-hobble is quite simple. The animal is first tied to a strong post or tree. A one-inch cotton rope (braided preferred over twisted) about 25 feet in length is placed around the neck and tied in a bowline knot at the shoulder on the side of the foot to be lifted. The rope is then threaded through the half-hobble ring, returned to and wrapped once around the anchor rope encircling the neck. By pulling on the leg rope, the foot can be raised to the desired height. The higher the foot is raised the more the animal is restricted; however, falling is more likely to occur when a foot is tied so high that it cannot touch the ground.

A horse that will not permit the half-hobble to be fastened to its hind leg is a much bigger problem. Tying the leg of such a horse is consid-erably more complicated and dangerous. It is done by the same type of rope anchored and tied around the neck as previously described. The handler must then take the rope at full length to a point directly behind the animal and let the rope lie slack on the ground. The horse can then be shoved over or otherwise encouraged to step over the rope with one leg, at which time rope slack is taken out with a jerk, which raises the rope high between the hind legs. Kicking usually occurs at this point; the handler should be positioned out of the danger zone. The handler then returns the rope, on the side with the bowline knot, to the shoul-der, while keeping the rope above the hock. He then runs the rope through the neck loop, taking most of the slack out of it. The rope is allowed to slip down over the hock until it engages the pastern joint. In this position the rope can be pulled to raise the foot. This action will usually cause a struggle and unless the rope is held very firmly it may slip and leave a rope burn. When the foot is at the desired height, the rope is then returned to the foot, where a wrap and reverse wrap

Hobble used to tie up a front leg.

are placed around the pastern. The excess rope is wrapped around the ropes leading back to the bowline knot, so that the animal cannot kick free, and is tied off on the anchor rope around the neck. The completed tie is stout and will not slip.

Both methods make the animal unsteady on his feet and frequently result in falls. For this reason a horse should be scotch hobbled only in an area with a soft footing and free from obstructions. When a fall is on soft ground, injuries rarely occur; the fallen horse should be left alone and allowed to regain its feet. The handler should always have a sharp knife with him for cutting the rope in an emergency.

Tying up a hind leg without the use of a half-hobble.

1. A 1-inch cotton rope 25 feet long is placed around the neck and tied in a bowline knot at the shoulder on the side of the foot to be lifted. The rope is placed so that the horse can be encouraged to step over it. The rope is then collected, raising it between the hind legs (kicking is likely at this point).

2. The rope is returned to the bowline knot at the shoulder while the rope remains above the hock. The rope is run through the neck loop, taking out most of the slack, then is allowed to slip down over the hock until it engages the pastern joint. The rope is then pulled to raise the foot.

3. The rope is returned to the foot and a wrap . . .

4. (completed first wrap)

5. and a reverse wrap are placed around the foot . . .

6. and tightened at the pastern joint.

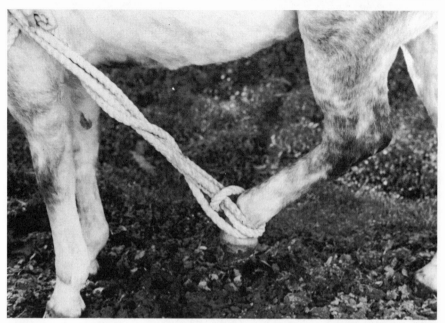

7. The excess rope is wrapped around the ropes leading back to the bowline knot . . .

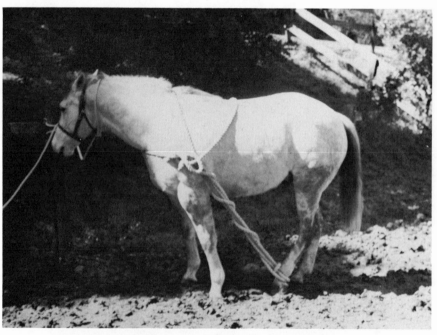

8. and is tied off on the anchor rope around the neck.

Hind leg tied up (side-lined). If the animal fights, the rope cannot slip and burn it.

3.4 Hobbles. Hobbling is a very effective method of restraint when used properly. No special training is needed for an animal that is hobbled to prevent its pawing or moving about when tied. However, when hobbles are used in place of tying, the animal must be taught to obey them. Any horse can quickly learn the knack of moving the hobbled feet together as one. By running in a series of jumps, a hobbled animal can attain considerable speed. The animal must be convinced that this trick should not be tried. To accomplish this the hobbles must provide a stable form of restraint during the training period. Hobble-breaking should be done where the ground is soft and the area is free from rocks, trees, fences and other structures and obstructions. The horse is probably going to fall at least once, perhaps many times, before the education is complete. A "dead man," such as a heavy log, buried fairly deeply in the ground, equipped with a strong three-foot length of chain, will serve as the anchor without endangering the animal when it falls. The chain should contain a swivel snap or link to prevent kinking and knotting, which otherwise is sure to occur as the horse maneuvers. When the chain is attached to the hobbles on the horse the scene is set for the training. After a few falls the disadvantage of moving when hobbled is understood and most horses will respect unanchored hobbles.

Hobble-breaking can also be done without the use of a dead-man anchor. One method requires a good rope-horse and a rider expert in handling ropes. The rope can be tied into the hobbles and dallied to the

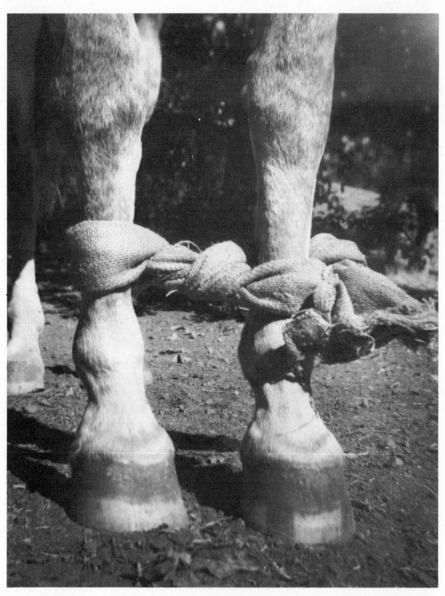

Sack hobbles are safe and inexpensive. All hobbles should be used in the cannon area of the forelegs. Horses hobbled by the pastern can travel easily and are more likely to dislocate one of the joints in the pastern area.

saddle horn. When the hobbled horse begins to plunge about, the slack can be taken out of the rope, pulling the hobbled feet toward the rear, which usually results in a fall and discourages further fighting. Another method is to put a half-hobble on a hind leg, then tie a stout rope between the half-hobble and the middle of the hobbles on the front feet.

Hobbles must be attached to the horse's cannon-bone area, never to the pastern. A pastern attachment allows much more freedom of movement and greatly increases the danger of serious injury to the several small bones and joints in the pastern area, which can be dislocated by a violent struggle.

Sack hobbles are very popular; they are inexpensive, easy to make from a feed sack, are soft and are unlikely to injure the horse.

Obviously there is an element of risk involved in hobble-breaking. For this reason the manager might decide not to hobble-break a valuable animal but will use another less dangerous method of restraint.

3.5 Twitching. A twitch is an instrument having a smooth handle, usually made of wood, 12 to 24 inches in length, to which is attached a small chain, rope or leather loop large enough to loosely encircle the

A half-hobble made from a burlap sack will prevent rope burns when a hind foot must be tied up.

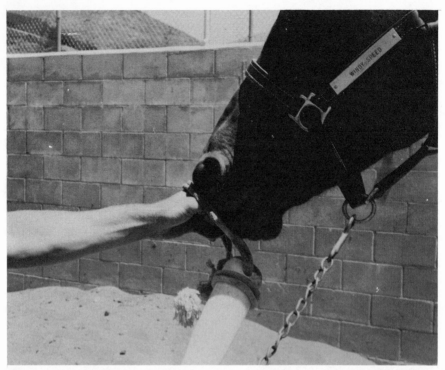

When applying the twitch, the handler stands at the side of the animal, reaches through the loop and grasps the upper lip. The handle is then twisted to firmly pinch the upper lip and inflict enough pain to keep the animal still for treatment.

horse's upper lip. The loop can be closed and drawn tight by twisting the twitch handle. Proper use of the twitch will not injure the animal but will inflict sufficient pain and discomfort to the area involved to draw attention from uncomfortable medication or treatment. The instrument can be used on the ear but usually causes ear shyness, which makes the animal very difficult to handle. If a twitch is used on the ear, the area should be rubbed after the instrument is removed, to help the animal overcome its fear.

The twitch can be quite easily applied to most horses. The handler, while positioned well to the side to avoid being struck, slips his hand through the loop and with the same hand grasps the upper lip between thumb and forefingers, and pulls it well forward. He then slides the loop off his wrist and onto the extended upper lip. With the other hand he twists the twitch handle until the slack is taken out of the loop, which tightens firmly around the upper lip. Throughout treatment the handler must attend the twitch and be prepared to slightly jiggle it or tighten it just before and during a treatment that will cause pain. Twitches that clamp onto the upper lip and need not be held are also available and are convenient when working alone.

The instrument works well when needed but causes severe discomfort and should not be used as a routine aid in treatment.

A properly applied twitch takes the animal's mind off unpleasant treatment but does not cause injury. The handler should be positioned well to the side, because some animals will strike when twitched.

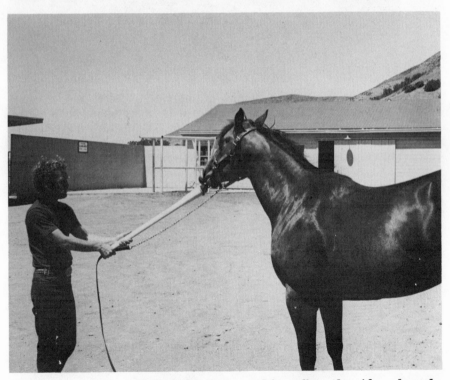

The handler has control of this mare and is well to the side, where he is in little danger should the mare suddenly strike.

This twitch made of light-weight aluminum is quick and easy to apply. The handles may be held together with a two-ended snap at the ends, which allows the handler to work alone. The disadvantage of this twitch over the long-handled loop type is that the handler must stand relatively close to the animal while holding the twitch.

3.6 Squeezing. Animals requiring treatment for sore shoulders or necks and those needing rectal or vaginal examination or medication can be immobilized by squeezing. If the ranch is not equipped with stocks for the purpose of immobilizing the horse in a standing position, it is sometimes possible to use a teasing chute or a horse trailer, but if none of these are available, it is possible to improvise by the use of a strong gate, sturdy fence or wall. When a gate is used it should be brought back snugly against the horse's side, pressing him against the fence with his head facing into the angle formed at the hinges. Facing the hinges is best because a horse will usually move forward rather than backward. Forward motion into the decreasing width of the angle will

A good set of stocks is invaluable for immobilizing animals so that they can be safely examined or medicated.

tend to wedge the animal at the shoulders. Once the horse is in position, the gate and the fence are tied together behind it with a stout rope. If a straight fence or wall is used, the horse can be secured tightly by a strong pole of sufficient length so that when in place it protrudes well in front of the shoulder and at least three feet past the tail. The front end is tied to the fence or wall, then the pole is closed tightly against the shoulder, side and stifle and is securely tied behind the animal. When in place the pole should be at shoulder and stifle height so there is no danger that the animal will jump over it. Improvisation is a poor substitute if the animal to be squeezed is likely to put up much of a struggle.

3.7 Throwing; Casting. When it becomes necessary to totally immobilize a horse, the animal should be tranquilized under the direction of a veterinarian. If the horse is heavily sedated, it may sometimes be downed simply by walking it in a circle; but if it has had only mild sedation, throwing an animal as large and strong as a full-grown horse is not an easy task. Throwing should be undertaken only in an area where the ground or other surface is very soft. Regardless of how the horse is taken down, its front and hind feet must be tied up or together immediately after it is on its side. As the horse struggles to rise its feet are extremely dangerous; the handler should always work from the side where the back will be when the horse is down. Once the animal is on the ground the handler should quickly place his own weight upon the horse's neck near the poll and twist the head up so the animal cannot rise. As soon as the horse is down to stay, a soft cushion should be placed under the head to protect against injuries and to keep dirt and other foreign matter from getting into the eye. All methods of throwing involve a means of destroying balance and impeding movement to regain balance. Following are a few methods used when full sedation is deemed unwise.

a) *Throwing by drawing the feet together.* Full hobbles can be placed on both front and hind legs, then a long rope anchored to the rear hobble and passed between the forelegs around the front hobble. The rope can then be pulled with force enough to work the forelegs and hind legs so close together that the horse will fall as he struggles. The same theory can be used with a half-hobble on each of the four legs. The rope is anchored to one half-hobble and passed through the rings of the remaining three. As the horse moves, his feet are pulled together until he loses his balance and falls.

b) *Throwing by tying up the hind legs.* A horse that kicks when the hind legs are handled is the most difficult. The procedure for this animal requires tying up first one hind leg by the method described in Section 3.3(b) and then pulling up the other hind leg. Of course, when the remaining hind leg is pulled forward, the horse must sit on its rump. From that position a strong pull on the lead rope going from the halter over the back will twist the animal's head and cause it to topple over on its side. When this method is used, padding should be provided for the hocks to avoid injury.

c) *Throwing by tying up a foreleg.* Horses are too smart to be downed more than once or twice by what would otherwise be the simplest method, accomplished by tying up a foreleg as described in Section 3.3(a). The halter rope is then run over the back, crossing from the side of the firm leg to the opposite side where the handler is standing. The handler quickly pulls the rope with force, turning the horse away from him. The unsuspecting horse will go down on the knee of the tied leg and quick action can roll it on its side. The drawback here is that the horse soon learns the trick and refuses to respond to the turning and may hop around on the firm leg.

3.8 Improvised Restraint. Occasionally a situation will arise when the handler is totally without equipment for use in restraining or controlling a horse. Such occurrences are rare, since a belt or other article of clothing can be quickly fashioned into a neck rope, but sometimes the occasion will arise so fast that there is no time to improvise. When it occurs, the hands can be used to best advantage by grasping the bridge of the nose with one hand, holding the top of the neck behind the ears with the other, and turning the animal's head toward the handler.

When one attempts to throw a horse with one front leg tied up, the lead rope is run along the side opposite of the tied leg, then over the back and pulled to make the horse turn. If it falls the handler will then be in the proper position to hold the animal down. The lead should be handled in this way in all three throwing methods discussed.

Occasionally instances arise in which a horse must be caught and held by hand.

A handler assuming such control must always be aware of the danger he is in if rearing or striking should ensue. If an animal is extremely unruly, an ear can be twisted by the hand previously holding the neck. Once the animal is quieted, the ear should be released and rubbed gently so that the horse will not be ear shy in the future.

Old-timers used to bite the animal on the ear, inflicting enough pain so that it would hold still. This technique is still used by some in dire emergencies; however, it is very painful for the horse and should be avoided if possible. Some of the very high-strung animals bred today cannot tolerate the pain of a twisted or bitten ear and will panic beyond all control. These methods must be used with good judgment.

4
VICES

(SYNOPSIS)

This chapter contains a description of bad habits often encountered in horses and suggests corrective action. The following synopsis by section numbers is provided as an aid for quick reference to specific subjects.

4.1 Vices Generally. Horses are not vicious by nature but some do acquire habits and behavior vices that are dangerous to themselves and their handlers. They do not reason before acting; instead, most of their actions are trained responses or merely reactions to discomfort, dissatisfaction or fear. The reactions that we call vices can usually be traced to mishandling, boredom or excessive nervous energy, but some undoubtedly are due to ill health or nutritional deficiencies. These vices are much more common in stabled horses that suffer from boredom or lack of exercise than in those having the free run of the pasture.

Whenever possible the vices should be overcome. Many can be cured by providing a companion such as a goat, to help prevent loneliness and boredom. If the source is apparent and can be eliminated by change of environment or equipment, it should be; but if the vice is a habit, the horse must be taught that the vice is not worth the consequences. The punishment must be so closely connected with the offense that the horse will recognize it as directly related to his own action, and if possible the punishment should be present with each offense in order to maintain the mental association of punishment to vice until the fault has been corrected. Since the handler is in the presence of the animal for only a small part of the day, mechanical aids are used and should be used to prevent the vices or to impose consequences so that the offenses committed out of the handler's sight do not escape punishment. Some of the acceptable aids are mentioned and described in this chapter along with specific vices.

A vicious, inhuman beating is not a corrective measure. The infliction of pain in a greater degree than necessary to make the point understood is the act of a vindictive person possessed of an uncontrollable temper, one who should not be permitted access to animals.

With the common vices set out below are suggested courses of action designed and recognized by horsemen as effective remedies.

4.2 Biting and Nipping. Of all vices, the stallion's inclination to bite comes as close to a vicious trait as any known. A biting stallion is dangerous and is apt to cause a severe injury. His attempt should be met immediately with a blow forceful enough for him to understand that such action results in severe pain, but the punishment inflicted should not be done in a place or manner that may result in permanent damage. The handler should never be off guard around a stallion, since a stallion can never be presumed cured of biting.

Unlike stallions, geldings and mares may occasionally attempt to nip, as distinguished from bite, and are more enthusiastic about learning from a sharp stinging rap than they are about nipping.

4.3 Cloth Tearing. Some horses, when blanketed or bandaged, are not content until they have ripped off the offending material. A common smooth but strong stick can be anchored to the halter ring extended between the forelegs and attached to a surcingle encircling the heart girth as a means of restricting neck movement. A commercially produced brace known as a neck cradle has the same restrictive function. Some horsemen prefer to fashion a heavy leather bib of sufficient width and length for attachment to the halter under the horse's chin to keep its mouth from the offending object.

4.4 Cribbing, Windsucking. These vices frequently cause colic or indigestion. Both actions result in the swallowing of air. The cribber places his upper front teeth firmly against a sturdy object, such as the rim

A horse that tears bandages or blankets can be fitted with a neck cradle so that it cannot reach the offending object.

of the manger; thus anchored, he pulls his neck into a rigid arch, which brings his muzzle straight back toward his body, and he swallows air. The windsucker, more accomplished in the art, can arch his neck without the use of an anchor. The arching in either case opens the esophagus sufficiently to allow the passage of air into the stomach. Unless caught in the act, a windsucker is not detectable; but the cribber of long standing can be recognized by worn teeth; the upper incisors appear to have been beveled. Unless these offenders are isolated, other horses will be quick to copy the stunt. The habit can be effectively controlled by a commercial cribbing strap or by the use of a wide common strap snugly fitted around the neck at the throat latch; this prevents neck arching, but does not impede eating, drinking or breathing.

4.5 Eating Bedding. Greedy horses and those lacking necessary nutrients in their diets are the most prone to practice the vice of eating their straw bedding. The sweetness of oat straw bedding encourages the habit and hence should not be used. A large amount of coarse straw in the stomach can cause colic. The best cure is bedding the animal in shavings, sawdust or other straw substitute. If a substitute is not available, spraying the bedding with a solution of one part moth crystals to

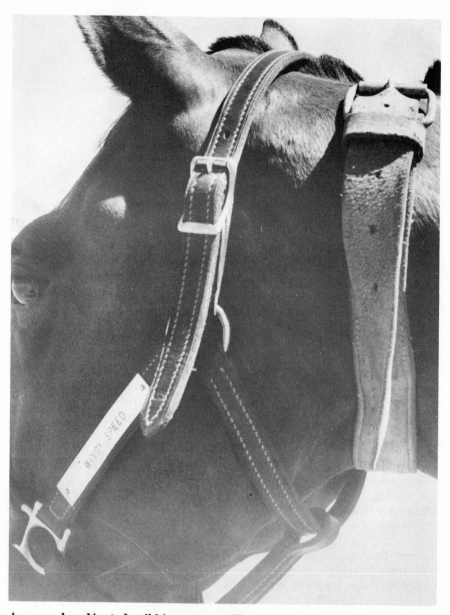

A properly adjusted cribbing strap will press on the larynx and prevent the swallowing of air.

five parts diesel or kerosene oil may stop this vice. Unless the horse is removed from the stall while the spray is being applied, his taste may become so saturated with the fumes that he cannot discern its taste in the straw. Fuel oil is flammable and the mixture may cause skin and eye irritation. For these reasons it should be used with caution.

4.6 Eating Dirt or Manure. This vice is very rarely practiced, but when in evidence it is an indication of nutritional deficiency or a symptom of colic. Punishment obviously is not a solution. If the vice persists even for a short while, its cause must be discovered and corrected; a veterinarian should be consulted. All horses at times do examine and smell droppings. This is not to be confused with eating. It is the distinct act of eating that should cause concern.

4.7 Eating Wood. Nearly all horses on occasion nibble or chew upon wood. Some, however, are confirmed wood-chewers or wood-eaters. Horsemen commonly refer to such horses as termites; but sometimes they are incorrectly referred to as cribbers. Cribbers are horses that suck wind—they are not wood chewers. Preferences seem to run in favor of redwood, with pine as second choice; but a confirmed wood-eater will accept any kind of wood available. Aside from the cost and labor involved in replacing boards, wood is not a very digestible substance; consumption of large quantities can cause colic. The habit may start from lack of something to do, or the want of bulk or nutrients in the diet. It has been noted that confined horses on complete pellet diets and those on a twice-daily feeding program are more prone to chew wood than pastured animals who can leisurely nibble throughout the day. Creosote and other commercially available bitter liquids can be applied to fence posts and stall wood as an effective deterrent. The animals should not be allowed access during the treatment of the area, or afterward until the treated material is thoroughly dry. The liquids if

Boards damaged by wood chewers can be very expensive to replace.

contacted while damp will cause severe skin irritation; and the liquid or its fumes, especially creosote, is irritating to the eyes and can cause serious eye damage. Pasturing a wood-eater after he has acquired the fault is not a cure. The termite may temporarily postpone his activity in preference to grazing, but will resume the habit when returned to the stall. He will chew or eat any wood not made distasteful. Wood-eating is another stunt that will be copied by other horses. Isolation from view is the rule for avoiding the development of a herd of termites.

4.8 Grain Bolting. Grain bolting is the vice of swallowing grain without chewing. Grain bolters are more prone to digestive disturbances because the grain is not adequately moistened with saliva for ease of passage through the digestive tract. In addition, a large amount of air may be swallowed with the grain. The food value is not completely extracted by the digestive tract because the indigestible hull is not cracked open by chewing to expose the nutritious center to the digestive juices. A few round stones the size of a baseball placed in the grain box presents enough of a problem to the horse to center his attention on the business of eating. Chopped hay, about an inch or so in length, mixed with the grain is a fairly sure control for grain bolting, since hay must be chewed and moistened before it can be swallowed.

4.9 Kicking. Kicking the stable wall is a vice sometimes found in the idle horse surrounded with boredom; mares are the principal offenders. These kicks break the silence and are not mere exercises intended to end in space; they are directed toward striking an object such as some part of the stall. However, aside from being irritating to the handler, the horse is in danger of damaging the foot. Turning the animal loose in a pasture and giving it plenty of exercise are the easiest solutions. Padding the target walls may reduce the irritating noise and perhaps lessen the danger of injury, but will not solve the problem. Often a burning barn light or a playing radio will provide a less boring atmosphere, but the habit usually cannot be broken unless the horse suffers from his act. A common practical way to associate punishment with kicking is by the use of a strap circling the leg above the hock with a short length of chain attached to it. When the horse kicks, he is immediately struck with the chain. The association of kick with pain will come through clearly and quickly. The chain device should not be used on a very high-strung horse because panic could result. Hobbling the hind legs together, while perhaps a slower remedy, can be used for the temperamental horse as well as others. A solution to the vice must be found, for fractures of the coffin bone are common in horses that kick stalls.

4.10 Pawing. Pawing with the forefeet while eating or anticipating food is a trait quite natural to the horse. No particular danger of injury

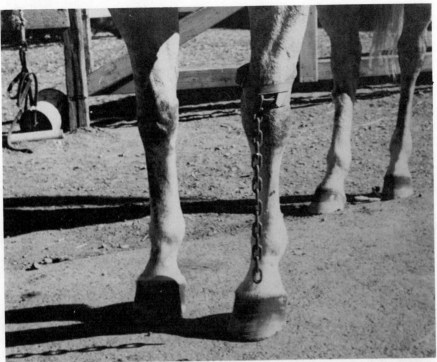

A length of chain attached to a strap above the knee can be effective against pawing. When this method is used the animal must be tied securely because some horses panic when they find they cannot escape the chain and may injure themselves or others.

is involved, but the habit will keep the barn help busy filling holes and adds nothing as a sales inducement. If the handler wishes to control pawing, he can do so by hobbling the forelegs together or by attaching a short length of chain to a strap fastened around a front leg above the knee. The horse will soon receive the obvious message. A very sensitive horse should not be subjected to the chain device.

An interesting point about pawing is that it is instinctive with horses when searching for food covered by snow. If the snow is not too deep the horse will survive when the bovine animal will starve. Cattle do not seem to have enough sense to use their feet, but they have been known to follow horses and forage upon the leavings of what the horses uncovered. Some authorities feel that the horse's trait of pawing indicates that it evolved in a cold climate.

4.11 Rearing and Striking.

a) *Instinctive Reaction.* Rearing and striking are instinctive reactions when the horse is cornered and frightened. These impulses cannot be trained out of a horse any more than a human can be trained to maintain an open eye when an object comes flying at it. Rearing and striking in self defense can hardly be described as vices. The best control over these natural but dangerous actions is through prevention of the incident. The handler should avoid situations that are likely to cause a horse to rear or strike from fear. When such a situation has arisen, the horse should be quickly moved away from the hazard; but, better than later safety, the situation should be anticipated and avoided. When controlling a frightened, rearing animal the handler should never be positioned in line with the forelegs. He should always move to the side at the first indication of rearing.

Horses getting acquainted for the first time have a tendency to squeal and strike at each other as they sniff noses. During occasions such as teasing or introducing a new horse to the herd, the handler should stand well to the side and out of the way of both animals' forelegs.

At times, treatment of the face, ears, eyes, mouth, teeth, neck or chest is required. Often the procedure causes pain or fear from which the animal cannot retreat. His natural inclination is to fight with his forelegs. The handler must be alert to the probability of such a defensive reaction and position himself as safely as the administration of the treatment will permit.

b) *Cultivated Action.* Some horses learn that rearing and striking have more uses than self preservation. These animals, fortunately few, are extremely dangerous and unpredictable. The trait is acquired by a few mares and geldings, and tends to linger in the gelding if castration occurs after stallion traits have developed. Animals who have this vice must be handled and treated as stallions, which are discussed in the following paragraph.

c) *Stallion Nature.* The stallion, as a means of asserting his masculinity, is predisposed to use his teeth and forelegs to alter a disagreeable situation. This paragraph is concerned with his strong tendency to rear and strike even when no provocation seems to be present. He has the same instinct to strike from fear that all other horses have, but concerns himself with many more things than self preservation. No assumption can ever be made that he will not strike, but he can be made more dependable by a competent, alert handler.

A stallion known for rearing or striking is more apt to exhibit this vice when led near other horses than at other times, but the possibility is always present. A bad actor should never be led without a metal nose band or strong chain attached to the halter, run over the bridge of the nose and secured to a stout leadrope long enough to permit the handler to move to a safe position without releasing his hold. The handler must be alert and if possible should jerk the leadrope sharply when a rear is indicated but before the horse rises off the ground. If the animal succeeds in leaving the ground, he must be permitted to complete the rear. A jerk while the horse is on the rise or at the top of the rear and in control of his strength may cause him to throw himself over backward or sidewise in an effort to avoid the nose pain. It might also provoke the horse to strike out at the handler. By the time the animal has reached the peak of the rear, the handler should be well to the side and prepared to apply severe additional punishment through jerking the leadrope as the animal starts down. The horse will then be solely concerned with landing and will not be able to strike. A jerk strong enough to turn the horse will impress him immeasurably. Following punishment, the horse should be asked to back for a few steps. If he responds, the episode is concluded; but if he still has rearing in mind he will attempt to rear again rather than back. Control requires split-second timing, an error in which can be a disaster to handler, animal or both. The handling of stallions is only for experts; the man and the animal are too valuable for experimentation.

4.12 Weaving. High-strung, nervous horses have an abundance of unused energy and when stabled for long periods may resort to oscillating motions while standing in place. This action is called *weaving.* Some horses add zest to the action by marking time with their feet in rhythm with the sway. Race horses and stallions are most likely to develop the habit. A nervous disorder is sometimes indicated, but usually weaving is just a way to pass time and is not particularly harmful. Stallions that have practiced the art for long periods have been known to suffer temporary infertility, and any horse may tire itself to the point of impairing usefulness.

Pasturing the offender may interrupt the practice, but once learned the horse will begin again when stabled.

Horses unacquainted with the activity will learn it by watching an offender. Isolation of the weaver is the sure way to avoid having an entire stable of weavers.

4.13 Crowding. Why do some horses delight in crowding people? Only the horses have the answer, but the people have the problem. This can be dangerous, not to the horses, but to the handlers. The large, heavy, strong animal can press a man against a wall with enough weight and might to cause serious injury, especially if the wall has the slightest protrusion.

The crowder's education in etiquette can be quickened by pressing, without jabbing, a rather sharp object, such as a hoof pick or spur, against its side during the crowding. It should soon learn that the consequences are too painful to make crowding fun. The handler should be equipped with the teaching instrument each time he enters the stall. Horses are very likely to cow-kick when punished in this way, so the handler should stand close to the animal's shoulder when applying painful pressure. If the habit is severe and the crowder is aggressive, its stall should be equipped with a rail or bar molding at stifle and shoulder height and far enough out from the wall to allow the handler to escape behind it.

4.14 Tail Rubbing. A horse that continually rubs its tail against an object is more likely to have a parasite than a vice. Examination is in order for signs of lice or fungus about the tail and for pinworms about the rectal area. If the areas are free from infestation, the tail, nevertheless, should be thoroughly shampooed. Dirty udders in mares and dirty sheaths in geldings and stallions will also cause tail rubbing. If tail rubbing is purely a habit, the problem can be remedied by the use of an electric hot wire along the face of its rubbing object or by a tailboard protruding from the stall wall that meets the buttocks at a point low enough to avoid the tail root and extending far enough from the wall to prevent wall contact.

If the offender is exceptionally high strung, the hot-wire device is likely to cause panic. The horse's temperament is of prime importance when considering the advisability of the hot wire.

4.15 Turning Tail

a) *Confined Horses.* Horses having the undignified habit of "turning tail" to the handler places the person in a dangerous position even though kicking is not the horse's usual idea. These animals are not necessarily anti-social; they have the desire not to be handled. Such an individualist can usually be broken of the habit in a small stout corral where it can be turned loose for the education. Every time the tail is turned to the trainer, one sharp sting of a whip across the rump should suggest that a different position is more desirable. After a moment or so, if no results are in evidence, another stinging rap will induce deeper thought. The horse will learn that tail away from a person is a much more comfortable position. This procedure repeated daily until the idea becomes permanent should solve the problem, but the sessions must be continued until the horse will face the trainer at all times even if the trainer walks in a circle around the animal. The trainer must be alert for a possible kick which, if it comes, requires further punishment. These sessions, however, should not be degraded into a form of beating; the horse's mental attitude can be changed for the worse unless the lesson is administered with reason. A horse trained always to face a person is sometimes referred to as *whip broke.*

b) *Unconfined Horses.* Pastured horses that refuse to come to the handler are time consuming and frustrating. This fault is similar to the vice of turning tail in close quarters, but sometimes develops into a sort of game. Even though closely resembling "turning tail," the fault cannot be trained out through the use of a whip. The size of the area makes it impossible for the trainer to enforce his demands. If punished, the offender will gallop away from the unpleasant situation. Horses that are whip broke are not so likely as other horses to have this pasture fault; however, the lesson learned in the small corral and in close proximity to the trainer may not be remembered when loose in the pasture. In any event, the horse does not associate facing the trainer with approaching the trainer from a distant point. This fault must be overcome by the association of some reward with response to the handler's call. One reward all horses understand is food. If the animal is called in the same way every time it is fed, the call and food will soon be associated and the horse will come when called. An animal that is difficult to catch should be haltered at each feeding. It will soon associate the halter with food rather than work and will no longer run away. The pastured horse should always be given a reward for coming, even if it is to be ridden.

Horses extremely difficult to catch may be educated by the use of a long dragging rope. The trick is to follow the horse; each time it is about to step on the rope, say "Whoa." It will soon develop the belief that the handler's command controls the impediment to its progress. A rope of one-inch cotton braid may be attached to the halter, but it is better to attach it with a half-hobble to a front foot where the horse will not readily see it and is more likely to step on it. A rewarding carrot or other tidbit offered each time the horse is caught will have a great influence over its desire to be caught.

4.16 Pull Back. Ordinarily the horses that set back and pull on the rope when tied were improperly broken to tie when young. The fault is particularly dangerous to the horse. The halter may break, the rope may snap, the post or other anchor may break or pull loose; any such mishap may cause a backward flip. A broken or dislodged anchor could result in a runaway, with the added danger of a fragment flying at the end of the rope.

A horse with this vice should never be tied directly to the halter. Halters, unless especially built for the purpose, cannot stand the strain.

The habit should be broken if possible. A one-inch cotton rope about 25 feet long should be tied in a bowline knot around the horse's body, encircling the withers and heart girth and its end passed between the forelegs and through the halter ring. The rope should then be wrapped a turn or two around a tree or exceedingly strong post and tied at a height a little above the withers. This will avert injury to the vertebral column of the neck just in front of the withers. A slipknot should be used at the anchor. A wrap or two around the anchor will prevent the knot from drawing too tightly or slipping down. The horse should be encouraged to fight, but not whipped or frightened. When the horse is actively working against the rope, it should not be encouraged, but when it is resting, sacking with a grain sack or a jacket will usually cause it to resume its efforts if it has not yet decided to give up. The animal should be left tied for several hours at a time.

The procedure should be a daily routine until the horse has forgotten the habit or until it becomes obvious that it is incurable. If the procedure is a failure, the horse should be hobble broke and never tied.

5
EQUINE HEALTH

(SYNOPSIS)

This chapter is a guide to basic sanitation, detection and treatment of health problems, prevention of and inoculation against serious equine diseases.

The following synopsis by section numbers is provided as an aid for quick reference to specific subjects.

5.1 Sanitation and Quarantine. Keeping the farm's horses in good health is perhaps the most important responsibility of the farm manager. The coming and going of large numbers of visiting mares increases the chances of introducing contagious diseases into the farm herd and greatly compounds the manager's responsibility. He must constantly be on the alert for signs of trouble and must maintain rigid, up-to-date health and sanitation programs. Such programs should consist of quarantine, daily inspection, routine tooth and hoof care, parasite control, and vaccination, with special emphasis upon rigid sanitation.

A complete health record should be kept up-to-date for each animal, so that general health status, hoof care, vaccinations, worming and treatments can be accurately reviewed at regular, short intervals. Such a health record, if well-kept and regularly referred to, is protection against overlooking necessary care and provides a medical history that is invaluable in diagnosing health problems as they arise.

The responsible manager will require that all horses visiting the farm or returning to it be quarantined for at least three weeks before being allowed on the main part of the ranch, and will quarantine all ranch horses that contact or are exposed to a communicable disease.

Good sanitation is the best prevention of most diseases and parasites; its practice should never be allowed to lapse.

A necessary part of disease control is disinfection of sick-animal quarters. If an animal contracts a highly contagious disease spread by indirect contact, such as distemper, all bedding and manure from that animal should be burned and the quarters thoroughly disinfected. Foaling stalls and breeding areas should be routinely disinfected. When a mare aborts, all aborted material should be sanitarily disposed of, the places it has lain should be thoroughly disinfected and samples of the aborted material sent to a laboratory for analysis to determine the cause of abortion. All breeding animals should be routinely inspected and cultured three or four days before breeding for signs of genital infection, which is certain to be spread by breeding.

Rats and mice must be controlled; they carry diseases, eat and foul the feed and destroy equipment. All grain containers and buildings should be made as rodent proof as possible. A few cats will control rats and mice if they are not too abundant; however, trapping and poisoning is required if the rodents are present in large numbers. Poisoning requires great care to avoid harm to livestock. The local Agricultural Extension Service can recommend the eradication program best suited to the particular situation.

Elimination of insect-breeding areas such as standing water and uncared-for manure piles is necessary to keep insect populations in check.

5.2 *Mosquitos*. Mosquitos are the vectors of many deadly blood-borne diseases such as sleeping sickness (*Encephalomyelitis*) and swamp fever (equine infectious anemia). They must be controlled as much as possible. Most successful control is achieved by elimination of breeding places. Swampy, boggy areas should be drained and filled with earth, and mosquito fish should be placed in the water troughs or other areas containing standing water. Most mosquito-abatement districts furnish mosquito-eating fish for troughs, ponds or irrigation canals as part of the mosquito-control programs.

5.3 *Screwworm Flies and Blow Flies*. Even though the national eradication program for screwworm has been very successful, the screwworm fly (*Callitroga americana*) is still a serious problem in the southern and southwestern parts of the United States, where the climate is most favorable for them. The female screwworm lays her eggs in open sores and occasionally in the moist secretion of the sheath of stallions and geldings. All open wounds and lacerations are potential egg-laying sites for the female screwworm fly. When the eggs hatch, the maggots produced literally eat the animal alive.

Blow flies lay their eggs primarily in decaying flesh, although several species also lay eggs in wounds. Unlike the screwworm fly, the blow fly maggot mainly eats dead tissue. The maggots cause continued irritation and enlargement of the wound. If left unattended the wound may become infected or so large it can kill or cripple the animal.

All animals should be examined daily for open sores that invite infestations. A suitable commercial repellent should be applied to all open wounds. Labels must be read carefully before using a repellent on the wound itself. Most pyrethrins are safe for use on the wound. Surgical operations such as castration should be postponed until the fly season is past and breeding should be scheduled so that the mare will foal in the spring ahead of the fly season. Large farms often maintain a screened fly-proof stall for horses recuperating from open injuries during the fly season.

Screwworm flies and blow flies are somewhat larger than the house fly and are blue-green or blue-black in color.

5.4 Bloodsucking Flies. Horse flies (*Tabanus sp.*), deer flies (*Crysops callida*), stable flies (*Stomoxys calcitrans*), horn flies and gnats are biting insects capable of transmitting blood-borne diseases such as anthrax, anaplasmosis, equine infectious anemia, encephalomyelitis, trypanosomiasis, tularemia and Corynebacterium pseudotuberculosis (pigeon fever). Both the horse fly and the deer fly go through an aquatic overwintering stage, therefore drainage of swampy areas is important for their control. The stable fly and horn fly lay eggs in decaying organic matter including manure and are therefore vectors of *Habronema microstoma*, the stomach worm larva responsible for summer sores. Gnats breed in running water such as streams and irrigation canals. Their breeding areas may be treated with pesticides to reduce the population of this irritating insect, which can debilitate horses by bothering them so much that they stop eating. Gnats often appear in such great numbers that the victim animal is in a black cloud of insects.

5.5 House Flies and Face Flies. House flies (*Musca domesticus*) and lesser house flies (*Fannia caniclaris*) ordinarily do not lay their eggs in open wounds, but they do feed on them and aggravate or infect the wounds, which may become persistent sores difficult to treat. In the summertime flies often irritate the animal's abdomen, causing sores that must be treated medically and protected from the flies. The female lays its eggs in decomposing organic matter wherever it can be found. Manure and soiled bedding are ideal places for nesting and should be sprayed heavily with a suitable repellent, insecticide or larvicide. Because the larvae feed on manure this fly is a mechanical vector of *Habronema* (summer sores). Animal carcasses are also breeding places and should be buried.

Most flies are now easily controlled by barn sprays applied after horses are removed and water troughs and feed containers are covered. Horses should not be returned to the stalls until the area is dry. If flies collect around the feed containers it is an indication that they are in need of cleaning. Fly bait may be spread about the barn or placed in shallow boxes covered with wide-mesh screen. Electric fly screens, fly traps, volatilizers or foggers are effective controls.

Animals may be dusted, sprayed or sponged with nontoxic insecticides. If an animal has sensitive skin, a water-dilution spray will be less irritating than oil-based solutions. No dusts, sprays or solutions should be used in or close to the eyes.

If extensive spraying of the farm is needed, it is advisable to use a large tank-type sprayer that can be mounted in the bed of a pick-up truck or similar vehicle and driven through the barns or pasture areas that are to be treated. A licensed exterminator may be contracted to come on a regular basis during the fly season. The reputable exterminator is thoroughly familiar with the chemicals he uses and is equipped with modern equipment. He is abreast of the new techniques of pest control and can provide protection that is effective for several weeks. Contracting the professional exterminator may be a money-saving practice for large ranches.

Flies winter in a dormant state either as pupa or as adults. Control is most effective and important in the beginning of the fly season to lessen propagation and at the end of the season to reduce the number continuing the cycle the following spring. Fly control should be on a continuing and diligent basis. The life cycle of various flies lasts from three days to three months and the female may lay from 10 to 1,000 eggs per batch several times during her life span. Each fly, unless controlled, will be responsible for thousands of additional flies during the season.

Face flies (*Musca autumnalis*) periodically invade various parts of the country and plague horses. This small, persistent and pesky fly lives on the fluids of the eye, the nose and any sore. It may spread contagious conjunctivitis, similar to pinkeye in humans. Treatment with antibiotic eye ointment may give relief to this condition. Face flies can cause such irritation to the animal that he may stop feeding, lose weight and become listless. Specific dusting powders or insecticides are effective in giving temporary relief but unfortunately these remedies cannot be applied directly to the eye where the face flies are particularly bothersome. Because it propagates in manure, the face fly is also a vector of *Habronema*. Dung beetles are introduced into some areas where the face fly is propagating, to infest dung heaps and feed on the larvae. Control by the use of nematodes (tiny parasitic worms) that lay eggs on the larvae of the face fly is in the experimental stage by the Department of Agriculture in hopes of preventing the insects from becoming a permanent problem.

5.6 Manure. Some ranches introduce red earth worms into manure compost piles to reduce the amount of manure requiring disposal and to provide fertilizer. The most convenient fly control is to place litter and manure in a compost pile or pit. The compost pile should be placed on a concrete floor to prevent contamination of the surrounding area and to retain the fertilizer nutrients. The pile should be covered to prevent flies from laying eggs in it, or treated with a suitable insecticide or fly repellent that will not harm livestock, alter the fertilizer value or affect worms or beetles that may have been introduced to aid in destroy-

The compost pile should be placed on cement and well away from the center of activity.

ing pests. The location of the manure pit should be far from the stables, convenient for pick-up but not close to the ranch population; it is unsightly, obnoxious and attracts flies. Many stables recycle manure and stall bedding by distributing it on idle pasture land for its fertilizing benefit and to aerate and condition heavy soil. The labor costs involved in use of the manure for fertilizing should be weighed against the cost of commercial fertilizers. All manure should be composted for at least two weeks to kill various parasite eggs and should not be spread where horses will be grazing for the following few weeks. Horse parasites usually do not affect other species of livestock so the manure can be used immediately and safely in pastures grazed by sheep or cattle. Manure can also be used to curtail erosion of slopes and banks, to reclaim unproductive areas or to "pad" rocky paths or trails.

5.7 Stalls and Paddocks. Any place where horses are confined, such as stalls or paddocks, should be kept clean. The alkaline content of manure is detrimental to hooves, and the ammonia in urine is irritating to the eyes and respiratory tract. A sprinkling of hydrated lime underneath the bedding where the horses will not come into direct contact with it will help keep the fumes and odor under control.

5.8 Stall Bedding and Cleaning. There are many different types of bedding suitable for horses. Kiln-dried wood shavings are generally considered most ideal for all bedding except for foaling mares. A bed of shavings is absorbent and the horse is not likely to eat it. Oak shavings, however, should be avoided as the tannic acid in it has a heating effect on the hooves. If shavings are used, large chunks of wood should be removed. While shavings are excellent, sawdust has serious drawbacks. Although it is very absorbent, it ferments rapidly when wet and produces a great deal of heat, which irritates the horse's skin. Maggots grow quickly in damp sawdust, eye irritations are more common because the small particles are easily picked up, and drains are often plugged with it. If sawdust is to be used, it should be from well-seasoned lumber to reduce heating and should be changed often.

Straw makes excellent bedding, especially for foaling mares. Rye straw has the longest stems and is toughest of all straws; however, it is expensive and sometimes hard to find. Wheat straw is often used but greedy horses are prone to consume it in large quantities. Barley straw is unsuitable; it contains awns that can severely irritate skin and damage eyes.

Unbaled straw is preferable for bedding; baling crushes the air space within the stems and usually breaks up the stem length. Straw bedding crossed layer by layer is less likely to separate and leave a bare space when the animal moves about.

Other material can be useful as bedding, such as dried leaves, peat moss or specially prepared bedding made from shredded dried sugar cane, corn stalks or cobs, peanut shells or other products that are absorbent but do not have sharp edges that could cause injury.

No matter what is used for bedding, it should be free from mold and dust and should have a clean appearance and odor.

Sand is very poor bedding; it is too chilly for cold, damp climates and it works into the cracks and crevices of hooves. If sand is used in the stall, it should not be sea sand, which contains salt. Horses are apt to eat it to get the salt. Sand colic, which can be fatal, will result.

Stalls should be cleaned at least once every 12 hours and more often, if possible, to prevent the horse from standing in damp, unhealthy surroundings. The bedding should be entirely removed and the stall allowed to dry at least once a week.

When cleaning the stall, all wet straw should be removed. Slightly soiled straw can be piled so it will air out and can be reused as the bottom layer of bedding, provided clean fresh straw is placed on top.

While the stall floor is being aired, the slightly soiled straw to be reused should be piled in the back of the stall where it will not be associated with the manger and the horse will not be likely to eat it. A very light sprinkling of hydrated lime on the bare floor will keep the odor under control, but it must be well covered with bedding to keep the animal's skin from direct contact with it.

5.9 Daily Animal Inspection and Grooming. Ideally, all horses should be closely inspected daily. Grooming is by far the best way to accomplish this. Unfortunately, the number of animals on a large farm often prohibits this method and daily grooming becomes limited to animals kept in stalls and small paddocks.

Good grooming is particularly important for the confined horse that is not receiving the circulatory stimulation of pasture activity.

Using the same grooming tools on many different animals is an unsanitary practice; they are notorious for transmitting diseases from horse to horse. Sponges are particularly offensive, for they are usually damp, have thousands of crevices that collect dirt and germs and are practically impossible to sterilize because heat will destroy the sponge. Chemicals that destroy germs remain in the sponge and can irritate the horse's skin and eyes. Soft rags can be used in place of the sponge. Rags are inexpensive, readily available and can be easily disinfected by boiling or even by spreading them out in the sun when not in use.

Unlike animals kept in large fields, the stabled horses cannot always select a clean place to lie down. Urine and manure must be continually cleaned off of them to prevent skin irritations. The stabled animals do not move about so much as pastured horses and lack the exercise needed to stimulate blood circulation and the oil glands of the skin. Such stimulation should be induced by vigorous daily brushing. Many ranches prefer vacuum cleaners made especially for this purpose.

The confined animals' hooves constantly pick up manure and urine-saturated bedding, even in the cleanest stables. The feet must be cleaned at least once or twice a day and treated occasionally to prevent thrush. Some bedding such as sawdust has a drying effect and horses bedded in this type of material should receive daily applications of a suitable hoof dressing. Horses kept in barns are the ones closely inspected by visitors and are usually the animals that will be shown soon, therefore they must be kept in the very peak of condition.

One important grooming point often overlooked is the cleaning of stallions' and geldings' sheaths. A great deal of dirt and body secretion collects in this area, providing a favorable environment for growth of bacteria that cause infection. The sheath should be cleaned at regular intervals, usually every month or so, although many animals require cleaning much more often. Scratching the horse over the kidneys will usually cause him to relax his penis enough to allow sheath cleaning. Mineral oil or glycerine introduced into the sheath will loosen the accumulation so it may be easily removed. An exceptionally dirty horse should have his penis and sheath washed with castile soap. Once clean, the area should be rinsed thoroughly and glycerine or mineral oil should be applied lightly to the area to lessen irritation. A dirty sheath often causes the animal to rub its tail.

5.10 Exercise. Planned exercise is essential for the confined horse. Daily exercise is the best health tonic known and is necessary for good circulation, muscle tone, digestion and prevention of constipation and colic. Exercise strengthens heart and lungs and builds muscles; lack of regular exercise is closely related to such diseases as lymphangitis and azoturia (Monday-morning sickness). Horses that are confined much of the time often stock up (develop swollen legs) due to poor circulation. If a horse stocks up often, permanent unsightly puffiness of the legs can result. Horses not receiving adequate exercise become bored, nervous, and difficult to work with and are apt to develop bad habits such as weaving, cribbing, pawing or kicking.

Adequate exercise can be provided in various ways. Animals may be ridden or turned out in a field or corral for a few hours each day. Some farms tie them on mechanical hot walkers for a specific length of time daily. Mechanical hot walkers may not prove beneficial to the animal's mental health if this is the only form of exercise offered over long periods; however, hot-walker exercise is far better than no exercise at all.

The best exercise is provided by driving, riding or ponying the animal. Stabled horses should be walked the first 10 or 15 minutes to allow them to limber up for more vigorous exercise. Failure to warm up a confined animal properly can result in strained muscles and pulled tendons.

A horse brought back to the barn must be walked until cooled to normal body temperature, after which a light, cool hosing is permissible in warm weather. Cold water must never be applied to hot muscles. Excess water should be removed with a sweat scraper and the horse

Mechanical hot-walkers are a convenience but should not be the only form of exercise offered over long periods of time. Horses can become very bored with only one kind of exercise, which is not beneficial to mental health.

kept out of drafts while damp. A hot horse should never be allowed to drink until he has cooled considerably and then should be allowed only a few swallows at several-minute intervals until completely cooled. A horse wet with old sweat is not hot if his temperature is normal.

5.11 Pastured Horse Inspection. Generally speaking, horses kept in a large field are much easier to care for than confined animals. They usually provide themselves with adequate exercise and the pasture offers a more sanitary environment. Sunlight dries and disinfects manure rapidly, thus reducing the internal parasite problem. The hooves of pastured animals remain in better condition because they pick up less manure and receive more exercising than do hooves of stabled horses. Pastured animals should nevertheless be inspected daily. The most convenient time for inspection is during feeding. By spending a minute at feeding time to observe each horse, the alert horseman can identify almost any serious problem immediately.

A little wet or cold weather will not bother a horse used to being turned out, but a wet blanket is much worse than no blanket at all. During rainy weather wet horses should never be brushed because brushing allows the water to reach the skin rather than "shed off" from the hair surface.

If the horse remains wet for extensively prolonged periods, hair from parts of the body such as mane and tail croup may fall out. The area should be washed with soap and water and the animal placed in a dry place periodically during the wet season. If the hooves remain constantly wet for extended lengths of time the outer hoof wall may become soft and tender. Drying dressings such as Kopertox or iodine painted on the sole may be helpful. The feet should be checked often for thrush and the horse should be placed in a dry area until the hooves become hard again.

Pastured horses occasionally get kicked, resulting in a lump or blood blister (*Hematoma*). Cold applications for 10 or 15 minutes morning and evening for several days will reduce the swelling. If the lump is still present after two weeks, it may be drained of a clear yellowish-colored fluid and warm, wet dressings applied to aid circulation in the injured area.

In extremely hot climates a pastured horse may suffer from heat stroke or exhaustion, especially if no shade is provided. This condition usually affects horses worked during hot weather, but may also affect young horses running in pasture. The animal appears very weak and may stagger. Sweating ceases and the hair coat is dry and hot. The horse should be cooled immediately by spraying with cold water or by applying ice packs to the head if the horse is staggering. The veterinarian should be summoned for treatment. This is a very traumatic condition and often proves fatal. If the body temperature is lowered by hosing or ice packs, the animal must subsequently be treated for shock and kept from chilling. Thumps (hiccoughs) often precedes heat exhaustion and should be heeded as a warning signal.

Areas of heavily concentrated manure generate considerable heat. New-born or young foals sleeping in such a location during the heat of the day may easily be overcome.

Horses suffer much more from dampness and wind than from cold. Heat and high humidity are not conducive to the comfort and well-being of horses. For this reason ranches in hot, humid climates often keep their animals in on hot days and turn them out at night.

Horses that have white or unpigmented areas, especially around the eyes, muzzle and ears, are often sensitive to sunlight and develop skin lesions. Such horses should be protected from the sun or bright light (kept in during the day and turned out at night) and antiseptic salve administered to the affected areas. Antihistamines and cortisone drugs often aid severe cases. Chemical reaction in the body caused by ingestion of certain toxic plants such as vetch, St. Johnswort, buckwheat, lantana or bur clover, or drugs such as phenothiazine used for worming can bring on this condition, referred to as photosensitization, derived from the photodynamic toxic action of plants.

Holding the tongue in this manner will keep the horse's mouth open for administration of medicine, examination of the teeth and gums or treatment of mouth ailments.

5.12 Teeth Care. Teeth problems are common in horses. The teeth of all animals should be examined by a veterinarian for sharp edges or other problems at least once a year. Undigested kernels of grain in the manure or slobbering while eating, refusing or fighting the bit, or bad breath are signs of mouth problems that should be corrected immediately. Teeth develop sharp edges and sometimes grow to uneven lengths and must be "floated" or filed occasionally so that they mesh for proper mastication of feed without irritating the cheeks.

5.13 Foot and Hoof Care. Every horse farm should have a competent farrier, preferably recommended by the veterinarian. Perhaps the surest way to tell the competence of a farrier is to examine the feet and legs of animals he has cared for during the past year or more. The farrier should make regular visits to the farm and trim the hooves of all ranch horses. The majority of the animals should be left unshod; shoeing is expensive and the blood circulation is usually better in unshod hooves because a shoe hampers frog pressure and the expansion and contraction of the hoof wall.

Animals with special foot problems such as severe quarter cracks will have to be kept shod. Horses being ridden to any extent, especially on rough, gravely ground, should wear shoes to prevent the hooves from wearing down too short and to prevent graveling, which causes severe lameness. The condition known as *graveling* occurs when particles of gravel work into the sensitive parts of the foot through the white line. Such a condition will cause painful inflammation, accompanied by heat and lameness. Eventually the particle works through the foot to the coronet band, where it emerges in an abscess.

A horse should be reshod every six to eight weeks, depending upon how rapidly its hooves grow and how quickly the shoes wear out. Horses that are allowed to go for extended periods of time between shoeings will often develop contracted heels and large, clumsy, unhealthy feet.

Unshod horses should have their hooves trimmed at eight- to twelve-week intervals. However, this period is quite variable, depending on individual differences in hoof growth and the type of ground upon which the animals are maintained. Horses on very soft ground or in a well-bedded stall need hoof trimmings more often than those running in large pastures with hard ground.

Certain points should be kept in mind concerning barefoot horses: 1) hooves should be trimmed whenever there is a half inch that can be removed; 2) hooves should be trimmed when they are wearing unevenly; 3) a barefoot horse should have more hoof left after trimming than one to be shod; 4) the edges of the hooves should be beveled so that they will not chip or crack easily; 5) if a small crack has developed, a groove should be rasped at the top of and at right angles to the crack to prevent further cracking. Horses with severe hoof cracks may have to be shod, perhaps fitted with hoof clamps to be worn until the cracks have grown out; cracks do not mend.

In all cases the hoof should be trimmed in such a way that no unnecessary strain is put upon the bones or tendons. When trimming properly conformed feet and legs, the farrier should trim the foot so that it is absolutely flat, with the side walls of equal length. Great care should be taken to see that the angle of the pastern continues in a straight line through the hoof. Under no circumstances should the toes be allowed to become too long and the heels too low, or the toes too short and heels too high. In either of these cases unnecessary strain will be placed upon the tendons and legs, greatly increasing the likelihood of unsoundness.

No matter how fine the breeding stock, the ranch will produce a number of young horses that will need corrective trimming. The younger the animal is, the easier it is to correct faulty leg conformation. Young horses start to develop permanent hooves at about three months of age. The hooves will then grow at the rate of a quarter to a half inch per month. Anytime after three months, if there is enough hoof to work with, corrective trimming can begin.

The two most common foot and leg faults found in horses are toeing-out (splay-foot) and toeing-in (pigeon-toed). When looking at the hooves of such horses from the front, it will be noted that one side is longer than the other. If toeing out, the hoof will be longer on the outside; if toeing in, the hoof will be longer on the inside. To correct either of these faults, the longer side should be cut down to the same length or a little shorter than the opposite side of the hoof (cut down the outside for the splay-foot animal and cut down the inside for the pigeon-toed animal). The greatest amount of hoof should be removed from the toe; the least from the heel. The heel should be trimmed so that it is level with the opposite heel but never lower. The long side of the toe can be trimmed slightly shorter than the opposite side to help correct the problem. If the condition is severe, small corrections should be made every few weeks, for if the direction of the foot is changed too rapidly, undue strain is put on the tendons, joints and bones.

Occasionally a young horse will have cocked ankles. In this case the heels are probably too long and should be trimmed; however, if the condition is severe, tendon and bone problems may be involved and a veterinarian should be consulted.

Contracted heels in barefoot horses can often be helped by lowering the heels so that the frog has more contact with the ground. Great care should be taken not to change the angle of the hoof too greatly in order to achieve frog-to-ground contact.

The frog should be trimmed carefully (but not excessively) to prevent ragged edges, which tend to collect filth that predisposes the hoof to thrush.

As stated in Section 5.9, all horses kept in stalls or small paddocks should have their hooves cleaned once or twice a day to remove accumulated debris that leads to thrush and to check for small stones that become lodged in the frog's cleft and the commissures alongside the frog. During wet winter months thrush may affect pastured horses as well as stabled animals. This disease usually begins in the cleft of the frog, spreads to other parts of the hoof and gives off an obnoxious odor and a black discharge. Excellent commercial preparations are available which, when used regularly, will control thrush. Proper frog-to-ground contact is important in controlling thrush by keeping the foot healthy; ground pressure causes a pumping action that aids blood circulation.

Brittle, dry feet often pose a problem to horse farms. Commercial hoof dressings applied according to directions are excellent for keeping feet pliable. Because of the large number of animals on a farm and the expense involved, it is usually not possible to inspect and treat the feet of pastured horses daily, in which case a beneficial and labor-saving technique is to allow the pasture water troughs to overflow so that animals must stand with all four feet in mud to drink. This is usually sufficient treatment for most animals' feet. Severely dried-out hooves and contracted heels respond well to dressings of pure neat's-foot oil applied with a small paint brush three or four times a week. Care should be taken to liberally apply the dressing to the coronet band, which is at the junction of the skin and hoof and is the area responsible for the growth of the horny wall, which must be kept healthy.

5.14 Parasites.

a) Parasites are generally classified as either external or internal. These terms refer to the outside or inside of the animal, whichever the parasites affect. Parasite infestation is a constant problem for the breeding farm and can result in heavy subclinical losses unless proper sanitation and worming schedules are followed.

The equine veterinarian will be up-to-date on the best methods of prevention, control and treatment for all types of parasites.

The appearance, performance and health of the animals and consequently the success of the farm can be greatly affected by parasites. Parasite control must be one of the farm's major concerns.

b) *External Parasites.* The most common, but by no means all of the external parasites are lice, mange, ringworm and ticks.

1. *Lice.* There are three kinds of lice that affect horses; however, for practical purposes they are usually classified as either bloodsucking lice or biting lice. Both types are very small, hardly detectable by the naked eye, but their presence may be indicated by the horses' irritability, restlessness and poor condition. An infected horse will scratch excessively, rubbing and biting the skin. The most infested parts of the horse are usually the mane, neck, shoulders, tail root and inner surface of the thighs. The skin irritation causes the hair to stand out and mat. In severe cases there is usually a loss of hair due to rubbing and scratching.

    Lice that are separated from the host die quickly. Their entire life cycle—living, dying and propagating—is spent on the host. They cannot be controlled by treating individual horses. The whole herd must be treated at the same time, otherwise the horses will reinfest each other. Commercial preparations, usually in the form of a dip or dust, are available for herd treatment. The initial treatment does not kill the eggs and must be administered again in three weeks in order to destroy the lice hatched after the first treatment.

2. *Mange (Scabies).* Today mange is a rare, reportable, communicable disease caused by the mange mite, which burrows into the skin, causing severe itching. The affected area becomes red and inflamed, the hair falls out and the skin thickens and scales. Mange is most troublesome during the winter months but is serious any time. If mange is suspected, the affected animals should be isolated and a veterinarian notified immediately. Lindane or toxaphene dips are effective.

3. *Ringworm.* Ringworm is a contagious skin disease caused by several different types of fungus that do not all respond to the same treatment. Ringworm is not confined to horses; it may be transmitted between man and horse, dog and horse or between other animals. The spores can live as long as a year and a half without a host and can be picked up from any contaminated object, such as grooming tools, bedding, fences, barns and tack.

The dry lesions appear as hairless, scaly or encrusted areas, grayish in color, usually beginning about the face or ears, upon the neck or at the tail root, but are also common in the back, flanks, croup and girth areas. The lesions begin with small centers and expand outwardly. They are more unsightly than debilitating unless they interfere with the saddle or bridle. Most types cause only mild itching, although some are very irritating to the horse. The affected horses should be isolated and treated by a veterinarian. Complete control requires disinfection of everything that could have become contaminated. Fungicide solutions and medications prescribed by the veterinarian will control the disease. Kopertox or iodine is sometimes effective in the early stages, while antibiotics may be needed for advanced cases.

4. *Ticks.* Ticks affect almost any warm-blooded animal and can be harmful, although not particularly painful, to the horse. Ticks cling to the animal by burying their heads in flesh, where they gorge themselves with blood. A heavily infested horse will suffer from anemia, which may become quite serious if the condition is allowed to continue.

Ticks are common vectors of many dangerous blood-borne diseases such as Piroplasmosis, Equine Infectious Anemia, commonly called Swamp Fever, and to a lesser extent Encephalomyelitis, also known as Sleeping Sickness. There are many different types of ticks, all of which may attach to any part of the body, although certain types will prefer particular areas. Ear ticks attach deep within the ear, causing severe discomfort. When infested with ear ticks the animal will hold his ears to the side, appearing lop-eared, and usually becomes quite head-shy and difficult to bridle. The bodies of affected horses can be treated effectively with commercial sprays, powders or dips, but ear treatment is more difficult. If low-pressure spraying within the ear is not effective, a solution of one-half chloroform and one-half light oil, such as salad or mineral oil, or a commercial liquid medication warmed to body temperature may be poured into the ear.

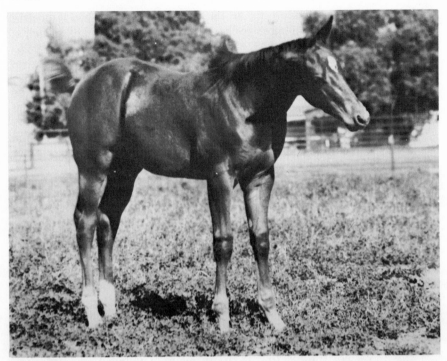

Colt with ear problem. Note that the head is tilted, with the affected ear held down.

Ticks are more prevalent in humid areas and are found in the tall grass and underbrush. It is almost impossible to exterminate them, but good land drainage to reduce humidity and removal of undergrowth from the pasture areas will do much to control them.

Occasionally the barns and weeds surrounding them become infested. If this happens the buildings and outside area should be sprayed with an effective insecticide. Oil (such as automobile crank-case oil) mixed with the insecticide will extend the effectiveness of the insecticide by clinging to the buildings and ground surface longer.

c) *Internal Parasites.* Internal parasites are generally more harmful and more difficult to identify, treat and control than are external parasites. Unfortunately almost all horses, even those appearing in good condition and excellent health, are the hosts of some internal parasites, which if not treated will soon become debilitating, resulting in substantial economic losses for the farm. Horses are often affected by several different parasites at the same time.

The general symptoms of severe internal parasite infestation are digestive disorders such as colic and diarrhea, loss of weight, retarded growth of young horses, unthrifty appearance, unhealthy hair coat, pot belly, lack of energy, listlessness, loss of appetite, swollen underline, coughing, anemia and many other indications of poor general health. The internal parasites of most concern to the horse farm are ascarids, bots, heel fly larvae, pinworms, strongyles and stomach worms.

1. *Ascarids (Ascaris [parascaris] equorum)*. Adult ascarids (large round worms) locate in the small intestine of the horse. They cause serious problems in young horses; however, animals over five years of age have usually acquired immunity to these worms. The larvae of the ascarid, ingested with contaminated feed and water, migrate through the organs and are especially damaging to the liver, lungs and heart. Large numbers are capable of causing severe tissue damage, producing toxins that make the animal sick, and can completely block the intestines, resulting in fatal colic. Pneumonia or a ruptured intestine are not uncommon complications of a heavy infestation.

In addition to the usual appearance of a wormy animal, a horse infested with these worms will cough in an effort to clear the migrating larvae from the trachea and lungs, but by coughing them up and swallowing, the animal actually assists the larvae in their travel into the intestine, where the parasites mature. The large nine- to twelve-inch white adult worm is often seen in the manure of infested animals, especially when the feed has been changed.

2. *Bots*. Bots are the larvae of the botfly (*Gasterophilus* sp.). It is the larva, not the fly, that is the problem. The bot fly resembles a small honey bee, except that the end of the abdomen curls under the body and is used for laying eggs, not for stinging. There are four different species, but all create essentially the same internal problem for the horse. Botflies are active during the warm part of the day from late spring to late fall. They do not sting, but their loud buzzing is a source of great annoyance to the horse. The white, yellow or black eggs are attached to the hairs on the horse's forelegs, neck, stomach, flanks, lips and lower jaw. Eggs and larvae gain entrance through the horse's mouth as it scratches with its teeth, and they migrate to the stomach and intestines where they attach to the lining and suck blood. The infestation may cause colic, anemia, loss of energy and, in severe cases, has been known to cause rupture of the alimentary canal.

To prevent infestation the eggs can be clipped or shaved off, the site of the egg attachment can be saturated with vinegar once a week to prevent hatching, or the area can be washed with warm, soapy water containing a disinfectant. The warm water stimulates hatching while the disinfectant kills the newly hatched larvae.

Horses should be wormed in the fall after the first killing frost if a bot infestation is suspected.

3. *Heel Fly Larvae*. There are two species of heel flies (also known as gadflies or warble flies), *Hypoderma lineatum* and *Hypoderma bovis*. Horses are not the natural hosts of the heel fly larvae, commonly known as "cattle grubs" or "warbles," but if horses are convenient, the heel fly will attach its eggs to the hair on their

heels, where maggots hatch and immediately bore through the skin, burrowing through the body to the internal organs where they are harbored until after the fall of the year. When it comes time for the maggots to mature into adult flies, they travel to the surface of the horse's back, just under the skin, where they form cysts while pupating into adult flies. The horses's make-up is not conducive to complete development (cattle are the natural hosts) and the larvae ordinarily die within the cysts. An experienced person may be able to remove the grubs by pinching the cysts, causing the grub to emerge through the air hole. However, pinching must be done at the right time and in the proper manner. If grubs are not removed the cyst may become infected and will not heal until opened surgically and cleaned out. The most harmful effects of the grubs are the unsightly appearance of lumps and sores and the temporary loss of the horse for saddle use until the sores have healed. To date, none of the systemic treatments used on cattle for control of the grubs have been registered for use on horses. Frequent use of fly repellent is about the only way to defend against the heel fly in heavily infested areas.

4. *Pinworms.* Pinworms are often referred to as "rectal worms" because they are seen around the anus. There are two types, large pinworms and small pinworms. The large worms live in the dorsal colon while the small worms complete the life cycle in the ventral colon. Large female pinworms (*Oxyuris equi*) lay eggs around the anal opening or in the manure, die and are then discharged with the manure. Ultimately the eggs drop from the anus or are expelled with the feces, grow to the infective stage outside of the horse and are later ingested as the horse eats and drinks.

Small pinworms (*Probstmyria vivipara*) live, produce live young and die without leaving the lower colon. The large pinworm is whitish colored and although fairly small—2 to three inches—can be readily detected in manure, but the small worms are so small that they may easily be overlooked. Anal irritation caused by large pinworms is very annoying to the horse and it will resort to vigorous tail rubbing for relief.

Both types of pinworms, while in the larval stage, will suck blood and, if present in great numbers, produce anemia and digestive disturbances.

5. *Strongyles* (*bloodworms*) (*Strongylus vulgaris*). There are approximately 60 different species of strongyles, the most difficult internal parasites to control. These worms cause bloodclots that can cause the death of the infected animal. Eggs are expelled with feces and pass through three larval stages on the ground. The third stage of the larvae crawl onto vegetation and are ingested by the horse. The larvae bore through the intestinal wall and migrate throughout the blood stream. Later the larvae return to the caecum and colon, where they attach to the lining and mature. The larvae remain in the blood stream up to five months, where they often cause clotting and where they cannot be successfully reached

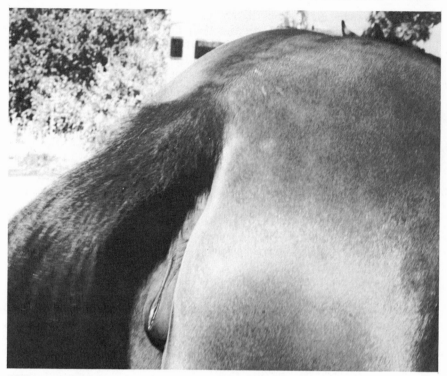

This mare has been rubbing her tail and holds it away from her body, indicating irritation from external parasites, pinworms or dirty udder (or sheath).

by medication. Besides causing clotting these parasites can produce severe anemia and other general symptoms of internal parasitic infection.

6. *Stomach Worms* (*Habronema spp.*) Three species of Habronema live in the stomach of the horse. Two of these, *Habronema muscae* and *Habronema microstoma,* may cause ulcers but usually do no great harm to the host; however, *Habronema megastoma* burrows into the stomach walls and forms nodules or small tumors in the stomach that can cause mechanical blockage.

The adult stomach worm lives in the stomach. Eggs pass out in the feces and are eaten by fly maggots. The stomach worm larva develops in the maggot. When the maggot pupates into a fly, the larvae migrate to the fly's mouth parts and are deposited around the mouth of the horse as the fly feeds on saliva. The larvae are then swallowed and mature in the stomach. Occasionally the larvae are accidentally deposited in a wound, causing a summer sore (see Section 5.19 (s) *Cutaneous Habronemiasis*).

Stomach worms are killed by copper sulfate, administered by a veterinarian.

5.15 Control of Internal Parasites. While it is practically impossible to completely rid the horse farm of internal parasites, much can be done to reduce the dangers and subclinical losses from heavy infestation. Aside from destroying the bot fly eggs by the methods mentioned in Section 5.14 (c) (2), the most effective control of internal parasites before they have gained entry into the horse's system is by carefully planned and rigidly enforced sanitation. It should be noted that most internal parasites are hatched in, mature in, or are carried by the animal's feces. This logically indicates that the best prevention centers around proper treatment of manure and sanitation of the places where it accumulates. Stalls, paddocks and pastures can become heavily infested with parasites, but proper cleaning, care and use of these areas can greatly reduce the hazard. Some basic preventative measures are listed below:

1. All feeding should be done in clean mangers, whether animals are kept in stalls or in pastures. Feed should never be allowed to contact the floor or bare ground, where it can become contaminated. Water troughs should be checked frequently for presence of manure and should be cleaned regularly. Since most of the internal parasites gain access through the horse's mouth, sanitary feeding and watering is extremely important.
2. Though stall cleaning should be done daily, frequent removal of spots of soiled bedding throughout the day will be beneficial.
3. Manure should be taken from paddocks and small corrals every few days, not exceeding weekly intervals.
4. Fresh manure should be placed in compost piles as discussed in Section 5.6. If the piles are not too large it is advisable to occasionally turn over and spread the manure so that the sunlight can destroy more of the parasites and other harmful organisms.
5. Do not fertilize pastures that will be grazed soon with manure that has been composted for less than three weeks. Fresh horse manure, if used, should be plowed under.
6. Avoid overcrowding pastures. Overcrowding concentrates contamination and increases exposure to parasites.
7. Rest pastures from horse grazing by rotation. The sun will destroy most of the parasites within two or three months. Grazing cattle or sheep is permissible in the meantime, since very few of their parasites affect horses.
8. Temporary pastures and pastures that are plowed under and reseeded every three or four years are much cleaner than permanent pastures.
9. Harrow or drag pastures frequently to break up and spread the manure for exposure to the sun and to speed drying.
10. Avoid exposing new pasture to horses that have not recently been wormed.
11. Routinely examine fecal samples under microscope for presence of parasites.

5.16 Maintaining a Worming Schedule. The internal parasite problem will vary greatly from area to area and even from farm to farm. Because of the great variation, a worming schedule adequate for one farm may not meet the needs of another. Each farm should establish its own worming schedule based upon fecal examinations and veterinarian recommendations. As conditions change the schedule must be reevaluated.

Many equine practitioners feel that frequent unnecessary worming is harmful to the horse's health. The substance that is poisonous to the worms can also be somewhat toxic to the horse, hence should be used as little as possible while still maintaining adequate parasite control. Ideally, fecal samples should be taken from all animals to determine which need worming; however, this method is impractical on large farms. Instead, routine fecal examination should be made over a period of at least one year to determine the best worming program for the individual farm. Random fecal samples should be taken periodically throughout the year to keep the management informed on the internal parasite situation.

A typical worming schedule often adopted, although perhaps not appropriate for all horse farms, is as follows:

1. Adult horses: Fall and spring, with a broad-spectrum drug and more often if parasites are prevalent.
2. Pregnant mares: Twice yearly for adults, plus the last month of pregnancy if microexamination indicates infestation. Very toxic worming medicines, especially those containing organic phosphates, should not be given to pregnant mares.
3. Horses 12 to 36 months of age: Every three to six months with a broad-spectrum drug. More often when kept in close quarters or where internal parasites are prevalent.
4. Foals: Beginning at two months, five times during the next twelve months if on irrigated pasture. If on nonirrigated pasture, worming should be at three-month intervals. Piperazine at the rate of one ounce per 200 pounds of body weight is safe for foals twelve weeks or older (Section 11.3).

5.17 Worming Procedures. There are several methods of worming horses. The one most favored by professional horsemen is "tube worming." This requires a soft rubber or plastic tube about nine feet long and five-eighths of an inch in diameter, which is inserted into the nostril, traveling down the esophagus into the stomach. The liquid worming medicine is then pumped through the tube. The advantage of this method is that all of the medicine is placed directly into the stomach at once and none can be refused by the horse, as is often the case when mixed and fed with grain. Most horses must be restrained and some must be mildly tranquilized in order to insert the hose. The animal is most likely to rear and strike as the tube is removed and the inexperienced handler is off guard. Tube worming should be done only by a qualified veterinarian; inexperienced persons may accidentally insert the tube into the horse's lungs and kill the animal by drowning it with worming medicine.

Liquid worming medication can be administered by a drenching syringe. However, this method is disagreeable to the horse and it may fight. Again, this method of worming should be attempted only by an expert.

A bolus (large pill) composed of worming medicine, given by a "balling gun" and forced down the horse's throat is not an acceptable method for worming, because there have been instances of the bolus becoming lodged in the throat for a time, resulting in severe ulceration of the esophagus. If boluses must be used, they should be dipped in mineral or salad oil to protect the throat, or the medicine should be given in gelatin capsules.

The safest worming method for the average individual to use is to put the worming medicine in the grain or water, or mix it with a sticky syrup or honey and place it far back on the animal's tongue. Because the mixture is sticky the animal cannot easily spit it out and swallows most of it.

Various other medications can also be administered by the last-mentioned procedure.

Another effective way to administer a medication is by use of a syringe. The tongue is pulled to the side and held firmly with the left hand, forcing the mouth to remain open. The syringe is then inserted and the contents placed well back in the mouth on top of the tongue.

No medications should be administered without first consulting the ranch veterinarian. He will specify the best treatment and indicate the proper method of administering it to the animals.

The following table indicates the apparent effectiveness of some antiparasitic drugs commonly used for horses. Although recommended dosages are given, the directions that come with the medication should always be read carefully to insure that proper dosage is given. It should be kept in mind that most medications that are toxic to parasites are also toxic to horses and should not be used indiscriminately. Some of these medications often produce illness in certain horses. However, piperazines and thiabendazole are usually quite safe to use.

ANTIPARISTIC DRUG CHART

Name of Drug	Dosage	Ascarids	Bots	Intestinal thread-worm	Pinworms		stomach worms	Strongyles			tape worms
					mature	immature		vulgaria	edentatus	small	
Carbon Disulfide	2.5 cc/cwt	good	excellent	—	none	none	—	none	none	none	—
Dichlorvos (Equigard)	1.6 gm/cwt	excellent	good	—	excellent	good	good	excellent	excellent	excellent	good
Di-phenthane-70 (Teniatol)	40 mg/kg	—	—	—	—	—	—	—	—	—	good
Parvex (Piperazine-Carbon disulfide complex)	4.0 gm/cwt	excellent	good	—	good	good	—	fair	poor	excellent	—
Piperazines	4.0 gm/cwt	excellent	none	—	good	good	—	fair	poor	excellent	—
Phenothiazine	2.5 gm/cwt	none	none	—	none	fair	—	good	fair	good	—
"	1.25 gm/cwt	none	none	—	none	none	—	fair	poor	fair	—
"	low-level	none	none	—	none	none	—	excellent	excellent	excellent	—
Thiabendazole (Equizole)	2.0 gm/cwt	good	—	good	good	good	good	excellent	excellent	excellent	—
Trichlorfon (Dyrex, Anthon, Neguvon)	3.6 gm/cwt	excellent	excellent	—	excellent	poor	good	good	fair	excellent	—
Dizan and Piperazines	2.0 gm Dizan /cwt 2.5 gm PPZ.	excellent	none	—	excellent	excellent	—	good	poor	excellent	—
Phenothiazine and Parvex	1.25 gm PTZ /cwt 7.5 gm Par.	excellent	good	—	good	fair	—	excellent	excellent	excellent	—
Phenothiazine and Piperazine	1.25 gm PTZ /cwt 4.0 gm PPZ.	excellent	none	—	good	good	—	excellent	poor	excellent	—

The internal parasites shown on this chart are the few that prevail and cause great concern; however, the possibility of infestation by other types should not be ignored.

The drugs mentioned above are not exclusive of other excellent anthelminitics.

The choice of medication should be made by the attending veterinarian.

5.18 Vaccination. An adequate vaccination program is important in preventing many serious diseases. Vaccines commonly used on the horse ranch provide immunization against: 1) tetanus, 2) encephalomyelitis, 3) influenza, 4) strangles, 5) rhinopneumonitis, and 6) various bacterial infections.

1. *Tetanus* (lockjaw). All animals should be vaccinated against tetanus. Horses vaccinated for the first time receive two intramuscular doses of tetanus toxoid vaccine one month apart and a booster of tetanus toxoid vaccine once a year thereafter.

2. *Encephalomyelitis* (sleeping sickness).

   a) All forms of encephalomyelitis are carried by mosquitos and sometimes by ticks. Horses should be vaccinated annually for this disease about 30 days prior to the mosquito season.

   b) There has been a vaccine against the Eastern and Western strains of this disease for several years. Two doses are given (7 to 10 days apart) and the dosage is repeated each year. There is now a vaccine combination for Eastern and Western encephalomyelitis and tetanus toxoid given intramuscularly. The older type of vaccine is administered intradermally.

   c) Venezuelan Equine Encephalomyelitis (V.E.E.) struck in the United States in the spring of 1971, causing serious nationwide concern before a vaccine could be made available. A vaccine that immunizes for many years or possibly for life has been developed and should be given to each horse. In some states this immunization is mandatory. Foals should be vaccinated for V.E.E. at six months of age.

3. *Influenza* (flu, Equine infectious arteritis, pinkeye, shipping fever, or epizootic cellulitis). This disease is caused by a virus and symptoms are similar to flu in humans. Young horses are especially susceptible but the virus may strike all ages and is highly contagious. The virus is most likely to be contracted where a large number of animals are gathered, such as at shows, sale yards or race tracks. All show animals should be vaccinated against virus infections at least two weeks before showing. The vaccines Equiflu II and Fluvac are given twice (4 to 12 weeks apart) followed with a booster yearly thereafter.

4. *Strangles* (distemper). Strangles affects horses between the ages of six months to five years, in particular. All show and racing animals should be vaccinated against this disease. Strangles bacterine is given in three doses (7 days apart) with one yearly booster thereafter.

5. *Rhinopneumonitis* (virus abortion, contagious abortion and colds). Rhinopneumonitis causes abortion in mares and colds in young horses. A modified live virus vaccine that cannot transmit the disease to unvaccinated horses has recently been developed. It can be used on foals 3 months and older and on mares after the second month of pregnancy with a follow-up dose 4 to 8 weeks later. The vaccine should not be used after the seventh month of pregnancy.

6. *Nonspecific Infections.* A broad-spectrum vaccine, Equibac (a mixed bacterin), is effective against staphococcus, E. coli and pasteurella infections. It gives protection against many reproductive tract infections, mastitis, navel-ill, diarrhea, pneumonia, shipping fever and other bacterial infections. Equibac is given in two doses (7 to 10 days apart) and is repeated annually.

The ranch veterinarian should decide which diseases to vaccinate against; in some areas he may choose to also vaccinate against diseases such as anthrax or leptospirosis.

Foals usually begin their vaccination program at three months of age. Earlier immunization may be administered by the veterinarian if a disease is prevalent in the area but this should not be undertaken without professional advice. If Venezuelan Equine Encephalomyelitis (V.E.E.) vaccine is given before six months of age, it should be repeated within the next six months to one year.

5.19 Health Problem Recognition and Treatment.

a) *General Discussion.* All ranch personnel should be well versed in proper horse care. They should be sufficiently familiar with horses to immediately recognize any indications of trouble and should be instructed to quickly report any injury, illness, abnormalities or suspicions of such to the manager no matter how minor the problem may appear. Anything out of the ordinary should be investigated immediately. For example, if a horse is seen standing alone some distance from the other animals, or if the other animals are romping and one is totally uninterested, the horseman should take a minute of his time to investigate the cause.

An attendant familiar with horses can tell immediately if the horse is not well by the facial expression of the animal. Even if there is no fever and no runny nose, the animal should be watched for several hours if it appears sluggish. Common signs of illness accompanied by a fever include a depressed appearance, decreased urination, constipation, diarrhea, increased thirst, indigestion or loss of appetite.

Incidences of respiratory disease usually increase when a sudden warm spell follows cold, damp weather. When this climatic condition exists ranch personnel should be especially alert for cold symptoms.

If the horse's temperature rises rapidly, his coat is standing erect, his skin is cold and his muscles are trembling he should be blanketed, quarantined immediately and a veterinarian should be called. These symptoms indicate an illness severe enough to require immediate professional attention.

Foot and leg problems are probably the most commonly encountered ailments. A standing horse usually rests one hind foot at a time, but ordinarily distributes its weight equally on both front feet directly beneath it. Any pointing of a front foot is an indication of soreness involving the foot farther forward. If the hind feet are advanced far under the body, thus relieving weight from the front feet, both forefeet are probably quite sore. If the horse's head bobs significantly when it

This foal appears very depressed and does not get up when approached. Its eyes are dull and runny and there is a thick discharge from the nose. The foal is definitely sick and needs immediate attention.

travels it is probably lame on a front foot. Viewed from the rear at the trot, if one hip rises higher than the other, the animal is lame on a hind leg.

If the horse is lame, the attendant should look for a hot area on the leg or in the hoof. If found, the leg should be submerged in or hosed with cold water or packed with ice. Under no circumstances should heat be applied to an already hot area; such treatment will aggravate the condition. If the hoof itself is hot, the foot should be searched for a foreign object or nail hole that could be causing the problem.

b) *Hoof Puncture Wounds.* When treating a puncture wound in the hoof the area around the wound is trimmed in order to expose the puncture, after which the wound is opened and cleaned well with hydrogen peroxide. If infection is present the foot should be soaked for 20 minutes in hot epsom salts. The puncture should then be packed with cotton soaked in very strong antiseptic (12% iodine is recommended). Even a slight infection requires a strong daily dosage of antibiotic until two days after the infection has disappeared. The possibility of tetanus must always be considered in the case of a puncture. When a valuable animal is injured, the manager should not rely on his own treatment but should phone the veterinarian immediately (see subparagraph (m) (7) below regarding tetanus).

c) *Skin.* The skin of the horse is normally quite supple, loose and easily moved in any direction. The coat should be sleek and glossy in warm weather, but in cold months it is rather rough and the hair stands out and is usually quite long on pastured horses. There should be no scaliness or dryness of the skin. A coarse, dull, rough coat may indicate worms, nutritional deficiencies or ill health. Itching and hair loss may be due to external parasites. A veterinarian should be consulted if the coat appears to be in poor condition.

d) *Temperature.* The normal temperature of the horse varies from 99° to 101° plus a few tenths of a degree; it is usually nearer 99° in the morning and higher in the afternoon. Foals may have a temperature of 102° on hot, humid days without being sick; however, a horse that has been resting and has a temperature of 102° is considered to have a low fever. A temperature of 104° indicates a definite fever and 106° or more is a serious fever. The temperature is most accurately and easily taken in the rectum by use of a mercury thermometer at least six inches long and having a Fahrenheit scale. A retrieval string with a clothespin or paper clip on the end to clip to the tail hair should be attached to a clean thermometer that is well lubricated with petroleum jelly or glycerine to aid in its insertion to almost its full length. It should be left in place a full three minutes for the reading to be considered accurate.

e) *Pulse.* The pulse of a horse can best be taken under the jaw bone on the submaxillary artery near the cheek. A normal pulse rate is from 35 to 55 beats per minute, but while the horse is vigorously exercising the rate may increase to slightly more than 120 beats per minute. A slow or irregular pulse is not normal; however, this condition is quite usual in a horse that is resting after hard work. The combination of fever with a pulse rate of over 80 beats per minute is good cause for immediate examination by a veterinarian.

f) *Running Nose and Common Cold.* Horses, like people, are subject to simple common colds. If a horse has no fever, is eating fairly well, is quite vigorous and is not dull or listless, his running nose is probably nothing more serious than a simple cold. He should be kept warm, dry and out of drafts and should be given plenty of water and a laxative feed such as warm mash. Proper rest is most important. Antibiotics should be administered if the animal is running a temperature. If the horse does not show a marked improvement in two or three days a veterinarian should be summoned.

Many deadly diseases begin much like the common cold. The animal must be watched closely and if there is any indication that the illness is something other than a simple cold the veterinarian should be notified.

g) *Coughs.* Coughs can indicate distemper, pneumonia, bronchitis, influenza, ascarids (roundworms) and many other ailments. In addition to indicating health problems, persistent or heavy coughing can result in permanent lung damage and requires the attention of a veterinarian.

h) *Eye Problems.* Colds and other diseases are often accompanied by runny eyes (both eyes). However, if the horse has runny eyes and does not appear to be sick, the eyes may be watering as a reaction to fine dust or pollen.

If only one eye is runny, it probably has been injured, contains a foreign object or is infected. Occasionally both eyes may be infected at the same time. Tear-duct blockage of either or both eyes is another possibility.

When no object is piercing the eye and no serious injury appears, the affected eyes should be rinsed with clean, luke-warm water, followed by a mild boric acid solution. Ophthalmic ointment may be applied to aid healing.

The eye can be rinsed thoroughly by saturating a piece of sterile cotton with the rinse solution, holding it above the eye and squeezing it so that the liquid trickles into the eye. Another method of rinsing is to detach the needle from a sterile syringe, fill the syringe with the rinse, anchor the hand above the eye and inject the solution into the outer

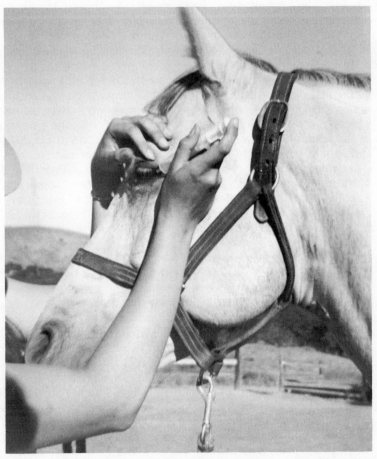

Proper use of the syringe (with needle removed) for flushing the eye.

corner of the eye so that it runs across the eyeball. The treatment should be administered every six or eight hours, but if the condition does not clear up within 36 hours a veterinarian should be consulted.

An animal with a seriously injured eye or one that contains a foreign object embedded within the eyeball itself needs the immediate attention of a veterinarian. No one other than a veterinarian should attempt to remove a foreign object piercing the eyeball. In the event the object has pierced the wall of the eye, removal will cause vitreous fluid loss and collapse the eyeball. A horse with such a wound should be cross-tied to prevent rubbing of the eye until the veterinarian arrives.

i) *Feces and Urine.* A normal horse will defecate about eight or nine times and pass from 35 to 40 pounds of fecal material during a 24 hour day. The quality of the droppings is a good indication of the condition of the horse's mouth, teeth and digestive organs.

The presence of whole grain or large pieces of grain in the manure is a sign that the horse is not chewing properly. It may be that failure to chew the grain is being caused by bad teeth or a sore mouth, either of which must be corrected. If, however, the teeth and mouth are in good condition, observation of the horse while it is eating may reveal that it is greedily bolting its grain. Ordinarily a large stone or two, baseball size, placed in the feed box will make the grain more difficult to pick up and will force the animal to eat more slowly.

Droppings of a healthy horse will be well rounded and moist enough to flatten a little upon hitting the ground. The manure should be free from obnoxious odor and slime. The color may vary from brownish-yellow to deep green, depending on the type of feed. The presence of blood in the feces is an indication of severe inflammation of the digestive tract and should be called to the attention of the veterinarian.

Dry droppings usually result from coarse, dry, indigestible feed, lack of water, or insufficient exercise. The horse needs some laxative feed such as bran, linseed meal, alfalfa and molasses meal or green grass, more water and an increase of regular daily exercise. If the condition does not improve, 16 ounces of milk of magnesia may be **given orally by syringe.**

Runny droppings are indicative of too much laxative feed (such as bran or green grass), fatigue, nervousness, overwork or a slight intestinal irritation. To correct this condition bran and green grasses should be eliminated or greatly reduced and the work load lightened. Adding oat hay or cottonseed meal to the ration may help. If the droppings are still watery, the horse should remain idle and should not be fed for a day. Any further treatment should be discussed with a veterinarian. Antibiotics are sometimes necessary to correct intestinal disorders.

Irritation to the digestive tract or heavily concentrated feed is usually indicated by foul-smelling, slimy droppings. The horse needs flushing with a laxative feed such as bran mash. It should be given plenty of water and concentrated feeds should be reduced.

Urine is normally a light-yellow-colored liquid that occasionally appears slightly clouded. The healthy horse urinates a quart or more of urine many times a day. If the color, quantity or intervals vary greatly a veterinarian should be consulted.

j) *Colic or Indigestion.* Colic is a broad general term describing almost any visceral pain. The most common type of colic is a digestive disturbance more accurately referred to as indigestion. Such colic is a primary killer of horses in any age group. Though there are many direct and indirect causes of indigestion, it is usually broadly classified as either gas (flatulent) colic, most frequently caused by fermentation of feed in the digestive tract producing a fairly constant pain, or spasmodic colic, in which pain usually occurs intermittently.

Colic may be either acute or subacute in nature. Acute colic is much more serious and calls for quick veterinary action. Great pain is caused when the stomach is severely distended by build-up of gas or too much feed. The tightly stretched stomach can be easily ruptured, resulting in the rapid death of the animal. The horse with acute colic is in a great deal of pain. Its face has an anxious expression, it will get up and down often, it will roll violently and even sit on its haunches like a dog or try to lie on its back to relieve the pain. Sometimes an animal in pain will even bite or eat itself. Often it will breathe rapidly and sweat profusely. Such colic is extremely dangerous and usually lasts only a few hours, after which time the horse will either die or recover. Obviously, an equine practitioner must be summoned immediately. The animal should be prevented from violently rolling or injuring itself. A mixture of steamed bran mash with ten aspirin tablets and a rounded tablespoon of Bicarbonate of Soda may be offered. If it refuses the mixture, aspirin may be mashed and mixed into a ball with honey and force fed. If the veterinarian has not arrived to give an analgesic or antispasmodic injection, the attendant may give 16 ounces of Milk of Magnesia by syringe placed far back in the mouth, if the animal can be handled. The handler should keep in mind that a suffering animal violently thrashing about is not aware of persons in its presence and may injure a careless attendant. The method of drenching with raw linseed oil is now seldom used because of the danger of the oil slipping into the lungs and killing the horse; however, oil can be effective when administered by stomach tube. If possible the animal may be walked very slowly.

Subacute colic usually occurs low in the digestive tract. It is generally spasmodic and may be caused by a number of things; most frequent causes are impactions resulting from eating very coarse feed and getting too little exercise, ascarids (large roundworms) blocking the intestines of young horses, sand or gravel accumulating in a low place in the intestinal tract (referred to as sand colic), irregular feeding or overfeeding. Other causes may be bloodworms blocking off the blood supply to part of the gut and resulting in gangrene, a twisted intestine or strangulation of the intestine by a hernia.

Symptoms of this type of colic are usually fairly mild during the first 24 to 36 hours, after which, if not relieved, they become severe. Going off feed is usually the first indication. The animal becomes listless and may lie down on its side, stretching out for a few minutes. It may hold its breath, grunt or moan, look back at its flanks, kick at its abdomen and paw the ground. Often the intestine is distended, pressing against the bladder and causing the animal to attempt to urinate frequently. One of the indications of impaction is that the animal will attempt to defecate often, passing only a few pellets. If not relieved, the condition will worsen. The animal will become anxious and demonstrate much more severe pain. Its lips and ears will be cold and cold sweat may appear between the fore and hind legs and on the belly. It may throw itself violently to the ground. When symptoms have progressed to this point, chances of recovery are very slight.

Although animals have been known to survive cases of subacute colic lasting up to five days, a valuable horse should not be gambled with; a veterinarian should be summoned as soon as indigestion is suspected. Sometimes the only outward sign will be that the animal is off feed or is listless, or is defecating with little results or urinating frequently. The handler can often determine if the animal has indigestion by listening to the abdomen. A tightly bloated or impacted digestive tract will become temporarily paralyzed and there will be no peristaltic movement, so the listener will not hear any normal digestive gurgles or rumbling.

As with acute colic, the animal should be made as comfortable as possible in a roomy, well-bedded stall while waiting for the veterinarian. The bran mash, aspirin and bicarbonate of soda mixture mentioned previously may be given, followed in a half hour by 16 ounces of milk of magnesia if necessary. The animal may be walked slowly to aid in the release of gas. A shallow enema of two or three gallons of mildly soapy luke warm water may be administered. Although the horse should not be bothered other than to prevent thrashing or violent rolling and should be slowly walked for ten minutes every half hour, it should never be left alone. Violent rolling must not be allowed, for the horse can easily twist an intestine or rupture the tightly stretched stomach or intestine, either of which will almost certainly result in death.

The veterinarian may administer a smooth-muscle relaxant along with large doses of water and enemas. In cases that do not respond to treatment, the veterinarian may have to resort to surgery.

The sick horse should have no feed until the day after recovery, and even then for a few days should receive only a light ration.

It should be pointed out that the horse is unable to vomit anything that has been swallowed. If a horse eats a toxic plant or spoiled feed, the material must pass completely through the digestive tract. For this reason, plus the fact that the horse's digestive tract is much more delicate than that of most other farm animals, special care must be taken to avoid spoiled, moldy feeds and toxic materials, to eradicate all harmful plants from the pastures and to check purchased hay carefully for any poisonous plants unwittingly baled with it. The farm should also strictly adhere to its worming schedule and/or fecal examination to avoid colic caused by ascarids in animals under five years of age and bloodworms in older horses.

k) *Abscesses.* Skin abscesses are localized areas of infection caused by a cut, scratch or bite that has become infected, or by an internal infection that has worked its way to the outside. Invariably the abscess is inflamed, filled with pus or fluid and swollen. While the cause of an abscess due to cuts or bites is relatively easy to determine because of its close proximity to the injured area, the cause of an abscess resulting from some sort of internal infection (commonly referred to as a boil) is much more difficult to diagnose. Several bacterial diseases produce abscesses; perhaps the most common of these is strangles (distemper), which produces abscesses in the head and throat area.

Treating abscesses: 1) A veterinarian should be consulted, especially when the cause is not known. 2) The abscess should not be lanced until it has come to a head; if it is incised too soon the infection may not be completely expelled. Hot water compresses alone or with epsom salts, or a commercially prepared poultice applied to the spot, will help bring the abscess to a head. When the hair falls off and the head of the abscess is soft and spongy, it should be lanced with a sterilized sharp instrument and allowed to drain completely. 3) When empty the cavity should be immediately flushed with an antiseptic such as hydrogen peroxide and a gauze plug inserted to provide good drainage and to prevent healing over. Repeated flushing for a day or two is advisable. Antibiotics may be given after the abscess has been drained, but *not* before, since injections too early may cause the infection to recess. If the animal is very ill the veterinarian may give antibiotics before the abscess heads, depending upon the situation. The decision to use antibiotics when abscesses are involved must be left to the veterinarian.

Pigeon Fever (*Corynebacterium pseudotuberculosis*) is a disease-causing bacterium appearing especially in California, and it manifests itself in large swellings and abscesses, sometimes reaching the size of a football. The abscesses are usually external and appear in the pectoral (chest) muscles, along the underline, in the mammary glands or the sheath, but sometimes may protrude inward, in which case the condition may be fatal. Pigeon fever requires treatment by the veterinarian.

l) *Bruises and Strains.* Bruises and strains are very common injuries. Cold applications can be used to reduce initial tissue inflammation and swelling. The cold constricts the blood vessels supplying the injured area and is helpful in reducing internal or external bleeding. Cold (*not heat*) should be used on injured tissue for the first day or two. Cold water compresses or ice packs are most beneficial. If ice packs are out of the question, cold compresses can usually be fastened in place and should be repeatedly saturated with cold water from a hose or bucket. The hose may be placed over the horse's back and loosely tied in place so that the cold water trickles over the injured area. Once the inflammation has subsided and all heat is gone (usually two days or more), the bruise or strain will benefit from an increased blood supply to facilitate healing of the damaged tissue. Heat applied at this time will greatly increase blood circulation. Hot water compresses or chemical poultices will do much to speed circulation, and liniment, which causes mild heat, is helpful in cases of minor bruises, strains and sprains. When inflammation and swelling are present, aspirin (50 grams for a 1,000-pound horse) may be given as an anti-inflammatory agent and to relieve the pain. Producing heat through blistering, firing or painting are very harsh methods of counter-irritation sometimes used by veterinarians in severe cases of ringbone or splints, but should never be used without a veterinarian's advice.

Cortisone drugs are sometimes injected into an inflamed joint, tendon or muscle to reduce inflammation and speed healing. However, cortisone should be administered only by a veterinarian and any surgery should be postponed for at least six weeks after injection of the drug so that the defense mechanisms of the body may build up again to fight infection and there is no danger of the body overreacting to the surgery. Broad-spectrum antibiotics should be administered along with treatment where cortisone is used. In many instances cortisone is very beneficial, but when it is administered this information should be prominently noted on the animal's health record.

m) *Wounds.* No matter how safe the fences and premises may be, horses have a way of getting hurt. Farm personnel must consistently be on the alert to spot injuries and must know how to give proper first-aid and to treat minor wounds.

Wounds are usually classified as: 1) abrasions (superficial or surface injuries such as nicks, scrapes and scratches), 2) incised wounds (such as those produced by sharp objects without damage to the surrounding tissue, 3) lacerations (rough, jagged tears or breaks that have much tissue damage), and 4) punctures (stab wounds that immediately close, such as those caused by nails, barbs, etc.).

All deep or severe wounds should be treated by a veterinarian to assure rapid recovery and to reduce the danger of a lasting blemish or subsequent unsoundness. However, horsemen should be familiar with routine procedures for treating all superficial wounds.

The procedures outlined below are generally considered to be the safest and most practical for treating fresh wounds.

1. A wound bleeding profusely must be tended immediately to stop the loss of blood. The bleeding can usually be controlled by direct pressure applied to a pad of cloth placed over the wound. If the wound is where it can be tightly bandaged over the pad, a pressure bandage that does not cut off the circulation may be applied. A leg can be bandaged quite tightly without impeding circulation if adequate padding is used. When the wound is where a tight bandage cannot be applied, it is necessary to hold the pad in place and apply hand pressure until the veterinarian arrives to take over treatment. After the pad is placed on the wound, it should not be removed; doing so will only hamper clotting and allow more dirt and debris into the wound. When a pad becomes saturated, more padding should be placed on top of it instead of removing the pad that is in direct contact with the wound.

2. If a big vein or artery has been cut and a pressure pad does not stop the bleeding, the vessel must be pinched off. Tweezers may be helpful if available, otherwise fingers will have to do. Fingers should be wrapped in cloth so that the slippery vein or artery can be grasped. The vessel will have to be held until the veterinarian arrives. It may be necessary to cut healthy tissue to expose the bleeding vessel, but speed is of greater importance than sanitation and tissue damage when there is imminent danger that the animal may bleed to death.

   Tourniquets are extremely dangerous to use and should be used only when hemorrhage cannot be controlled any other way. When blood circulation is totally cut off or almost completely restricted, gangrene may set in. If bleeding cannot otherwise be stopped, the tourniquet of course must be used but it must be tightened no more than necessary to control the bleeding and must be loosened for a minute every 30 to 40 minutes to permit a fresh supply of blood to pass to the tissue. The tourniquet must not be forgotten. The arterial blood that is coming from the heart is pumped in spurts and is bright red, whereas the venous blood is on the return trip, flows steadily and is quite dark red in color. The tourniquet used to stop arterial blood must be placed between the heart and the wound, but venous bleeding must be controlled by pressure on the side of the wound farther from the heart.

3. A gaping wound should be immobilized as much as possible and not allowed to dry out until the veterinarian can arrive and suture it closed. It may be moistened with water but absolutely no medication should be used on a wound that is to be sutured.

4. If bleeding does not present a problem, the area around the wound should first be clipped free of hair, dirt and debris, then the wound should be rinsed with soapy water. If treatment is to be concluded, the superficial wound should be washed out with hydrogen peroxide, treated with an appropriate dressing (see subparagraph 9 below) and well bandaged.

   Water tends to slow the healing process and therefore should not be used on a wound after the first cleansing. Sterile gauze or cloth should be used to remove debris from the wound. Dry cotton should never be used on a wound; its particles will adhere to the wound and impede healing. If cotton is to be used it should be very well saturated.

5. A wound that is to receive the prompt attention of a veterinarian should not under any circumstances be medicated by the handler. Prior medication could greatly interfere with veterinary treatment, especially if the wound may require suturing. However, if the handler does treat the wound, he must tell the veterinarian immediately upon his arrival exactly what was done and what medication was used.

   In cases that are not severe enough to require veterinarian attention, the handler should apply antiseptic ointment or powder directly to the wound. If much tissue is damaged or exposed, or if the wound is very dirty and contaminated, systemic antibiotics should also be administered. Nonirritating antibiotic medications for deep wounds and bandaging should be used wherever possible.

6. Puncture or stab wounds are very common on the horse farm. Nails, sharp instruments, splinters and barbed wire often result in puncture wounds. This type of flesh wound is sometimes extremely difficult to treat. Such a wound is hard to find if the piercing instrument does not remain. The wound ordinarily closes immediately without bleeding sufficiently to clean the tissue or expel the debris from the pit. Infective germs have a made-to-order nest

unless the hole is opened, flushed well with hydrogen peroxide and made to heal from the inside outward. The surface opening of a large hole should be packed with sterile gauze saturated with an antibiotic ointment and bandaged to keep out dirt. Caustic tissue-damaging medication must not be used because dead tissue within the wound will hamper healing. The wound should be treated daily with an antibiotic ointment. If it does not respond to treatment within a few days the veterinarian should be advised. This type of injury is apt to develop a deep-seated, unobservable abscess. If the horse goes off feed or otherwise appears ill within a few days after the wound occurs, the veterinarian must be told of the condition.

Puncture wounds create an extreme hazard of tetanus (lockjaw). (See subparagraph (7) below.)

7. Tetanus protection should be foremost in mind for all wounds. Horses are prime targets for this deadly disease. A tetanus antitoxin injection should be administered within 24 hours of injury, especially if the animal has not completed its routine immunization inoculations. Puncture wounds present the greatest danger because they do not bleed enough to cleanse the deep wound and the top closes over, providing an ideal anaerobic (no air) environment for the tetanus bacteria to grow in. A tetanus antitoxin shot is indispensable when treating a puncture wound, whether or not the horse has received routine immunization.

8. Deep wounds should be bandaged to protect against dirt, dust, contamination, flies and other irritants. Good bandaging may also help prevent movements of the injured area, thus facilitating healing. Bandages should be changed every three days until a healing base is established, after which the open air is very beneficial and the bandage can be removed. Occasionally the handler may detect infection in the wound as he changes the dressing. The symptoms of infection are the presence of pus, heat, soreness and swelling. The wound should be soaked in warm epsom salts for 20 minutes or so, repeating two or three times a day until healed. If the wound does not improve rapidly, a veterinarian should examine the horse and prescribe further treatment. In most cases of infected wounds antibiotic injections are advisable.

A leg bandage should never be applied without sufficient padding; it must be tight to stay in place, but if applied tightly without padding, it may restrict blood circulation.

Bandages for clean, slight, surface wounds in the form of plastic sprayed over the injury are quite useful if the wound is difficult to bandage properly.

9. Wound dressings. All dressings are not for all wounds; some types may be very injurious if applied to the wrong type of wound. If the handler is not absolutely sure of what dressing to apply he should telephone the veterinarian for the advice he needs. Never use a medication that is unfamiliar without professional advice.

The following subparagraphs indicate the types of wounds and some of the dressings usually considered safe for the particular type of wound:

a) Types of wound: Almost any open wound, although especially appropriate for deep wounds, those reaching close to a joint and those to be closed by suture.

   Dressing: Antibiotic ointments; Nitrofurazone (these dressings are antiseptic, nonirritating and have no drying properties).

b) Types of wound: Shallow surface cuts; abrasions (such as superficial nicks, scrapes, scratches and surface burns).

   Dressing: Antiseptic; gentian violet; furazolidone; Merthiolate; tincture of iodine. Follow with ointment or salve when wound is dried. (These dressings are irritants and produce drying, which is not desirable for wounds having deep penetration, those to be sutured or wounds near joints where movement will cause a dry wound to reopen).

c) Types of wound: Deep or open wounds needing growth of fill-in muscle and skin.

   Dressing: Enzyme sprays; scarlet oil; sulfa, penicillin and tetracycline powders (these dressings allow the growth of proud flesh, uncontrolled tissue growth). Limit use on or below the knee and hock, where healing is sometimes difficult and proud flesh is produced.

There are some types of dressing that should not be applied except under veterinary supervision. The caustic irritants such as butter of antimony, copper sulfate, silver nitrate, some specific acids and lime are sometimes used to prevent or remove unwanted tissue growth (proud flesh), but these can be very damaging unless expertly administered.

Surface abrasions may have an application of a light healing agent such as vaseline or other petroleum jelly without restriction.

When a wound is to be left exposed, it is wise to cover it with a gall salve to discourage flies.

n) *Broken Bones.* Unfortunately, broken bones occur occasionally. If a fractured leg is suspected, the leg should be padded with sheet cotton, then splinted with a light board, stick or heavy cardboard and wrapped in a heavy blanket. Some ranches stock inflatable plastic splints. The horse should not be tranquilized because an animal in shock is very apt to fall unconscious if tranquilized. No pain killers should be administered; if the leg is painful the animal will not try to use it. This is as it should be, to prevent greater separation of the bone and additional damage. An equine practitioner should be summoned immediately.

o) *Artificial Respiration.* Occasionally an accident, such as electric shock, or a disease such as tetanus will cause a horse to stop breathing. If this should occur, the handler can try to revive the horse by driving his knee or the heels of both hands very forcefully into the abdomen just below the rib cage in an attempt to stimulate diaphragm movement. This should be continued rhythmically as in breathing until the horse recovers or until it is determined to be futile. The veterinarian may attempt heart stimulation by manipulation or by an injection of adrenalin. Artificial respiration for new-born foals is discussed under Section 10.6.

p) *Tracheotomy.* There have been cases of animals choking to death because of blockage of the trachea, the presence of a strangles (distemper) abscess involving the epiglottis, or swellings due to snake bites closing off the air passages. If it becomes obvious that the animal is going to die before the veterinarian arrives, as evidenced by a blue color of the gums, a tracheotomy can be attempted. Haste is more important than proper sterilization techniques; infection can be coped with if the animal survives. A common sharp pocket knife is suitable for the operation. The emergency operation requires that an incision be made into the trachea (windpipe) about 5 inches below the larynx (adams apple). The cut should follow the length of the trachea and should sever two or three of the cartilage rings. The opening can be spread by the fingers, a piece of tube or any other hollow object available until the veterinarian arrives.

q) *Snake Bite.* There are four poisonous snakes native to the United States. They are the coral snake, water moccasin or cottonmouth, copperhead, and 26 kinds of rattlesnakes. Coral snake bites are very rare due to the snake's relatively small distribution, its docile disposition and its short fangs placed on the sides of the jaw. The coral snake produces a neurotoxin that attacks the nervous system, making such a bite extremely dangerous and usually resulting in death by paralysis of the diaphragm. The cottonmouth, copperhead and rattlesnake are all pit vipers and produce a hematoxin that attacks the red blood cells and causes tissue damage. Rattlesnake bites are by far the most common. Because horses are curious, a great many bites are on the head; the majority of the rest are on the lower legs. If a horse is bitten, a veterinarian must be called immediately. He can administer venom antiserum, steroids, antibiotics, and tetanus antitoxin to combat bacteria carried into the wound by the fangs. The greatest danger from pit viper bites is that of suffocation. A six-to-eight-inch piece of hose can be inserted into the nostrils to keep them open as the head swells. In severe cases a tracheotomy may be necessary. Cold water or ice packs applied to the bitten area will slow venom absorption, reducing the body's reaction to the venom.

r) *Injections or Shots.* There may be instances when the ranch personnel will be called upon to give shots to horses. Unless the ranch has a full-time veterinarian employed, the routine vaccines are usually obtained from the regular veterinarian and given by ranch employees. A supply of antibiotics and tetanus antitoxin should be kept on hand for use in emergencies.

Since all ranches should have regularly scheduled visits from a veterinarian familiar with the ranch animals, all injections should be given only with his approval.

Should an emergency arise and a veterinarian cannot be contacted, the manager may elect to administer a shot or give medication. He should, however, inform the ranch veterinarian of his actions at the earliest possible time.

When antibiotics are called for, proper administration is of the utmost importance. Treatment should continue for three or four days and at least until thirty-six hours after the last symptom was noted. Doses should be substantial to guarantee the destruction of all disease-causing organisms lest a resistant strain be developed that will defy treatment. Antibiotics should not be used indiscriminately. The body builds up a tolerance to them over a period of time and perhaps, when really needed, they will offer little protection.

The labels of all medicines should be read carefully, especially those dealing with injectable substances. Many vaccines, bacterins and antibiotics must be kept in a cool place or refrigerated to prevent deactivation. All have an expiration date printed on the label and all vials that have passed the expiration date should be discarded immediately.

Special care must be taken that the proper dosage is administered. Dosage will vary greatly between the same vaccines produced by different companies, due to the difference in the concentration of the fluid; for example, a dose of tetanus toxoid may vary from 1 cc to 5 cc. An overdose of certain medications can have very serious consequences.

Disposable sterilized needles and syringes can be obtained from the veterinarian or through farm supply stores. They should be discarded immediately after use, since old needles are a common method of transmitting serious blood-borne diseases from one animal to another. Most ranches prefer not to use mechanical sterilizers, because the disposable needles and syringes are inexpensive.

The site or location of the injection must be thoroughly cleaned by a suitable solvent such as alcohol or soap and warm water to remove all disease-carrying dirt and oil. (Alcohol and soap do not sterilize but they do cleanse.)

The most common types of injections are:
1. Intradermal—into the skin
2. Subcutaneous—under the skin
3. Intramuscular—into a muscle
4. Intravenous—into a vein

Intravenous will not be discussed here, because these injections should be left strictly to the veterinarian.

The label on the vial will indicate the proper method of injection and should be followed carefully. The method of inoculation is determined to a certain extent by the liquid carrier in which the medication is suspended. Therefore it is possible that different vaccines for the same disease must be given by different types of injections to be effective.

Intradermal shots may be given in any location, but the skin of the neck is most commonly used. A lump at the site of the injection usually results; therefore, the vaccine is often administered where the mane will hide it. A very fine needle is used, no. 20–25 gauge, ½″–¾″ in length. The skin is grasped between the thumb and forefinger and

Proper method of administering an intradermal injection.

pinched up. The needle is slipped in at an oblique angle and the medicine is slowly injected. Fluid injecting easily indicates the needle is through the skin and the injection is subcutaneous instead of intradermal, and correction must be made.

A subcutaneous injection is administered in the same area by grasping the skin between the thumb and forefinger, pulling it away from the body, piercing the skin below the thumb into the space created between the skin and the muscle. A no. 22–25 gauge needle ¾″–1″ long is used.

An intramuscular injection may be given in any large muscle. The most commonly used sites are the top of the rump in the gluteus muscle, in the lateral neck muscle, in the tricepts muscle above the forearm and in the pectoral muscle between the front legs. The pectoral muscle is the most convenient site for a novice to begin, because the place is less painful and there are no bones or nerves close to the skin that are likely to be struck. Occasionally a small bloodvessel is ruptured, producing an unsightly swelling, therefore this area is not suitable for a horse that is to be shown soon. The novice should have the exact locations carefully pointed out by the veterinarian before attempting to give injections himself.

The best procedures for giving an intramuscular injection is to detach the needle from the syringe and grasp it firmly between the thumb and forefingers. The animal is then patted vigorously with two or three swift striking motions with the side of the hand. Patting deadens the nerves to a certain extent, making the shot less painful. The hand is then turned so that the needle is stuck quickly into the muscle on the next pat; the rhythm should not be broken and if done properly the animal is often unaware that the needle has been introduced. If blood runs out of

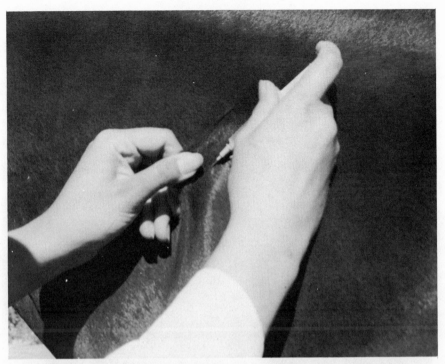

Correct procedure for administering a subcutaneous injection.

the needle a vein or artery has been struck and the needle must be moved before the injection is made. The syringe is then attached and the contents slowly injected.

Care should be taken to remove as many air bubbles as possible from the material before it is injected. This is accomplished by holding the syringe with the needle straight up, allowing the air to rise to the top and pushing the plunger slowly until all air is expelled. When removing the needle the thumb and forefinger should be placed at the sides of the needle and the skin should be held in place to prevent pain when the needle is withdrawn.

In preparing for the injection it sometimes becomes difficult to withdraw the vaccine from the vial. In this case air may be injected into the bottle to release the vacuum. Germs are introduced with the air so the vial should not be kept long after introduction of air. Vitamin shots are more likely to become contaminated than are antibiotics.

If the needle breaks off in the muscle or skin, it must be removed immediately, even if a veterinarian has to be called to perform surgery.

s) *Disease and Ailment Chart.* The following chart gives pertinent information for many common diseases and ailments.

Note: An animal that has received a sedative type of medication should not be worked or ridden until the medication has worn off. Pain relievers and relaxants often temporarily cause the animal to appear well, but the condition can be aggravated by use of the animal while it is under the influence of medication. The veterinarian should recommend the period of idleness for a horse that has received a systemic treatment or is impaired by a severe injury.

1. When giving an intramuscular injection, the needle is separated from the syringe and held while the animal is given two or three rhythmical pats.

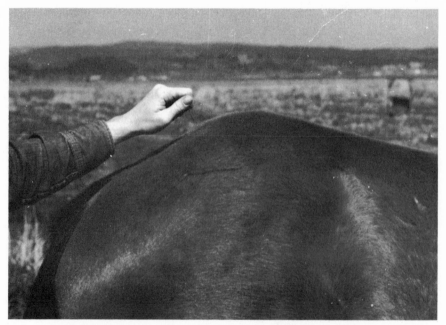

2. Without breaking the rhythm the needle is extended . . .

3. and driven in on the next pat.

4. Once the needle is inserted and no blood comes from the needle, the syringe is attached and the contents slowly injected.

DISEASE AND AILMENT CHART

Name	Comments	Symptoms
Abortion	See Equine Viral Rhinopneumonitis	
Allergic dermatitis	Allergic reaction to feed, medication, organism or organism toxin. Some cases commonly called "feed bumps" or "pasture bumps." Sometimes caused by a Vitamin A deficiency.	Sometimes splotchy loss of hair, especially around face. Itchy rash with lesions. Mostly on neck, face, shoulders, chest and forelegs. Can be accompanied by edema and difficulty in breathing. Sometimes swollen udder or sheath.
Anemia	Noninfectious. Lack of red blood cells and hemoglobin due to dietary deficiency of iron and occasionally copper, cobalt or certain vitamins. Also caused by large numbers of bloodsucking parasites.	Listlessness, loss of appetite, weakness, emaciation. Low red blood cell count.
Anthrax (Splenic fever, Charbon)	Usually fatal. Caused by the Bacillus anthracis bacteria. Usually not transmitted by direct contact, but by bacterial spores that escape from a dead animal and are ingested by other animals. Spores remain virulent in soil indefinitely. Once an area becomes contaminated, the only protection is vaccination to prevent spread. Carcasses must not be opened. Should be burned (not buried). Horses less susceptible than cattle and sheep.	First signs similar to colic, followed by high fever, large hot swellings in throat, neck and chest areas, bloody discharge from rectum, sudden death.
Azoturia (Paralytic myoglobinuria, Monday Morning disease, Tying-up syndrome, Lumbar Myositis)	Noncontagious, nutrition-related. Usually occurs in well-conditioned, hard-working animals that are given one to three days rest without reduced ration. Occurs when animal is worked again. Sometimes aggravated by vitamin deficiency and/or calcium-phosphorous imbalance.	Appears suddenly, usually within ½ hour of when exercise begins. Stiffness and tenseness of muscles, usually those of back, loin, croup and hindquarters but may affect all legs and shoulders. In acute cases involved muscles are hard and hot. Animal often falls and cannot rise, sweats heavily and breathes hard. Urine reddish-brown color. The same symptoms often appear in mild form and are referred to as tying-up. Cause, treatment and prevention are the same for tying-up as for azoturia.
Biliary fever	See Piroplasmosis	
Broken Wind	See Heaves	

DISEASE AND AILMENT CHART

Treatment	Prevention
Contact veterinarian. Give 8-10 antihistamine or other anti-allergic medication in feed twice daily, accompanied by medicated baths. Insecticides if indicated. Cortisone on advice of veterinarian if chronic.	Avoid feeds and medications the animal is allergic to, i.e., prickly plants, vetch hays, linseed meal, phenothiazine, etc. External parasite control (water-diluted insect repellents are less irritating than oil-based solutions).
Remove all blood-sucking parasites. Provide adequate iron, copper, cobalt and vitamins in diet. Veterinarian may give iron injection for severe cases.	Prevent parasite infestation. Provide adequate nutrients in diet.
Contact veterinarian at once. Early treatment with antianthrax serum and massive doses of penicillin can save some animals if disease is detected early enough.	Vaccination. Vaccine will occasionally produce severe reaction, sometimes death. Should not be used unless area is known to be contaminated with anthrax.
Complete rest. Do not move animal unless absolutely necessary. If animal must be moved, lead VERY slowly to stall. Until vet arrives, warm applications to affected muscles will make animal more comfortable.	Reduce feed when animals are not being used.

Name	Comments	Symptoms
Colic (Indigestion)	Complete discussion in section 5.19 (j).	Abdominal pain, anxious expression, rolling, kicking at abdomen, looking at flanks, listlessness, loss of appetite, frequent attempts to defecate with little success. Symptoms vary greatly according to cause and location of pain.
Conjunctivitis	Inflammation of conjunctiva of eye. May be caused by germs, allergies, foreign object in eye, dust, mud, pollen or other irritants.	Heavy flow of tears, spasmodic blinking. Watery discharge may later become thick and sometimes foul smelling.
Coughs and Colds	Common mild, viral, respiratory infection similar to human cold. Contagious.	Listlessness, slight temperature, nasal discharge, coughs dry at first then become moist.
Cutaneous Habronemiasis (Summer Sores, Granular dermatitis)	Summer Sores are result of wounds accidentally infected by larva of stomach worm (*Habronema*) carried by common housefly. Because wound is not normal environment for larvae they cannot complete life cycle so migrate throughout wound, causing severe irritation. During winter months larval activity slows down and wound appears to heal, but when weather becomes warm larvae again move within wound, causing new inflammation and proud flesh formation.	Wounds slow to heal have red, angry appearance and develop a great deal of proud flesh. Wound appears to heal during winter but becomes active again during warm weather.
Diarrhea (Scours)	Indicates digestive disturbance. May be caused by illness, poor feed, change in feed, nervousness, physical overactivity or other causes affecting digestive tract. Observe animal for signs of illness. Diarrhea in foals may be caused by mare being in heat.	Droppings are of a semifluid consistency.
Distemper	See Strangles	
Encephalo-myelitis	See Equine Encephalomyelitis	

Treatment	Prevention
Contact veterinarian immediately. Steamed bran mash with 10 aspirin for 1,000-lb. horse and 4 Tbsp. baking soda. Make comfortable. Prevent violent rolling. Lead very slowly or leave alone. Give enema. If no improvement before vet arrives give 16 ounces of milk of magnesia by syringe. Aspirin. Vet will give analgesic or antispasmodic injection.	Feed only high-quality feeds free from dust or mold. Follow a regular feeding schedule. Do not feed on sand or gravel. Follow rigid parasite-control program.
Consult veterinarian. Rinse with lukewarm water or mild boric acid solution. Antibiotic ointment on advice of veterinarian.	Good sanitation and dust control.
Cough medicine, decongestants, antibiotics on advice of veterinarian. Aspirin. Rest, mild laxative diet, protect from cold, drafts and overheating. Cold symptoms often precede many very serious diseases, therefore horses with colds should be watched. If condition does not improve or worsens, contact veterinarian immediately.	Good general health. Proper feeding, sanitation and parasite control. Quarantine all incoming animals.
Excess proud flesh (granulation tissue) must often be removed surgically and bandaged with a caustic powder. Organic phosphate compounds may be given orally to destroy Habronema larvae. Cortisone or antibiotic ointment to control inflammation.	Sanitation. Fly repellents and barn sprays. Manure disposal. Rigid parasite-control program. Clean wounds with soap and water. Cover for 3–4 days. Apply fly repellent around wound to discourage flies.
Withhold food for 24 to 48 hours. If due to washy, laxative feed, change diet. If condition persists or if there is blood in feces, consult veterinarian for antidiarrheal medicine. Broad-spectrum antibiotic. Aspirin (10 tablets for 1,000-lb. horse).	Proper feeding methods and use of high-quality feeds.

Name	Comments	Symptoms
Enteritis	Inflammation of intestine, often complication of other diseases such as influenza. May be caused by bacterial infection, poisons, parasites, bad feed, overwork, chills and other causes affecting the digestive tract.	Loss of appetite, rapid, hard pulse. Frequent defecation. Colic symptoms, mild temperature. Diarrhea.
Equine Encephalo-myelitis (Sleeping Sickness)	Viral infection of central nervous system. 4 different viruses produce the disease. In the U.S., Eastern and Western Encephalomyelitis and VEE (Venezuelan Encephalomyelitis). All transmitted by bloodsucking insects, principally mosquito. Birds harbor E. and W. strains, acting as carriers. While E. and W. cannot be transmitted directly VEE can be. Insect carries E. and W. from bird to animal. Because of short time virus is in the blood, E. and W. strains are not often transmitted from animal to animal by a vector. 30% mortality for W. strain, 90% for E., mortality rate for VEE not yet available. Disease is transmitted to man. Usually permanent damage to brain and spinal cord if animal recovers.	Drowsiness, depression, excitability, walks in circles, staggering gait, drooling, difficulty swallowing, grinding teeth, unable to stand, walks into objects, paralysis. Fever 102° to 107°. If fatal, death within 2–4 days.
Equine Infectious Anemia (Swamp Fever, EIA.)	Incurable disease caused by virus attacking red blood cells. Most prevalent in damp, poorly drained, marshy areas with large numbers of biting insects. Can be transmitted by contaminated needles or tattoo instruments, biting insects and ingestion of contaminated feed or water. Infected brood mares transmit disease to unborn foal. Disease usually appears in May or June, declines in late fall. Chronic cases may occur throughout the year. Affected animal remains a carrier for life. EIA certificate required for travel to Hawaii, Canada and other countries. Disease may be acute or chronic.	Attack accompanied by high fever, 105° to 108°. Weight loss, depression, weakness, swelling of underline, sheath and/or legs, diarrhea, frequent urination. Animal hangs head, stands with hind legs forward under body. Becomes anemic. Mucous membranes of eyes and mouth pale or yellowish.

Treatment	Prevention
Call veterinarian. Keep animal warm and comfortable.	Proper feeding and parasite control.
Call veterinarian immediately. Isolate animal in a quiet area. Provide palatable feed, water. Keep animal on feet. Protect from mosquitos. Watch for complication of pneumonia. Aspirin (50 grams for 1,000-pound horse).	Vaccinate all livestock. Combination vaccine for Eastern and Western strains given yearly prior to mosquito season. Venezuelan encephalomyelitis vaccine is thought to immunize for several years. Mosquito eradication.
No cure at present time.	Suspect animals should be isolated immediately, biting insects must be eliminated. Blood test. If positive, animal should be destroyed and carcass burned or buried very deep with quicklime. Area thoroughly disinfected. Boil instruments 15 minutes.

Name	Comment	Symptoms
Equine Influenza (Equine Infectious Arteritis, Shipping Fever or Epizootic Cellulitis)	Highly contagious. Caused by filtrable virus. Spread by direct and indirect contact. Affects young animals primarily (except young foals nursing mares and receiving passive immunity through milk). Immunity for variable lengths of time after recovery. Low death rate. Can cause abortion. Stallions may carry virus in semen from few weeks to several years. Should not be used for breeding until semen has been checked for virus presence.	3 to 14 day incubation period. Symptoms sudden. High temperatures 103° to 106°. Depression, loss of appetite, weakness, watery discharge from eyes and nostrils, eyes may be clouded. Cough, rapid breathing, swelling in ventral parts of body and legs.
Equine Virus Abortion	See Equine Viral Rhinopneumonitis	
Equine Viral Rhinopneumonitis (EVR)	Primarily a viral respiratory disease of young horses, it causes abortion in pregnant mares. Spread by direct contact with infected animals or aborted fetus.	Young horses demonstrate cold symptoms. Mild fever, nasal discharge. Abortion in mares late in pregnancy, 8th–10th month, or foal may be born dead or die shortly thereafter.
Farcy	See Glanders	
Founder	See Laminitis	
Frostbite	Rare in animals that are well fed and have winter coats. May occur in cold weather when animal must stand in mud or if worked in areas where salt has been used to melt snow.	Similar to a burn. Surface cells die, fluid oozes through skin. Area becomes swollen and painful.
Glanders (Farcy)	Rare in U.S. Suspect animals given mallein test. Infected animals destroyed. Chronic, incurable, highly contagious. Caused by bacteria.	Glanders form: Nasal discharge, ulcerations of nasal membranes, hard enlargement of lymph glands between jaws, wheezy respiration, head may swell. Farcy form: Lymph glands and vessels swell and stand out. Ulcers appear on swellings and discharge yellow pus.
Grease Heel	See Scratches	

Treatment	Prevention
Consult veterinarian. Absolute rest. Keep warm and out of drafts. Provide adequate water. Danger of fatal complications such as pneumonia. Antibiotics. Cough medicine, aspirin (50 grams for 1,000-pound horse). Steroids and antihistamines may be administered by the veterinarian.	Vaccinate all animals that will be exposed to nonfarm animals (show and race horses). Isolate infected animals, burn bedding and discharges, disinfect stalls and equipment. Quarantine incoming animals.
Antibiotics and sulphonamide compounds can be given to foals. Cough medicine. Aspirin (25 grams for 500-pound foal, 50 grams for 1,000-pound horse).	Vaccination, but it does not give permanent immunity. If used, animals should be vaccinated in late fall or early winter. Once begun, vaccination program must be continued, since the virus has been introduced to the farm with live vaccine. Good sanitation, isolation of mares that abort. Lab analysis of aborted fetus. Disinfect area. Separate mares from young stock, quarantine incoming animals.
Consult veterinarian. Keep animal dry. Apply antibiotic ointment to affected area.	Never wash animal's legs in very cold weather, avoid wet, boggy areas, provide good-quality feed (breakdown of roughage in caecum generates a great deal of heat).
None. Consult veterinarian for mallein test.	Destroy all infected animals and burn carcasses. Isolate all suspected animals. Give mallein test.

See Scratches

Name	Comments	Symptoms
Heaves (Broken Wind, Chronic pulmonary emphysema)	More common in confined animals. Often an allergic reaction to dusty, moldy feed, pollen, dust. May be complication of chronic cough or another respiratory disease such as strangles.	Difficult breathing, jerking of flanks, coughing, nasal discharge, appearance of "heave line" along lower side of barrel.
Hydrophobia (Rabies)	All warm-blooded animals are susceptible. 100% fatality. Death in 3–4 days. Incubation from 9 days to many months. Virus in saliva of infected animals. Transmitted by bites of wild animals, dogs, etc.	First sign is usually itching where bitten. Rabid horse often becomes vicious. Attacks people, other animals, solid objects. Usually drools. If horse does not become vicious it will become lethargic and act much like an animal with sleeping sickness.
Infectious Anemia	See Equine Infectious Anemia	
Influenza	See Equine Influenza	
Laminitis (Founder)	Inflammation of sensitive laminae of horse's foot. Caused by overfeeding of high quality nutrients, retained afterbirth, continued work on hard ground and certain diseases. Fat horses with heavy crests are more susceptible from overeating. Animals recovered are more susceptible to further attacks.	Extreme pain in forefeet, hind feet are well under body to relieve weight from front feet. Fever 105°–106° is common. Hot feet, rapid pulse. In chronic cases rings form on hoof wall and sole will drop.
Lockjaw	See Tetanus	
Mastitis	Bacterial infection of mammary tissue of udder. Infectious or noninfectious. Most likely to occur in lactating mares. Not so common in horses as in other livestock.	Heat and hard, painful swelling in affected quarter. Milk may be thick and stringy with pus or blood in it. Abscesses may develop in severe or neglected cases. May require surgical drainage or removal of infected quarter.
Melanoma (Neoplasm)	Skin tumor. Common in gray horses, rare in other color. Almost always occurs in animals over 15 years of age. Perhaps 80% of gray horses are affected. Not contagious.	Tumors most frequent in dock area, under tail and around anus. Also around head or udder. Occasionally on internal organs.
Monday Morning Disease	See Azoturia	
Moon Blindness	See Periodic Ophthalmia	

Treatment	Prevention
Avoid damaged, dusty or moldy hay or grass. Sprinkle water on hay to reduce dust just prior to feeding. Feed pelleted ration. Place in clean, grassy pasture. 8–10 antihistamine tablets for a 1,000-pound horse.	Same as treatment.
None. Call veterinarian immediately for positive diagnosis. Infected animals extremely dangerous. If the animal must be shot to avoid injury to other animals or to people, care must be taken not to injure the animal's brain. The complete brain will be needed for positive diagnosis.	Vaccinate all ranch dogs and control stray dogs. Vaccinate all animals exposed to rabid animals. Disinfect all water troughs, feed boxes and other areas where infected animal may have deposited saliva.
Summon veterinarian at once. Stand animal in cold water. Bran mash with 8 antihistamine or 10 aspirin twice daily. Groove hooves at veterinarian's recommendation or use special shoes for protection.	Proper feeding methods. Examination of afterbirth to determine that none is retained. Do not work animals on hard ground. If animal is predisposed to laminitis let hooves grow somewhat longer.
Call veterinarian. Antibotics administered systemically or directly to site of infection.	Good sanitation. Wash mare's udder before foal nurses.
Early treatment by veterinarian is usually successful. Advanced cases may require surgery if tumors interfere with riding equipment. If growth is malignant surgery must be in early stages. Later surgery will spread the tumors.	None.

Name	Comments	Symptoms
Mud Fever	See Scratches	
Night Blindness	See Vitamin-A deficiency	
Periodic Ophthalmia (Moon Blindness)	Thought to be caused by lack of the vitamin riboflavin in diet. Also associated with leptospiral bacteria. May also be caused by eye parasites such as Thelazia sp. (eye worms) or Onchocerca, and streptococcal eye infections.	Eye appears cloudy. May clear up but reoccurs resulting in blindness eventually.
Piroplasmosis (Biliary fever)	Caused by protozoan parasite that attacks red corpuscles. Transmitted by biting insects, primarily the tick. Recovered animals less susceptible to further infection. Recovered animals act as carriers for indefinite period.	High, irregular fever 105°–106°. Membranes of eye and mouth pale, then darken to orange or reddish-brown. Purple blotches eventually appear. Attacks of piroplasmosis are often preceded by colic. Urine dark, constipation with brown feces coated with mucus. Unsteady on feet. Degree of secondary infection. 8–10 day duration.
Pneumonia	Inflammation of lungs. May be caused by 1) bacteria or virus as complication of other diseases such as strangles or influenza, 2) injury to lungs, 3) foreign material in lungs such as drench accidentally given into windpipe, 4) worm larva in lungs. Mortality if untreated 50%–75%. Damp, drafty barns and changeable weather predisposes animals to pneumonia.	Severe chill followed by high temperature 103°–107°. Mucus membranes are red. Reddish discharge from nose, may later thicken and become grayish. Rapid, difficult breathing. Soft, short cough. Rapid, irregular pulse. Ears and legs cold.
Rabies	See Hydrophobia	
Scours	See Diarrhea	
Scratches (Grease heel, Seborrhea, Mud Fever)	Both scratches and mud fever are closely associated with wet, dirty surroundings. Long hair on fetlocks holds moisture and germs.	Scratches: Inflammation of fetlock and heel area. Heat, swelling, runny exudate. Severe cases cause lameness, proud flesh ("grapes") usually on hind legs. Mud Fever: Similar to scratches but occurs on parts of limbs where mud splashes. May affect entire leg.
Shipping Fever	See Equine Influenza	

Treatment	Prevention

Contact veterinarian. Antibotics usually helpful. Put on high Riboflavin B$_2$ diet. | Provide adequate riboflavin in diet, good parasite control, proper sanitation.

Contact veterinarian. Isolate animal, make comfortable, protect from biting insects and sun. Feed easily digested, laxative diet. Full recovery will take at least one month. | Control ticks and other biting insects.

Consult veterinarian immediately. Keep animal warm, provide plenty of water, easily digested feed, fresh air; avoid drafts. Antibiotics and sulfa drugs are usually administered for two weeks. Aspirin and decongestants. | Careful observation and good care of animals with minor illnesses to prevent secondary pneumonia infection. Good sanitation and parasite control. Proper administration of oral medications.

Clean area thoroughly and keep area dry and clean. Clip hair from area and apply antibiotic ointment. | Good sanitation. Do not allow animal to stand in muddy or marshy areas. If muddy, allow to dry and brush off instead of washing. Water will aggravate condition.

Name	Comments	Symptoms
Sleeping Sickness	See Equine Encephalomyelitis	
Sporadic Lymphangitis (Weed)	Probably caused by lack of exercise and excess high-quality feed. Tendency to reoccur.	Hind legs usually affected. One or both (or all) legs become painfully swollen and hot. Often fever of 104°–105°. Shivers and sweats, usually remains standing.
Stones	See Urinary Calculi	
Strangles (Distemper)	Highly contagious bacterial disease involving upper respiratory tract and lymph glands of head. Most common in horses 2–5 years of age. Life-long immunity when recovered. Mortality rate low, 1%–4%. Most troublesome during wet seasons. Spread by direct or indirect contact. Some horsemen do not try to prevent the spread of the disease but feel that the animals must catch distemper once to be immune to it.	High temperature 103°–105°. Listlessness, loss of appetite. Thick yellowish-white nasal discharge, moist cough. Hot, painful swelling between jaws. May have difficulty swallowing. Abscesses must rupture and drain before animal recovers.
Summer Sores	See Cutaneous Habronemiasis	
Swamp Fever	See Equine Infectious Anemia	
Tetanus (Lockjaw)	Caused by bacteria Clostridium tetani, which gains entrance through wound. Puncture wounds are most apt to result in this disease because they close over quickly, providing the anaerobic (airless) environment needed. The tetanus bacteria is normal inhabitant of horse's digestive tract and is constantly present in manure. Mortality is 50%–60%. Recovery in 8–10 weeks.	Stiffness and spasms of voluntary muscles. First symptom will be stiffness of gait. Difficulty eating. Face muscles affected. Extreme excitability. 3rd eyelid partly covers eye. Usually no temperature until disease is far advanced. Saliva may drip from mouth. Convulsions may occur.
Thrush	Disease of the frog of foot. Associated with damp, unsanitary stable conditions. Healthy feet are less susceptible to disease. Good frog-to-ground pressure is necessary to prevent thrush.	Dark, foul-smelling discharge from cleft of frog and commissures of the hoof. Lameness in severe cases.

Treatment   Prevention

Consult veterinarian. Force animal to
move. Feed laxative diet. Withhold
grain. Keep susceptible animals
turned out where they can move
around. Veterinarian may recommend
sulfa drugs and antibiotics.

Reduce rations for idle horses.

Consult veterinarian. Rest, keep
warm. Do not lance swelling until
well pointed, flush abscess cavity 2 or
3 times daily with hydogen peroxide.
Counterirritants or hot compresses
can be used on abscesses to bring them
to a head. Antibiotics after abscess
heads but not before, because if given
too soon will cause abscess to recess.
Consult vet on advisability of antibiotics.

Vaccinate animals that often leave
ranch to go to shows, races, etc.
Quarantine incoming animals. Isolate infected animals. Burn bedding and dis-
charges to prevent spread by indirect
contact. Do not handle other horses
after caring for sick animal until clothes
are changed and hands thoroughly
washed. Disinfect quarters after animal
is recovered.

Call veterinarian immediately. Isolate
animal in darkened stall. Veterinarian
will administer tetanus antitoxin, antibiotics and muscle relaxants. Provide
plenty of water, soft laxative feed or
mash. Animals have been known to
choke on hay when throat muscles are
affected. Soft or liquid gruel or mash
is best. Intravenous or stomach tube
feeding may be necessary. Enemas
and catherization may be required.
Support critically ill horse in a sling.

Vaccinate annually.

Keep feet very clean and dry. Trim
frog. Apply strong drying antiseptic
preparation. Many good commercial
preparations available.

Proper sanitation, regular hoof cleaning, good hoof health. Proper frog
pressure of utmost importance.

Name	Comments	Symptoms
Urinary Calculi (stones, water-belly, gravel)	Presence of stones in urinary tract blocking passage of urine. Exact cause is unknown but probably associated with nutrition. Incidence is higher in chalky soil areas. Urinary Calculi is not a problem in mares.	Frequent attempts to urinate without success, pain, colic symptoms. If blockage is not remedied bladder may rupture or uremic poisoning set in.
Virus Warts (Papillomatosis)	Caused by contagious virus infection. Most prevalent in young horses. Unsightly but not serious.	Appearance of warts on muzzle, underparts of body and/or sometimes on penis of stallions and geldings and on mare's clitoris. Can become as large as a baseball but most remain quite small and resemble cauliflower.
Vitamin-A Deficiency (Night Blindness, Xeropthalmin)	Caused by lack of vitamin A, necessary component of visual purple in rods of eye, and necessary for epithelium tissue health.	Faulty vision at night. Runny eyes. Eventual blindness. Poor hoof development. Increased incidence of respiratory disease.
Weed	See Sporadic lymphangitis	

Treatment	Prevention
Consult veterinarian. Muscle relaxants often ease passage of stones. If relaxants fail surgery usually becomes necessary.	Proper feed management. 1%–3% salt added to ration may be of value in areas where incidence of urinary calculi is high.
Warts usually disappear in two or three months with no treatment. Occasionally may have to be removed surgically. Ointments that may speed disappearance are available.	Vaccine available.
Add synthetic Vitamin A and carotene to diet and treat secondary infections.	Feed adequate amounts of Vitamin A (carotene).

5.20 Nursing Care. A horse with symptoms of a contagious disease should be separated from the other horses to prevent spread of the disease. Some horses become extremely upset when by themselves, increasing the danger of complications. The manager must use his judgment when dealing with such a horse. A radio or companion such as a goat in the stall is often helpful in calming this type of horse.

Generally speaking the animal should be cared for much the same as a person. He should be made as comfortable as possible, protected from cold or wet weather and kept out of drafts. He should be stabled in a well-bedded stall of sufficient size to allow ample room in which to get up and lie down easily without danger of injury or becoming cast. The bedding must be kept very dry and clean. Aspirin (10 tablets or 50 grams for a 1,000-pound horse) often relieves minor discomfort of various minor illnesses.

Ventilation is important in keeping the air fresh and aiding in keeping the area dry. Sunlight and fresh air are most beneficial to the sick animal and if the weather is mild, and depending upon the seriousness of the illness, taking the animal out of doors for an hour or so in the middle of the day may hasten recovery. A corral or paddock where it will not have to be tied is preferred. If the weather is chilly, damp or windy it is best to keep the animal indoors.

The sick horse should be groomed thoroughly. Grooming aids the blood circulation, eases sore muscles and makes the animal feel better. A sick horse appreciates having its mouth rinsed with water and its face, nostrils and corners of the eyes gently wiped with a damp cloth several times daily.

If the weather is cold the horse should be blanketed. If the legs are cold they should be wrapped. In warm weather, if the horse is outside or if the stable is not insect-proof, a fly sheet will help prevent insects from irritating the animal.

The horse should not be forced to eat, although it may be tempted with its favorite feed or with special tid-bits such as carrots, apples or other choice morsels, as suggested in Section 6.15. If it refuses its feed the ration should be removed after fifteen or twenty minutes and fresh feed offered an hour or two later. Clean water (with the chill removed) should be available at all times. If an animal is suffering from a respiratory ailment, the feed may be dampened slightly to reduce dust.

The veterinarian may prescribe a special diet to conform to the needs of a horse with a particular ailment, but generally as the horse begins to recover it should receive a laxative ration to relieve constipation resulting from inactivity and fever.

Many people underestimate the value of extra attention to sick animals, but like people they seem to recover more rapidly if given special consideration and made comfortable. A fine line must be drawn between helpful attention and overhandling that may distress an already sick animal. Horses that like people may benefit from extra attention, while horses that are suspicious of humans should be left alone as much as possible.

5.21 Accident Hazards. Many horses appear to be accident-prone; if there is anything to get hurt on, they will find it. The ranch employees must be constantly on the alert for protruding sharp objects, loose wire, nails, tools, and the like.

Leaving halters on horses is a dangerous practice; they can be death-traps. Horses often catch them on hooks, nails or snags outside the stall as they scratch over the Dutch door or a barrier. They have been known to get a foot caught in the halter while scratching their heads and choke to death.

Rope halters can shrink in the rain and cause great discomfort. The halter parts constantly in contact with an area of the face will eventually cause irritation or produce a groove on the animal's nose. Halters left on horses indicate a lazy attitude on the part of the stable help and, while more convenient, they are not conducive to the animals' safety.

5.22 First-Aid Kit. The manager and hired help should be able to treat minor ailments and should be well acquainted with first-aid measures to be employed before the veterinarian arrives. Quick action often means the difference between life and death.

Every farm should have at least one first-aid kit easily available in case of emergency, and should maintain a good supply of equipment and medication that may be needed from time to time. No medication should ever be used on a horse just because it happens to be handy. No treatment at all is a thousand times better than the wrong treatment, which can kill or cripple an animal.

Most farms stock the following materials for emergency first aid:
first-aid book or chart
veterinarian rectal thermometer
very sharp surgical knife
bandage scissors
tweezers
hypodermic syringes—2, 10, 20, 35 cc
hypodermic needles—16, 18, 20, 22 gauge
large, blunt hypodermic needle for flushing abscesses
roll of sterile cotton
sterile cotton balls
sheet cotton
gauze bandages and pads
adhesive tape
antibiotics (check expiration date)
antihistamines
aspirin
tetanus antitoxin (check expiration date)
antiseptic dusting powder
antibiotic ointment
ophthalmic ointment
boric acid
sterile lubricant (KY jelly)

vaseline
Vick's vaporub
Epsom salts
medicated clay poultice (Numotizine or Antiphlogistine)
Nitrofurazone
gentian violet
tincture of iodine
colic medicine
cough medicine
liniment or leg brace for sore muscles and tendons
leg wraps
elastic bandages
alcohol
disinfectant
wound fly repellent
fly spray
ear tick medicine
fungicide
external parasite solutions for ringworm, lice, ticks, etc.
thrush remedy
milk of magnesia for constipation
antidiarrheal for diarrhea
bucket
twitch
flashlight
25 feet of ¾" or 1" cotton rope
wire cutters
hoof nippers
hoof knife
hoof pick
hoof rasp

This list is by no means complete; some farms may choose to delete some items or add others to meet their own particular needs. No matter what materials are stocked, care must be taken that the supply is always kept adequate and that broken equipment is repaired immediately. Accidents and illnesses can occur at any time.

Towels, tail wraps and other such items not damaged by heat may be sterilized by sealing them in a heavy paper bag and placing them in the oven at 225° for half an hour. A pressure cooker may be used to sterilize needles and other small metal or glass items if a laboratory autoclave is not available. Boiling for 20 to 30 minutes will also kill most germs.

5.23 The Veterinarian. The reader may feel that there has been over-emphasis on the importance of calling the veterinarian promptly. In a great many cases veterinary attention may prove to have been unnecessary; however, when expensive, registered animals are involved one cannot be too cautious. The loss of one animal is usually the monetary equivalent of many years of emergency veterinary calls. Gambles are not worth the possible disastrous and expensive consequences. Veterinarians are the breeding farm's best friends.

5.24 Dead Horse Carcass Disposal. At some time during the ranch operation it will be necessary to dispose of the carcass of a dead horse. In every case without exception, the cause of death must be known or determined by the veterinarian. If the veterinarian determines that the cause of death was from a noncontagious cause and presents no danger to the other livestock, the carcass may be disposed of to a tallow or rendering company. If local laws do not prohibit, the carcass may be buried on the ranch, but if it is to be buried, the site should be away from where water could be polluted or where the carcass or its contaminants might be washed out. The body should be placed in a pit, sprinkled generously with quicklime and buried at least three feet beneath the surface of the ground. Carcasses can also be burned. This is accomplished by piling a large amount of flammable material on the carcass—especially old tires, which make a hot fire—sprinkling with gasoline and igniting. Of course fire safety and pollution regulations must be observed. Burning is recommended in all cases where animals have died from dangerous communicable diseases.

6
NUTRITION

(SYNOPSIS)

This chapter is a practical, nonscientific approach to equine feeding. It contains much information and many nutrition points of basic concern to the horse farm and its manager. The following synopsis by section numbers is provided as an aid for quick reference to specific subjects.

6.1 General Statement. Quality breeding stock and adequate facilities are a necessary foundation for a successful operation, but unless the animals are properly fed and cared for, profitable production cannot be maintained.

Sufficiency of feed is not enough if the animals are to compete favorably with the quality of horses found on today's market. The days are gone when the professional horse breeder could turn a band of good mares and a suitable stallion out to pasture and return a few years later to round up a herd of young horses ready for training. The horses then were hardy and able to fend for themselves but would not meet the needs of today's market. The present trend is to show and race horses as young as possible. The demand, especially in the racing industry, is for quick return on investment, and whether the breeder likes the idea of training young horses or not, he is almost forced to follow the demand or fold up his operation. If the young horse is to remain sound and make a good showing, he must be healthy and must have received plenty of top-quality feed from birth.

The ability of the manager to select the best feeds at the right price and to feed in the way that the animals can get the most from the ration can mean the difference between a profitable business and a tax write-off.

Improper feeding can result in nutritional ailments, unsoundness, lameness, digestive disorders, lowered fertility in both stallions and mares, decrease in quantity and quality of the mares' milk production, loss of vigor and increased cost of care plus decrease of profits.

Horses cannot be fed properly on the basis of guess. The nutritional requirements for all classes must be understood. While a standard feeding program should be developed for each class of animal on the farm, the feeder must make adjustments to meet the needs of individual horses. Some animals will become obese on diets other horses do poorly on. Aside from the normal body demand, a horse exercised heavily will burn a great deal of energy, which must be replaced. Some experts, more particularly the National Research Council and the commercial feed companies, have tried to classify horses by age, weight, size, sex, lactation, required labor—such as idle, heavy, medium, light. While such classifications are valid, each term is so vague in itself that feeding charts cannot be geared to meet the needs of a particular horse, nor do feed charts or nutrition tables consider the difference in metabolic rates between animals. These experts are the first to acknowledge that all feeding charts are only general guidelines and must be modified by the sound judgment of the man who knows the horse and its particular requirements. While a scientific approach to feeding by the use of nutrition tables is essential, a complete understanding of food values, keen observation of each horse, attention to the expenditure of energy, long experience in horse feeding, weather conditions and good feeding sense all must combine in determining the best rations for individual animals. It is not suggested that feeding charts can be dispensed with; on the contrary they are quite necessary and extremely helpful and informative.

All nutrition tables do not consider all nutrients but, due to the importance of protein, calcium, phosphorus and Vitamin A, most nutrition tables show the amounts of these nutrients found in common feeds as well as the total percentage of all nutrients that can be digested from the feed. The abbreviation T.D.N. accompanied with a percentage figure indicates the total digestible nutrients contained in the feed. It represents the maximum of protein, carbohydrates, fats and energy that can be derived from the feed. The digestible protein usually designated as D.P. or Dig.Prot. is also shown by a percentage amount.

When considering protein the digestible protein is the most important figure, since not all of the crude protein can be used by the animal. A serious mistake is made if the figure for crude protein is relied upon as the digestible content. In some instances only about 60% of crude protein can be digested. If the crude protein percentage is far greater than the digestible protein percentage, the excess nitrogen could prove toxic for the horse. Rations prepared for ruminant animals are often very high in a form of crude protein known as urea. However, the horse's digestive system, especially that of the young animal, is not very well adapted to convert free nitrogen to protein. There may be a possibility of nitrogen poisoning from feeding too much urea, especially to young horses; however, this point is disputed. It is agreed that urea is not used well by the horse, therefore, because of possible nitrogen poisoning, urea should not be used in horse feed.

The content of calcium (Ca) and phosphorus (P) is usually shown for each as a percentage; however, it is not uncommon for these two minerals to carry values by weight shown in grams. The Vitamin-A component is referred to by whole numbers and decimal fractions and is measured by 1,000 international units (I.U.); but sometimes the percentage of Carotene from which Vitamin A is derived is the factor shown. The amounts of Carotene are so small that, when converted to weight, they will result in measurement by milligrams. The nutritive ratio (Nut.Rat.), if expressed, is shown in whole numbers and decimal fractions; for example 1:3.6. The ratio then is 1 to 3.6 and is arrived at by dividing the Digestible Protein figure into the sum of the digestible carbohydrates plus 2.25 times the digestible fats contained in the ration.

All nutrition tables do not show exactly the same percentages and ratios, but are relatively close. The values have been determined in the past by analysis through practical experimentation. What values remain in the fecal material and urine after digestion are subtracted from the values of the feed input. (Today the studies are much more refined and take into consideration energy lost through heat, gas and other minor factors). Obviously conditions and feed quality cannot be identical in all programs and slight variations result.

The older nutrition tables tend to intermix weight systems; heavy weights are shown in pounds equal to 16 ounces while the light weights, such as minerals, are shown in grams. The double system adds to the difficulty of formulating a balanced ration. Currently the trend is to show all weights in metric figures as well as pounds or percentages, which is a great mathematical help in adjusting the ration components.

When formulating rations it is often helpful to know that 454 grams equal one pound and that one milligram is 1/1000th of a gram. The National Research Council Encyclopedia of Feed Composition now shows the complete feed analysis in the metric system or percentages, except the Vitamin-A content, which is by I.U.s (International Units of Potency). A few years ago the N.R.C. decided also to include in its feed analysis the caloric system of measuring energy values. The N.R.C. Encyclopedia of Feed Composition contains highly refined analytical data for some seven thousand animal feeds. Feeds fed to horses are but a small part of that work. It is an indispensable tool for nutritionists but is, perhaps, too extensive as a practical guide for the horse farm. The N.R.C. bulletins on horse feeding and nutrition are less complex and deal exclusively with horses. These bulletins contain very reliable feeding information and are guides that all horse farms should have and use.

6.2 Body Requirements, Digestive System, Feed Palatability, Water.
a) *Body Requirements.* Like all animals, the horse requires a diet that supplies the proper amounts of protein, energy (carbohydrates and fats), minerals and vitamins. Unlike ruminants, horses have only one stomach and a comparatively small overall digestive tract. They cannot physically handle excessive quantities of bulky roughage. Even if roughage could meet the needs of a working horse, the quantity that would be needed would greatly overstretch the digestive tract and distend the underline to the extent that comfort and mobility would be impaired. However, the horse needs enough roughage to keep feed moving easily through the digestive tract, thus lessening the chance of constipation and colic. Wood eating and fence chewing often result when horses are maintained on pelleted complete feeds. Even though the pellets expand several times their dry size when eaten and mixed with digestive juices, they are not so bulky as roughage and are consumed too quickly to satisfy the animal.

Nature designed the horse as a grazing animal. When free to graze at will the horse nibbles leisurely all day, selecting the most desirable feed in the area. By confining the horse, man controls not only what kind and how much feed the animal will consume but also the time at which it will eat it. The feeder should keep the horse's natural feeding habits in mind. Horses do much better and utilize their feed more efficiently when fed small quantities often and at the same time each day. The digestive tract is designed to assimilate the amount of feed it can comfortably accept; more than that is not effectively used. Excessive bulk is tiring to the horse and interferes with its lung expansion and capacity for breathing. It is not good judgment to feed too much roughage to a horse that will soon be exercised, since that will drastically hinder its ability to perform.

b) *Digestive System.* Digestion begins in the mouth, where the feed is masticated and mixed with saliva. Saliva not only moistens the food for easier passage through the esophagus but also provides the enzyme ptyalin, which begins to convert starches into sugars. The chewed feed passes through the esophagus to the stomach, which has a maximum capacity of four gallons. The gastric juices secreted by the stomach lining begin breaking down proteins and fats. From there the feed passes into the small intestine, which is about 70 feet long, has a diameter of about three inches and a twelve gallon capacity, where starch, protein and fat are further broken down for absorption. The mixture then continues into the first section of the large intestine, called the caecum. The caecum, also known as the water gut or blind gut, is over three feet long and one foot in diameter and is analogous to the appendix of the human digestive system. This organ is the site of fermentation of cellulose by bacteria. The bacteria not only convert the cellulose into sugar but also produce a certain amount of protein and synthesize limited amounts of Vitamin B and K. Unfortunately for the horse, the caecum is located at the wrong end of the digestive tract. Unlike the ruminant, whose rumen acts like the caecum and is located in front of the stomach, the horse cannot fully utilize the fermented product because the feed has already passed through much of the digestive tract. In addition, the capacity of the caecum is only about 30 quarts, while the capacity of a cow's rumen is about 160 quarts. For this reason horses cannot handle so much roughage and must have higher-quality feed than ruminant animals. The final processing and assimilation of nutrients takes place in the great colon, where there is still some bacterial action, digestion and absorption of nutrients. The great colon is about 12 feet long and 10 inches in diameter. Waste material is solidified in the small colon and is discharged through the rectum in the form of dung balls.

The delicate digestive tract of the horse is no more equipped to handle feeds tainted with mold, dust, dirt or germs than that of the human. Such feeds can and do cause severe colic and other digestive disorders and must be avoided.

Except for Vitamin D and possibly some Vitamins B and K synthesized in the caecum, all other nutrients, minerals and vitamins must be provided by the feed. Aside from dirt-free, unspoiled foodstuffs and clean water, the ration should be properly balanced to furnish the horse's needs without causing constipation or undue looseness of the bowels.

c) *Feed Palatability.* Like people, horses too have their likes and dislikes and may refuse to eat feeds having an offensive odor or taste. The palatability of a ration refers to its attractiveness to the animal. A ration that horses eat well and seem to like would be considered palatable. If the feed is not palatable or the water is tainted, the horse will reject it, or if very hungry or thirsty, it may take a small quantity to tide it over in hopes that the next ration will show a better understanding on the part of the feeder.

The competent feeder is the one who can formulate a palatable ration and at the same time maintain the proper balance of nutrients. The horse has a sweet tooth; a little molasses mixed with less palatable concentrates or supplements will greatly increase the attractiveness of the feed. There is no economy in feeding what will not be eaten or what, when eaten, is inadequate. The single most expensive operational cost is feed, but unwarranted economy in this area is far more costly in terms of lost production, income and farm reputation for quality.

d) *Water.* Clear, clean, pollution-free water is of course absolutely essential to the horse, and is necessary for proper digestion. A horse will drink 8 to 10 gallons or more during a normal day. Lactating mares require much more water for milk production. Water requirements are increased by dry feed, sweating, high temperature and low humidity. Older horses' digestive tracts sometimes become sluggish and are more prone to impact. Extra salt can be added to the ration to increase consumption of water and reduce chance of indigestion.

A horse should never be allowed to drink its fill when overheated. It should be walked and allowed a swallow or two of water every few minutes until body temperature has returned to normal. Horses should have water free choice except when overheated or when water has been withheld for a long period of time.

6.3 Vitamins.

a) *Vitamins generally.* Because the average ration usually meets the vitamin requirements of mature horses, the importance of vitamins to the horse was not fully recognized until recently. Many broodmares and stallions were labeled genetically inferior and poor producers when they were actually suffering from vitamin deficiencies.

Vitamin deficiencies are slow to appear. The early symptoms are not too noticeable and progress slowly until, when finally recognized, the animal is suffering from a severe deficiency.

Most losses due to vitamin deficiencies are subclinical, showing up as poor production and inadequate growth. However, serious bone disorders may develop in young animals deprived of sufficient vitamins. The body is able to store a certain amount of vitamins, but this store will be quickly depleted if the animal does not receive adequate amounts in its diet.

Horses are known to need Vitamin A, some of the B group (riboflavin, niacin, thiamine, pantothenic acid and B_{12}), Vitamin D and Vitamin E.

b) *Vitamin A.* Vitamin A is synthesized from a plant chemical, carotene, which is present in large quantities in young green pasture. Alfalfa hay is one of the best sources of carotene, but there is a rapid loss during storage, especially during high temperatures. Alfalfa that has been well protected in storage will lose half of its Vitamin A value in much less than a year's time. Sun-cured grasses and alfalfa are lower in carotene than those cured by dehydration.

Deficiencies are often encountered in association with confined animals on a diet lacking in green or yellow feeds, or animals fed bleached hays that lack the nutrients provided in green forage, or animals fed parched feeds following lengthy drought. Vitamin A is thought to be responsible for epithelium tissue health. Most symptoms are closely associated with improper function of that tissue. Horses that are deprived of sufficient Vitamin A, after they are fully grown and developed, may be affected with night blindness, sometimes total blindness, lower fertility in both male and female, lowered resistance to lung and intestinal diseases or infection, poor hoof growth and poor muscular coordination. Although the young horse suffers from the same ailments as the adult horse, its problems are much more acute. Before birth the foal depends exclusively upon the mare, but even if the mare supplies sufficient Vitamin A, very little is stockpiled by the foal. When it is born the foal's epithelium tissue is exposed to foreign substances for the first time. Dirt, dust and germs must be offset by healthy tissue. Lack of Vitamin A has a decided retarding effect upon growth and development of the young horse. The bone and hoof structure is very dependent upon it. When a horse has been deprived of carotene for a long period, the system loses the ability to convert it into Vitamin A. It takes a long time, three years or more, before full conversion power is restored. This means that although sufficient carotene may have been added to the ration, the body will still not receive its needed Vitamin A. The problem can be helped by adding synthetic Vitamin A to the concentrates. This needs no conversion as does carotene, and is readily accepted by the body. The substitute brings about a more rapid recovery from the deficiency and is an inexpensive means of preventing a deficiency.

c) *Vitamin B*. It is not known if the entire group of Vitamin B complex vitamins is essential for horse health; however, riboflavin (B_2), thiamine, niacin, pantothenic acid and B_{12} all have a definite influence upon health maintenance. Vitamin B deficiencies usually occur when green feed is not available and a poor grade of hay and grain is being fed. Lack of the vitamin shows up as impaired night vision, sore back, anemia, nervousness and lack of appetite. Lack of riboflavin (B_2) is the specific vitamin deficiency believed to cause periodic ophthalmia (moon blindness). When neither green pasture nor high quality roughage is available, Vitamin B can be added to the ration by feeding a small amount of brewer's yeast or a commercially prepared supplement.

d) *Vitamin D*. The absorption of calcium from the gastrointestinal tract requires the help of Vitamin D. Calcium is vital as a component of bone. Without Vitamin D a calcium deficiency results and serious bone conditions arise. If the shortage is critical the young growing horses may develop rickets, and older horses, especially lactating mares, may suffer from osteomalacia, characterized by softness and brittleness of the bones and loss of bone density. Horses are able to make Vitamin D from the action of the ultraviolet rays of sunlight on the skin and can extract it from good-quality sun-cured hays or pasture grown on well-fertilized soil. A Vitamin D supplement can be given in the form of cod-liver oil (which is not very palatable) or irradiated yeast.

e) *Vitamin E.* Vitamin E (formerly known as Vitamin X) is the anti-sterility or reproductive vitamin found in abundance in the leaves of many plants, wheat germ and all seed oils. Lack of it causes low fertility in both stallions and mares and tends to cause muscular atrophy. Experiments conducted in Canada by veterinarians Darlington and Chassels indicated that Vitamin E increased conception rates and stallion sex drive, improved the quality of semen and improved the stamina and performance of race horses.

Sufficient Vitamin E is present in most rations, but if a deficiency exists it can be supplied by alphatocopheral succinate added to the ration. A Vitamin E supplement should be considered for breeding stock a month or so before the breeding season starts and for race horses at the time they begin training. Advice of the veterinarian or nutritionist should be sought before adding a Vitamin E supplement.

6.4 Minerals.

a) *Minerals Generally.* Horses require minerals as do all other animals. Except for sodium, chlorine and perhaps iodine, ordinary feeds of good quality, grown in well-fertilized soils, generally contain the needed minerals in sufficient quantities, but mineral problems can occur at times. Rations should be adjusted with the mineral requirements of each class of livestock in mind. Mineral supplements can be and usually are fed with the ration to balance the minerals that may be lacking in the feed. Although the body will store excessive minerals, they must be replaced as used to avoid ultimate deficiency.

Cobalt, copper, iron, manganese, magnesium, molybdenum, sulfur and zinc are of little concern in feeding, since almost all rations meet the horse's body requirements for these minerals. Selenium is not required, but if the feed abounds in it, poisoning can result; thus the danger of selenium is not in its absence but in its excess. Deficiencies of iron are seldom encountered, but should be considered in cases of anemia.

Other necessary minerals are given special attention in the following subparagraphs.

b) *Salt* (sodium chloride). Most plants contain little if any salt, therefore salt must be supplemented in all situations. If available free choice, the horse will take only what he needs, from 1½ to 3 ounces a day (seldom as much as 1½ pounds per week). If fed measured quantities with the ration, the horse must accept what it gets; if too little, it may develop a craving for salt; if too much, it will have to eat the salt to get the complete ration and will drink excessively, which causes irritation of the kidneys and thins the blood, causing performance and nutrition loss. Much salt is lost in urine. During hot weather or heavy exercise horses may lose a great deal by sweating. A horse in need of salt will show physical signs of fatigue. Lack of salt affects digestion, retards growth, decreases milk production, causes stiff muscles and rough hair coat and increases the possibility of heat stroke. The abnormal craving for it may appear as chewing fences, licking plaster, drinking muddy water by preference or eating dirt. Salt can be provided in block or crystal form and should be protected from the weather. Each salt station in the pasture should serve about 12 to 15 horses and should be moved regularly to prevent overgrazing in the area surrounding it. Some horsemen maintain that because a horse's tongue is not so rough as a cow's, the horse cannot easily lick enough salt from a solid block to meet its body needs and should be fed salt crystals.

c) *Calcium and Phosphorus.* Calcium and phosphorus are discussed together because they interact, and improper ratios in relation to each other can have serious consequences. It is necessary to control the quantity regularly consumed as well as the ratio. It is far better for the horse to have a little too much calcium than not enough. The calcium must at least equal the phosphorus and for safety should be a little higher, perhaps 1.1:1. The ration should consistently furnish a little more calcium than phosphorus. Bone problems are likely to appear when the phosphorus content exceeds the calcium. The animals are destined to unsoundness when calcium or phosphorus are deficient or the calcium-phosphorus ratio is top-heavy with phosphorus. When this deficiency or imbalance is suspected, a blood test to diagnose the problem should be made. The ratio can then be adjusted to compensate for the problem. The competent feeder will be very careful to see that the ration will meet the calcium and phosphorus requirements in the proper ratio. Nutritional charts that show the percentages of calcium and phosphorus contained in specific feeds are useful in computing the amounts and ratios. Commercially prepared feed mixes take into account the Ca:P requirements; however, breeding farms that formulate their own rations must consider the Ca:P problem.

Everything the animal eats must be evaluated when figuring the Ca:P ratio. Normally the grains and roughages contain ample phosphorus but are deficient in calcium. An imbalance of calcium to phosphorus is likely to be encountered when feeding large quantities of grain and protein supplements. Green feed and legumes are high in calcium and should make up the roughage in the diet of animals receiving highly concentrated rations. Animals that have been pushed for maximum growth in order to be shown, raced or sold as yearlings are the most likely to suffer from soft bones due to a calcium-phosphorus imbalance. An excess of phosphorus actually results in a calcium deficiency. The horses should be allowed free access to a carefully prepared mineral mixture that provides the proper ratio of calcium and phosphorus.

If horses do not receive calcium and phosphorus in adequate amounts in the proper ratio, bone diseases such as ringbone or sidebone will appear in adult horses and rickets will afflict the younger animals. Sore back will also result from calcium-phosphorus imbalance.

d) *Iodine.* Feeds grown in areas where the soil is deficient in iodine, such as the Great Lakes region, will not supply the needs of brood-mares, stallions and foals. Mares deficient in iodine during pregnancy are very apt to produce a weak and sickly foal that may be afflicted with goiter and therefore be more prone to navel-ill.

Stabilized iodized salt containing from 0.01 to 0.05 percent potassium iodide fed free choice is the supplement most used.

6.5 Roughages, Concentrates and Supplements. The two main feed categories are roughages and concentrates. Roughages are those feeds such as hay, pasture and silage which are high in crude fiber content and low in energy value. These feeds usually do not have a total digestible nutrient (TDN) value over 50%. The legume hays such as alfalfa, clover, and the like are much higher in protein than other roughages and, if of good quality, may maintain an idle mature horse. Roughages alone are inadequate for lactating mares, growing colts and fillies, stallions and heavily worked horses. Hays with heavy, coarse stalks and stems are inadequate in calcium and protein for horses of any category. Fine hay with an abundance of leaf foliage or grain head is much more desirable.

Concentrates are feeds that are high in energy and low in fiber, containing less than 18% crude fiber by dry weight. Their total digestible nutrient content is generally between 70 and 80 percent.

Supplements are feeds or feed mixtures used to improve the nutritive content of the ration and are fed in addition to other feeds. Plant source protein supplements fed to horses usually contain more than 30% protein and are high in certain vitamins and/or minerals. The high protein content may be because the natural basic ingredient is high in protein, but more usually it is the result of protein concentration after the carbohydrates and fats have been extracted from the basic feed by a milling operation. Animal source protein supplements are not palatable to horses and should not be used.

6.6 Hays (roughages). Before the manager can decide on good balanced rations for the horses on the farm he must have a thorough knowledge of the feeds available. He should know which are the best sources of proteins, carbohydrates, fats, minerals and vitamins as well as which feeds horses find most palatable.

Some of the more popular horse feeds are discussed in the following paragraphs.

a) *Alfalfa Hay.* Alfalfa hay is perhaps the most popular hay used on breeding farms. It is more difficult to cure properly than grain hays and should be checked carefully for signs of mold or sweating. Good alfalfa will have fine stems, much leafiness and will have been cut prior to full blossoming. It is mildly laxative, contains high-grade protein and provides carotene (Vitamin A), Vitamin B$_2$, D and nicotinic acid (niacin). It is very palatable and should not be fed free choice to idle animals, for they tend to overeat it. Alfalfa is excellent for young animal growth and for lactating mares; it is often the only hay fed to young stock and broodmares. Although excellent in most respects, it does not produce hard fat. Horses fed only alfalfa tend to sweat heavily; therefore, it should not make up more than half of the hay ration for show or race horses. A hay ration consisting of ⅓ to ½ alfalfa combined with high-quality hay produces a hard finish while maintaining the laxative quality of alfalfa, significantly reducing digestive disturbances. Alfalfa-grass hay mixture, grown and baled together, makes high-grade horse hay.

b) *Barley Hay.* Analysis indicates that barley hay is similar to oat hay in value, being somewhat lower in protein; however, bearded barley hay should not be fed to horses because it causes severe mouth wounds.

c) *Grass Hay.* Grass hay is made from grasses commonly used for pasture. Grass hay is very popular in the northern United States and is similar to timothy hay in feed value.

d) *Oat Hay.* Oat hay is one of the best grain hays for horses. Mature idle horses can be maintained on high-quality oat hay alone. It is low in protein and should be fed with a protein supplement whenever any kind of production is required of the animals consuming it. Oat hay is also low in calcium and does not meet the mineral requirements of most horses, especially those of lactating mares and growing foals. Oat hay must be cut and cured while in the "dough" stage to insure maximum nutritional quality. When properly cured it retains some green color and a large amount of grain. Red oat hay is one of the best varieties for horses; the stems are shorter and finer than those of most other varieties. The yield per acre is low, making red oat hay difficult to purchase, consequently it must usually be contracted for in advance of planting or be grown by the ranch.

e) *Timothy Hay.* Timothy hay is very popular in some areas. It is cured easily into a shiny, crisp, clean hay that is quite free from mold. As with grain hays, it is low in protein and minerals and should be supplemented. It will suffice for mature idle horses, but if fed alone is inadequate for lactating mares, young growth or for horses used for any type of production.

f) *Volunteer Hay.* Volunteer hay usually refers to a grain hay crop coming up from seed lost from shattered heads during harvest the previous season. The quality of such hay varies greatly from area to area and within the field itself. Volunteer hay must be carefully examined for poisonous plants and weeds that can cause mouth injuries.

g) *Wheat Hay.* The unawned variety of wheat hay is excellent for horses. It has a very sweet flavor and animals will devour it even if the stalks are rather coarse. The feed value is similar to oat hay. Bearded wheat hay produces mouth wounds and is not recommended.

h) *Wild Hay.* Wild hay refers to hay harvested from an uncultivated field that has produced enough excess forage to warrant baling. This random crop varies greatly and may consist of any grass or legume species. It is usually low-quality hay, lacking in protein, and is frequently intermixed greatly with weeds. Foxtails or other awned grasses are found in wild hay and cause mouth injuries. Quality often varies throughout the same field and all bales must be checked to insure uniformity. Most horse farms avoid the use of wild hay.

i) *Green Pasture.* There is no finer or more nutritious feed available for horses than good-quality green grass. It is extremely palatable, a tonic for horses, and is high in quality vitamins, especially carotene. Animals on good green pasture gain weight and have higher fertility and fewer digestive disturbances than animals on other feeds. Very young pasture has a high moisture content, which makes it impossible for the animals to eat enough to supply all of their needs; they must be supplemented with hay and perhaps grain. As pastures mature and begin to dry, protein, digestibility and palatability all decline. The T.D.N. in young pastures can be expected to be about 65%; however, this figure will vary greatly due to plant variety and maturity.

6.7 Grains (concentrates).

a) *Cracking, Crimping and Steam-Rolling Grains.* For most efficient feed use, all grains and seeds should be opened by cracking, crimping

There is no finer or more nutritious feed than quality green pasture. This particular field is Piper Sudangrass.

or rolling. Opening the kernel insures that the highly digestible part of the grain will be exposed to the digestive juices. This does not mean that grains or seeds should be finely ground; on the contrary, the dust from such milling can irritate the lungs and may cause heaves. In addition, finely ground grains tend to "ball up" when mixed with saliva, making them hard to swallow. This can result in impactions. During the crushing process, fanning the material will blow away most of the dust and dirt. Steam rolling (the crushing of the kernel in the presence of steam) also greatly reduces dust. Dust irritates the horse's nostrils and thus reduces the palatability of the feed. If dusty feed must be used, it should be dampened with water and steamed or sprinkled with a solution of half water and half molasses. The preparation of the grains and seeds in this fashion helps old horses with bad teeth and the youngsters under 10 to 12 months of age to chew and digest the feed better. The cracking, crimping or rolling should not be done far in advance of use. When exposed to the air, the kernel loses much of the aroma and has a tendency to deteriorate somewhat in feed value. The opened hull allows the evaporation of natural moisture and the feed becomes dustier. The processing should be limited to a six- or eight-week supply.

b) *Barley*. Whole-grain barley is hard and extremely difficult for horses to chew well enough to break open for easy digestion. Much food value is lost when the grain is swallowed whole. Steam-rolled barley, however, is very digestible, high in energy and is often substituted for oats, especially on the Pacific coast. It is a heavy feed and, unless mixed with a bulky, lighter feed such as oats or bran, can cause colic. Although more concentrated, its overall value is close to that of oats. Some horsemen avoid its use on the theory that the particular protein found in barley is difficult for the horse to digest, but if steam rolled and mixed with other grains, little problem is encountered.

c) *Corn*. Corn is high in energy and very palatable, but is low in protein and calcium. It has a high fat content and provides more caretene (Vitamin A) than other grains, but provides no bulk. It should be rolled or cracked and mixed with a bulkier feed to avoid colic. The cob and grain ground very coarsely or chopped provides a good grain with plenty of bulk. The high energy value makes it extremely good for horses performing strenuous labor but produces body heat and consequently sweat. The lack of protein must be made up by alfalfa, legume hay or a protein supplement. This is especially true for lactating mares and growing youngsters.

d) *Milo*. Whole-grain milo is hard and difficult to chew. It should not be fed as whole grain; cracked or steam rolled, it is a good concentrate but, having no bulk, it should be mixed with feeds such as oats, to provide the bulk needed to avoid impaction and colic.

e) *Oats*. Oats, a light grain, is by far the best grain for horses. The hull (30% or more of the kernel) provides bulk, enabling consumption of large quantities unmixed, without much danger of colic. As with other grains it is low in digestible protein and calcium and due to the bulk is lower in total digestible nutrients but can be consumed in greater amounts. Oats is the preferred grain for all horses. Rolling or crimping increases the nutritional value and is recommended, especially when feeding very old or very young horses.

f) *Wheat and Rye.* Wheat and rye grains are fairly good in digestible protein, high in total digestible nutrients, and have energy values close to those of corn. They are difficult for the horse to chew and should never be fed unless rolled. They form a sticky mass that tends to ball up in the digestive tract, which causes severe colic. For this reason, and because they are usually very expensive, they should make up no more than 1/4th of the grain ration if used at all.

6.8 Protein Supplements.

a) *Linseed Meal.* Linseed (flax) meal is the preferred high-protein supplement for horses due to its palatability; it contains about 34% digestible protein and is a good source of phosphorus but contains no Carotene (Vitamin A). It is quite laxative and acts as a conditioner for horses, causing them to shed earlier and grow a very healthy coat. Since linseed meal is both laxative and expensive, no more than a pound a day should be fed. Flax grown in cold climates may be high in toxic hydrocyanic acid. For this reason it should be fed sparingly and carefully.

b) *Cottonseed Meal.* Cottonseed meal is an excellent source of protein (40%), and is high in phosphorus. Although lacking in Carotene (Vitamin A) and Vitamin D, it does contain some Vitamin B. Because it is a very heavy feed, it can cause colic if overused and has been known to cause a toxic condition when used excessively. Quantities in excess of one pound a day have been fed with no ill effects, but much less is a safer level. This meal produces a fairly solid fat and results in a glossy coat.

c) *Soybean Meal.* Soybean meal contains 43% protein and, like other seed meals, is high in phosphorus. It contains Vitamin B, particularly niacin, but is lacking in carotene and Vitamin D. As with cottonseed meal, soybean meal is very heavy and can cause digestive disturbances; one pound should be the limit in the daily feed mixture. The adding of this supplement like other protein supplements produces a healthy coat with a glossy finish.

d) *Meat or Fish Meals.* Meat or fish meals are not suitable as protein supplements, primarily because they are not palatable to the horse.

6.9 Other Supplements.

a) *Wheat Bran.* Wheat bran is made from the outer layer (aleurone) of the wheat kernel. It is a popular laxative feed for horses and is highly palatable. It is high in phosphorus but low in calcium. This feed has been associated with bladder stones (urinary calculi). Bran mashes need not be routinely fed, but if so fed, a pound and a half per day should be the limit. An occasional feeding of 2 or 3 pounds of mash is desirable when the condition of the horse or the nature of the feces indicates a need for a laxative. A bran mash is easily made by dampening 2 to 4 pounds of wheat bran with boiling water, adding ⅔ tablespoon salt, covering and allowing to steam for a half hour.

b) *Beet Pulp.* Beet pulp is quite digestible, although bulky due to its high fiber content. Calcium is high but phosphorus is low. It contains carbohydrates and is high in energy but quite low in protein, carotene (Vitamin A) and Vitamin D. Though not particularly palatable to horses, they will nevertheless accept beet pulp.

c) *Molasses.* Molasses is exceptionally palatable due to sweetness and is added to feeds (especially grains) to increase consumption and reduce dustiness. It is high in energy and is often laxative. Animals should not consume more than one or one and a half pounds of molasses a day; larger amounts not only can be too laxative but can cause excessive sweating, which reduces the amount of hard work the animal can do in hot weather. In addition to providing energy, molasses is high in iron and calcium but contains almost no protein and carotene.

d) *Alfalfa Molasses Meal.* Alfalfa molasses meal is a mixture of finely chopped alfalfa hay and molasses. Its feed value is only slightly higher than that of alfalfa hay; however, it is very palatable, easily chewed and can stimulate old horses with poor teeth and young stock to consume more feed. It is a good source of iron and energy and is fairly high in carotene.

e) *Yeast.* Yeast seems to be a tonic to horses. When used, two tablespoons per day are usually fed. It is high in the Vitamin B complex, protein and phosphorus.

6.10 Pelleted Complete Feed. In recent years commercial feed and milling companies have recognized the need for a feed mixture having all components needed by horses and measured in quantities proper for general use. These mixtures contain chopped roughage, grain, supplements, minerals and vitamins pressed together in pellet or wafer form and packaged in feed sacks or boxes convenient in size and weight for easy handling. This feed is easily stored in a small area and requires only distribution to the horses in proper amounts at feeding time. The feed is far more bulky in the digestive tract than in the package. The saliva, water and digestive juices cause it to swell to several times its dry size, but it nevertheless does not produce the bulk of nonpelleted feeds. It is consumed quickly; some horsemen believe too quickly, since the horse prefers to devote much time to eating. Some horses, because they finish the ration quickly, may resort to chewing on the manger or other wood at hand. Lack of roughage quantity is thought by many horsemen to be the primary cause of wood chewing or wood eating. It has been noted that animals used heavily and fed strictly on pellets take on a gaunt, drawn appearance, even when fed in excess of recommended amounts.

For the owner with very few horses, the pelleted complete feed may be the best approach, especially if he does not want to spend time studying equine nutrition; but for the sizable horse farm that uses much feed, pellets can become quite a bit more expensive. Although hay takes up more storage space and there is some waste when hay is fed in bulk, the savings should be considered against the additional cost of pelleted complete feeds if used extensively. Like all businesses, the horse farm is ultimately concerned with cost and gross income. Pellets are a relatively new innovation in horse feeding. The advantages and disadvantages should become more apparent as more research is done regarding the pelleted rations.

6.11 Feeding Classification Table. All horses do not have the same nutritional requirements. Each class needs special consideration. Age, development, type of work—all bear upon nutritional needs. Foals are a special feeding class and gestating mares and stallions in production, although generally classed broadly as mature horses, nevertheless must be singled out as distinct subclasses. In connection with this chapter on nutrition, sections in other chapters should be consulted for special attention to stallions, mares and foals and for other points of general concern, as follows:

Sections 2.16 through 2.20—Pastures

Section 4.8—Grain bolting

Sections 8.4 and 8.5—Stallion feeding

Sections 9.1, 9.2, 9.3 and 9.4—Mare feeding, diet and pasturing

Section 11.2—Foal feeding

No charts or tables can be expected to unalterably classify or prescribe rations for all horses, but the following tables suffice very well until some special situation comes to the manager's attention that indicates that a change of ration is in order.

FEEDING CLASSIFICATION TABLE

LIGHT HORSE RATION (1000 lb. mature horse)

STATUS	Crude Protein (% C.P.)	Lbs. Dig. Prot. (D.P.)	lbs. Cons.	lbs. Rough.	gm. Ca.	gm. P.	mg. Carotene
Mature							
Maintain	9	.7-0.9	0	23-27	13.7	15.4	50
Light Work	9	.7-1	4-5	20-23	13.7	15.4	50
Medium Work	9	1.0-1.2	8-10	20-23	13.7	15.4	50
Heavy Work	9	1.2-1.4	13-14	20-23	13.7	15.4	50
Stallions							
Breeding Season	14	1.4	8-15	18-25	15	15	60
Idle	9	.7-0.9	0	23-27	13.7	13.7	50
Mares							
First 2/3 gest.	9	.7-0.9	0	23-27	13.7	15.4	50
Last 1/3 gest.	12-14	1.2-1.4	10	25	15	15.4	60
Lactating	14-16	1.4-1.7	10-15	18-25	24.4	22.2	60
Creep fed Foals	20	2	5-8	0	12	10.6	12
Weanlings to 12 mos.	15	1.2	6-9	Free C.	13.4	12.5	24
Yearlings							
12 mos. to 18 mos.	11	1.1	8-13	Free C.	13.4	13.4	36
18 mos. to 24 mos.	11	1.1	5-10	Free C.	11.6	11.6	48
Two-Year-Olds	11	1.1	5-10	Free C.	12	12	50

Table abbreviations: Ca.—Calcium; Carotene—produces Vitamin A; Cons.—Concentrates; C.P.—Crude Protein; D.P.—Digestible Protein; Free C.—Free Choice; gest.—Gestation period; gm.—gram; lbs.—pounds; mg.—milligrams; P—Phosphorus; Rough.—Roughage.

Note: 1) The table indicates suggested minimum requirements and does not consider the ratio of calcium to phosphorus, as discussed in Section 6.4 (c). When using the table it must be kept in mind that calcium must be supplemented in some cases in order that the ratio is not less than 1:1 and preferably Ca:P should be 1.1:1.

2) Maintain means requirements for mere maintenance of a mature idle horse (a pastured horse exercises too much to be classified as idle).

   Working denotes being ridden or the expenditure of equivalent energy, generally classified as follows: light work—2 or 3 hours of pleasure riding; medium work—3 to 5 hours of pleasure riding; heavy work—5 to 8 hours of pleasure riding.

3) Cold weather requires more feed for heat to maintain an animal.

Feeding, as suggested by the above table, is a good starting place, but there is no substitute for diligent daily attention to the individual horse's general condition and attitude toward its ration.

6.12 Feeds and Feeding Check List. Several miscellaneous points concerning feeds and feeding are mentioned here as a quick check list that may be of use:

a) Equipment.

   Low mangers or racks for feeding youngsters or groups of horses are accident hazards. Hay racks open on all sides will prevent the cornering of a horse while it is eating. Heavy rubber tubs spaced about 20 feet apart in the pasture or paddock are excellent for feeding grain and separating the horses to avoid fighting; they are portable and can be moved to clean areas each day.

   Feed boxes must be kept clean and fresh at all times; they should be scrubbed often and sun-dried. Metal, rubber and plastic material greatly facilitate cleaning.

   Feeds should never be fed upon the ground, especially in confined areas; if so, the horses are bound to become reinfested with worms and ingest dirt and filth, which can lead to digestive disturbances.

   Nose-bags, if used, should have a number of small holes near the bottom for drainage, to prevent drowning in the event the horse tries to drink while wearing it. The nose-bag should be removed immediately when eating is finished and should not be used except when an attendant is nearby.

b) Grain (Concentrates).

   Overfeeding of concentrates can cause founder. Grains should never make up more than 50% of the ration's T.D.N.

   Grains fed often in small amounts contribute to more thorough digestion.

   Horses closely confined for long periods lack conditioning exercise and should be fed $\frac{1}{2}$ to $1\frac{1}{2}$ pounds of bran mixed with oats daily to avoid constipation and colic.

When practicable, a small feeding of hay just prior to feeding grain is preferable.

Steam-rolled grains are less dusty than dry-rolled or cracked grains and are therefore preferable.

Grains should never be finely ground; this causes extensive dustiness.

Good-quality grain is free from heat, mold and ground damage, has not sprouted, contains no excess moisture and cannot be pressed into clumps.

Horses that bolt grain need corrective control such as thin spreading of the grain, mixing the grain with straw or placing a large smooth stone in the grain box.

c) Hay (Roughage).

An all-roughage diet lacks energy for horses required to perform production or work, limits the capacity for breath and distends the underline into a "hay belly," sometimes referred to as "pot belly," which has an unsightly appearance as well as being cumbersome and tiring to the horse.

Horses need roughage, at least 50% of the T.D.N. of the ration, to avoid colic and constipation.

During cold weather, animals require more roughage; the breakdown of roughage in the caecum releases a great deal of heat, which the animal uses for body warmth.

Alfalfa hay cut too soon is very laxative and often causes diarrhea, also referred to as scours.

Good-quality hay is free from dirt, dust and mold, has a sweet, clean aroma, is light green in color and has fine stems. Grain hays have considerable grain intact and nongrain hays have an abundance of leaves. If properly cured and baled, the bales will be springy and will bounce when dropped, will not sweat and will not be hot.

d) Pasture

Good pasture is the best roughage a horse can get. Free grazing should be permitted whenever possible, but if not possible some freshly cut green feed should be furnished. Green feed is especially important for young growing animals. When the grass is short, pastured horses need more feed to make up for lack of grazing. Pastured horses are not truly idle horses and need more feed than charts indicate for maintenance.

A hungry horse should be fed hay before it is turned out to a lush green pasture. A satisfied animal will be less apt to gorge itself and develop colic.

e) Storage.

Moist feeds lose much of their nutritional value when in storage. The drier the feed when stored the better the nutrient retention. Tightly packed moist feeds, when stored where ventilation is restricted, have a tendency to mold and sometimes cause fires due to spontaneous combustion.

f) Ration Changes and Feed Variety

All ration changes should be made gradually. New feeds should be blended with the customary feeds so that the changes will be gradual rather than abrupt. Abrupt diet changes sometimes upset the digestive system.

A variety of feeds should be available for three reasons: horses, like people, tire of the same ration; deficiencies in one feed are made up by other feeds; and variety provides a means of occasionally cutting costs.

g) Minerals and Vitamins

Free access to salt should be provided. Trace minerals can be included with the salt. Commercial blocks containing minerals are excellent sources when placed for free-choice use. Supplements can be purchased or home-mixed to cover specific vitamin and mineral requirements.

h) Water

Horses will drink from 8 to 10 gallons or more of water daily; lactating mares need much more.

Water should be available free choice at all times, with the exception of horses hot from work or exercise. Such animals should be allowed only a few swallows every three or four minutes and should be walked until cooled to normal body temperature. Health depends upon the presence of sufficient water in the body at all times.

i) Miscellaneous Notes.

Horses on good hay as the sole feed, consume about 2.5% of their body weight every day.

Horses generally eat better when competing with others for their share.

Careful observation of horses when eating will show which are the "shy" feeders and which the greedy. The ration of each horse must, if necessary, be adjusted on an individual basis.

Horses eat better if required to clean up each ration; they should not have feed before them constantly.

Hay-belly (pot-belly) can be reduced by feeding pelleted complete feeds or high protein concentrates and smaller amounts of high-quality hay.

Hay sprinkled with salt or sweetened molasses water (using equal parts of molasses and water) often tempts a sick horse or one off his feed.

Dirty, moldy or damaged foodstuffs must be avoided to prevent colic and other digestive disturbances.

Dusty foodstuffs must be avoided to prevent respiratory ailments such as heaves.

Routine inspection of manure will indicate how well the feed consumed is being utilized by the individual horse.

Green feed must be fed immediately after cutting and uneaten green feed must be thrown away, since harmful bacteria and mold will grow in it fast enough to cause serious illness within a few hours.

Bad teeth or parasite infestation will impair the horse's ability to use its feed. Regular dental inspections and examinations for parasite infestation should be made to prevent loss of feed efficiency.

6.13 Grain-Ration Formulas. The farm may choose to make up its own grain mixture formula to compensate for deficiencies in its particular pastures or to utilize a feed that can be obtained economically or easily, or the farm may choose to make use of one of the many well-balanced grain mixtures commercially marketed. Whatever the origin of the mixture, a typical well-balanced grain-ration formula consists of:

Components (100 lb. mixture)	Percentage	Weight in lbs.
Grains (rolled corn, oats, barley or milo)	82.0	82.0
Protein Supplement (linseed meal)	5.8	5.8 (5 lbs. 13 oz.)
Molasses (blackstrap)	10.0	10.0
Minerals (calcium and phosphorus 0.5 each)	1.0	1.0
Salt	1.0	1.0
Trace Minerals (Cobalt, copper sulfate, iron, manganese sulfate, potassium iodide, zinc)	0.2	0.2 (3+ oz.)

An example of a popular grain-ration formula is marketed on the West Coast under the brand name "Frontier Mix"; per ton it consists of:

Components	Pounds
Red Oats	1000
Rolled Barley	200
Cracked Corn	200
Wheat Bran	100
Linseed Meal	100
Dehydrated Alfalfa Leaf Pellets	300
Molasses	100
Plus	
Salt	20
Oyster shell flour	20
Dicalcium phosphate	10

One pound of this formula supplies .113 lbs. D.P. (digestible protein), .712 lbs. T.D.N. (total digestible nutrients), 4.36 gm. Ca (calcium) and 2.87 gm. P (phosphorus) and provides adequate carotene through the dehydrated alfalfa-leaf pellets. This mix is meant to be fed with oat, timothy or grass hay. When five pounds of this mix is fed with 20 pounds of oat hay, the horse is supplied with 1.57 pounds D.P., 12.96 pounds T.D.N., 40.8 gm. Ca, and 31.6 gm. phosphate. This is adequate for maintaining any adult animal and the calcium-phosphorus ratio is quite adequate.

6.14 Nutrition Experts. Determining the nutritional requirements of the ranch horses is an art that takes years to develop. While books can assist in many respects, knowledge acquired through daily observation of the eating habits of the individual animals is essential for a true understanding of the feeding requirements. It may be wise for the ranch to consult a nutrition expert when developing the rations to be used on the farm. Such an expert will be able to analyze the soil, water and pasture grasses. He will be quick to recognize deficiencies in the ration which must be corrected. He can greatly assist in the development of good pastures. Such specialists may be located through the various agricultural colleges, the Farm Bureau or the Department of Agriculture.

6.15 Finicky Eaters. Pastured animals are not usually finicky eaters. Companionship, physical activity, variety of feed and competition from other horses all encourage healthy eating habits. However, young horses and stallions confined to stalls or separated from other animals often become finicky eaters. Confinement and isolation furnish an unnatural environment that often depresses the horse and causes nervousness. Such animals sometimes reflect a disinterest in their feed in spite of having nothing to do but eat. Of course, the cause of finicky eating is not always the mental attitude; such an animal may be physically ill and should be examined by the veterinarian. If it is in good health, many tricks can be used in an attempt to stimulate its appetite.

a) Vigorous physical activity will often create a natural good appetite.
b) If the horse is being fed alfalfa, a small amount of grain hay may be added, or if it is on one type of hay, a mixture may be fed, such as one-half red oat and one-half alfalfa.
c) Another type or mixture of grain may interest it. This of course must be of similar content so that the animal's digestive system will not be upset by an abrupt change.
d) Feed can be withheld for a day and then given sparingly. As the horse cleans up the food, more can be added, but when it stops eating, the remainder should be removed. Leftover feed often reduces appetite. The point is to create an anxiety for food so that the animal will eat more when it is fed.
e) A tidbit the horse likes added to its feed may stimulate eating. As suggestions, carrots, apples or other fruit, watermelon, dried bread, oatmeal, sweets such as honey, sugar (white or brown) or molasses diluted with water. (Many tempters are quite laxative, especially sweets.) Hand feeding may coax eating.

f) Competition will often increase appetite. A greedy horse can be placed next to the finicky eater and feed placed in close proximity to the other horse. The greedy animal's attempts to get the food may stimulate the finicky eater to consume more.

g) Feeding grain in a nosebag will sometimes encourage the horse to eat more. (Holes should always be punched near the bottom of the nosebag to allow water to drain out should the horse submerge it in an effort to drink. Nosebags should be used only when an attendant is nearby and should be removed immediately when the horse finishes eating.)

h) When even slight eating has been reestablished, a tonic added to the grain may further increase the appetite. Such an additive may be any of a number commercially marketed or one prescribed by the veterinarian.

i) In the event the refusal of sufficient feed is prolonged, the veterinarian should be kept informed. He may want to make a further examination or administer a medication; Vitamin B shots are sometimes used to increase appetite. Aspirin is effective in some instances.

6.16 Hay Procurement. Many ranches purchase hay standing in the field before it is harvested; others contract farmers to raise it for them or grow it themselves. These practices increase the likelihood that the horse farm will get the exact type of hay desired. Red Oat hay has a low yield per acre and is rarely grown for sale purposes; however, it can be obtained when the breeding ranch maintains its own hay-raising operation or when that type is contracted for in advance of planting. By the purchase of standing hay or hay to be grown, the manager has the advantage of supervising its cutting and curing to ranch satisfaction. He can take delivery in bulk and can designate light-weight bales or loose baling for ease of handling and less crushing. Hay pressed very tightly in the bale is more difficult for the horses to eat.

When buying hay, the manager should know the quality produced in the area where it is grown. There is great variation in hays grown in different areas, due mainly to soil properties and rainfall. The differences in nutritive value and palatability in hay from different localities are quite substantial. The manager must know the agricultural conditions under which it is grown; otherwise he will not know what he is getting until he begins feeding it.

If it is possible to obtain and feed some of the hay before buying all of it, the horse's attitude toward it can be a good indication of its quality. Hay that is high in nutrients and beneficial minerals will be relished and well cleaned up by the horses, even though a little coarse. Before contracting in advance of planting, it is highly advisable to test the horses with hay grown on the hay farm the previous season.

Where the breeding farm has the farm land available or can lease it nearby, haymaking can be a very profitable sideline. Fluctuating market price, cost of transportation, control in times of scarcity and saving the farming profit all add up to good business for the horse-breeding farm.

6.17 Grain and Molasses Procurement and Cost of Total Digestible Nutrients. Some farms purchase whole grain in large bulk lots and

Rodent-proof and moisture-proof grain storage bins should be situated so that delivery can be made conveniently.

crack, crimp or steam roll it themselves in six-or-eight-week supplies. This timely use of the opened kernels avoids the loss of natural moisture and deterioration of feed value. Quite a saving can be had in purchasing grain by the carload or truck-trailer load through a reputable grain broker and by buying molasses by the tankload. However, the large investment in equipment necessary for processing the grain makes this method of grain procurement impractical for smaller farms.

Ready-mixed grains can be purchased wholesale if the order is large enough, but most farms prefer to formulate their own rations and pay the local feed mill to mix it in one-ton lots. In any case the grain must be protected from the weather and rodents, preferably stored in moisture-proof and rodent-proof bins, which can be purchased as manufactured or built to specification. The farm-storage bins for bulk grain should be situated so that the delivery vehicle can make direct transfer of the grain into storage by the use of a chute, which the vehicle carries for the purpose.

In calculating the cost of various grains, the farm manager's attention should be focused upon the cost of the Total Digestible Nutrients. It is the feed value that counts; the higher the nutrient content of the feed, the less quantity is needed. The comparative cost for grain-feeding values can be easily determined by establishing the cost of the nutrient value. Since grain is usually sold by the hundredweight, this is a convenient measurement to use. Assume only for the sake of example that the total cost of oats is $4 per hundred pounds, and the oats have a TDN percentage of 71 and that the total cost of barley is $3.40 per hundred pounds and the barley has a TDN percentage of 76. This means that in buying oats, 71 pounds of TDN costs $4 or a little over 5.6 cents per pound, while in buying barley, 76 pounds of TDN cost $3.40 or a little over 4.4 cents per pound. Obviously, on this hypothesis, barley is about 1.2 cents per pound of TDN cheaper.

All feeds can be compared for cost in this way if the current price and the TDN are known.

The cost of mixtures, although requiring more extensive calculations, can be compared in the same fashion; however, the cost of additives such as minerals, vitamins and molasses must be included in the total cost.

7
BREEDING

(SYNOPSIS)

This chapter considers the breeding season, breeding methods, artificial insemination, mare classes, service to outside mares, physiological aspects, determination of the mare heat cycle, covering procedure and determination of pregnancy. The following synopsis by section numbers is provided as an aid for quick reference to specific subjects.

Section 7.1   Breeding Season (early breeding for early development; official birthday for performance horses; breeding methods)

Section 7.2   Pasture Breeding (only one stallion in the pasture; problems with registration of foals; hazard)

Section 7.3   Corral Breeding (hazard; not recommended for general use)

Section 7.4   Hand Breeding (controlled breeding; recommended for general breeding program)

Section 7.5   Artificial Insemination (procedure; advantages and disadvantages; foal registration problems; utilization of crippled and unsound stock)

Section 7.6   Outside Mares (mares on short visits; mares visiting for the entire season; isolation to avoid communicable diseases; identification numbers; mare's records)

Section 7.7   Ranch Mares (separation of barren, maiden and foaling mares; records)

     a   Barren Mares (causes of barrenness; examination of breeding health; rectal and vaginal examination)

     b   Maiden Mares (examination; opening hymen; vaginal septum; infantile genitalia; irregular estrus; jumping)

     c   Foaling Mares (breeding on foal-heat, or post-parturient estrus; advantages and disadvantages)

     d   Sutured Mare (vaginal health; opening at breeding and foaling time)

Section 7.8   Breeding Time (life of egg, ovum; life of sperm; ovulation; heat cycles; indications of heat; breeding intervals; reproductive tract examination)

7.1 Breeding Season. Compared to other animals the horse is not especially prolific. The national foal crop averages about 60%; however,
farms with efficient breeding programs and skillful management quite
consistently achieve foal production of close to 85%. The success or
failure of a breeding operation is dependent upon the farm's production
of as many quality horses as possible. The manager's ability to organize
and maintain an effective breeding program is the very foundation
of a successful breeding farm.

The time of the year the foals are wanted determines when the mares
must be bred. The date the foal is born is of great concern to those who
want to race horses, show young horses at halter, or enter them in
futurities. These horses in particular need as much growth as possible
by the time they are expected to compete in their age groups. For the
purpose of competition January 1 has been set as the birth date of all
horses, regardless of the day of parturition. The closer after January 1
the foals are born, the more time they will have to grow and develop
before reaching their first recognized birthday and the greater their
advantage will be over competing animals foaled later but whose annual
birthday is also January 1. The weather conditions and farm facilities
will have a great deal of bearing on how early the mares should foal.
Generally the farther north the ranch is situated, the later the regional
foaling season begins, due to the cold, wet weather conditions in the
early months of the year.

Normal gestation of a mare is approximately 337 days. For this
reason many farms begin breeding around the middle of February,
anticipating that many foals will arrive in the first two months of the
following calendar year. This is an unnaturally early time for horses
to breed and only a small percentage of mares are likely to conceive.
The rate of ovulation increases with the number of daylight hours and
the amount of green pasture grass consumed. Mares are often referred
to as "long-day" breeders because they are most fertile in the time of

year the days are longest. The most fertile period for breeding mares is when their winter coats are about half shed out. The most natural months for horses to breed are from April through June and these are the months with the highest rates of mares settled. In the northern states the season is a little later. Mares can be settled year round, but foals born later than June will have more difficulty competing with earlier foals in their age class. If the young horse is not to be used for racing, showing or sale, the mare may be bred to foal later in the season and the foal will benefit from milder weather.

Several methods of breeding are employed by the various breeding farms. Pasture breeding, corral breeding, hand breeding or artificial insemination can all be used.

7.2 Pasture Breeding. Pasture breeding is accomplished by turning the stallion out with a band of broodmares and allowing him to remain with them throughout the breeding season. For the foals to meet most associations' registration requirements the mare must not be exposed to more than one stallion of breeding age, and if the stallion pastured with the mares is changed, a period of at least 30 days must elapse before a different stallion is given access to the band of mares. Although the conception rate in pasture breeding is very impressive, usually higher than that of other methods, it has its hazards. When expensive animals are being bred, pasture breeding is not a sound economic practice; turning a valuable stud loose with a band of mares greatly increases the likelihood of injury over the slight danger involved when only the stallion and the one mare are bred under the control and supervision of the studman. Another disadvantage is that it is difficult to establish in advance parturition dates for mares that are pasture bred. Also, some stallions running with mares become very aggressive and will attack another horse or a person that invades his domain.

7.3 Corral Breeding. When corral breeding, the two animals to be bred are turned loose together in a small pen. The corral should be empty except for the mare in heat and the stallion to which she is to be bred. They can be inconspicuously observed for assurance that all is going well. This method is often employed with young stallions that are not yet sure of their roles. If the stallion is young and unsure, it is wise to use an older, gentle mare that will not kick. Left to themselves the young horse will eventually figure out what is expected of him. This method is not recommended for regular use with valuable stallions. A cantankerous mare can be the ruination of a stallion if he is kicked. A young, shy, breeding stallion can be placed over the fence from a mare in heat for a day or two so that he can have time to become acquainted and more aggressive.

7.4 Hand Breeding. The preferred and safest method of breeding used on most farms is hand breeding. With this method both the stallion and the mare are under control at all times and precautions against injury and infection can be and are taken. This method is highly recommended and will be discussed in detail later in this chapter under Sections 7.8, 7.9 and 7.10.

7.5 Artificial Insemination. Artificial insemination for horse breeding is not used extensively for the following reasons: 1) the semen does not live long outside the vagina (from 10 to 30 minutes after ejaculation) without careful preservation. (A preserving method has been developed so that semen may be frozen and shipped.) 2) Collecting semen, preparing it for use and impregnating the mare are very complicated and difficult procedures. (The inseminator must be well trained in this field. Many colleges and insemination companies conduct special insemination classes, usually dealing with cattle only, but the principles are much the same and are of value to the horse breeder.) 3) Most breed registration associations recognize foals gotten by insemination only if the stallion and mare are both on the ranch premises at the time of impregnation.

The most sanitary method of collecting a stallion is by letting him mount a mare in heat and diverting the penis into a specially designed collecting tube known as an artificial vagina. The semen is diluted and divided into portions equal to the number of mares to be bred. The inseminator inserts a small speculum into the vagina of the mare. The interior of the vagina is then lighted by means of a flashlight or examination lamp so that the relaxed cervix can be easily located. A pipet is threaded into the opening of the cervix as far as possible and the semen is then injected into the cervix through the pipet.

The impregnation can otherwise be accomplished by the same use of the speculum and light to implant a gelatin capsule containing semen by pressing it deeply into the cervix, where it will dissolve and release the sperm. This method has a slightly higher rate of infection due to handling and is not generally used.

All equipment used must be thoroughly sterilized. Disposable pipets and plastic gloves are commercially available for use in insemination procedures.

The advantages of artificial insemination are: 1) several mares (perhaps 10 to 15) can be bred by the sperm of one stallion ejaculation; 2) mares suspected of having an infection may be bred in this way with no danger of infecting the stallion; 3) a small or crippled mare may be bred with no danger of physical injury. Likewise, a lame or crippled stallion can be bred more easily to a larger number of mares; 4) a localized infection in the mare's vagina can be by-passed, thus reducing the possibility of spreading the infection to the uterus.

7.6 Outside Mares. Procedures for stud service to off-ranch mares may vary depending on distance and the preference of the farm manager and the owner of the mare. The owner of the mare may determine when the mare is in heat and bring her, along with her records, to the farm for the duration of her estrus cycle. This procedure is convenient for the owner of the mare that is stabled nearby but impractical for those coming from considerable distance. The procedure most often preferred is for the owner of the mare to deliver her to the farm prior to the breeding season and board her there until she is determined to be in foal or until the breeding season is over.

A short visit creates some objectionable conditions, which can be overcome when the stay is to be for the whole season.

The chance of introducing a communicable disease into the ranch herd is increased with the influx of outside mares. As a precaution against this threat, newly arrived mares should be kept at the isolation unit for three weeks prior to being moved to the main portion of the ranch. If a mare is brought to the farm to be bred during only one heat cycle, she should remain at the isolation unit for the duration of her stay. Communicable diseases are too costly to risk for the price of a stud fee. Short-visiting mares usually have a lower conception rate, often because the owners are perhaps not too knowledgeable in determining the proper time to breed, or because most owners without a stallion cannot tease to establish the mare's heat cycle accurately. As a result, the number of successful breedings to such mares is quite low. In addition, the mare will probably be upset by the change of surroundings; such distress can greatly reduce fertility. Mares arriving at various times create a difficult scheduling problem for the manager; he must maintain a breeding program that will not overtax the stallion. If the manager is familiar with a mare he will know her estrus cycle and how long it will last so that he can place her in a workable breeding schedule and have her covered at the proper time. Many breeders will not breed a mare if there is doubt that she is in heat, preferring to skip that period. Because of these complications many farms do not accept a mare unless she is to remain throughout the breeding season.

Almost all farms prefer the long-term arrangement. Under the full-season visit plan the outside mares are usually received about the middle of December or the early part of January (one or two months prior to the beginning of the breeding season) so that they can be teased to establish their heat cycles prior to the actual breeding period. The mares are kept at the farm isolation unit for three weeks before being brought onto the main ranch, and are teased at those facilities during the quarantine. When released from isolation the outside mares can either be kept separated from the farm mares or mixed with them, depending on the facilities of the farm and the preference of the manager.

Usually the breeding farm will ask for a case history for each visiting mare, which should include a health record, a treatment record (such as worming dates and inoculations), past breeding complications and any other information that may be relevant to that particular mare, including any exposure to a communicable disease outbreak. The owner of each mare should be kept informed of the breeding progress of his mare—when the mare is determined to be in foal, probable foaling date, complications encountered and all other matters of concern to the mare's owner. Postcards are a simple way of informing the various owners, unless an emergency arises that requires an immediate decision. Such postcard reports may be printed in form style and kept on hand for convenience during the breeding season. A postcard may avoid future hard feelings and, perhaps, legal entanglements.

Most farms require all visiting mares to be accompanied by a recent certificate of good health from an equine veterinarian, but nevertheless a cautious manager will provide for another examination of the mares at the ranch in order to detect and prevent the spread of costly infections and to assure that the visiting mares will conceive with the least possible delay.

Except for the special necessary considerations mentioned above, the entire breeding procedure is the same for visiting mares as it is for the ranch mares.

Both visiting mares and ranch mares should be marked for quick, easy identification. A safe, practical method is to shave an identifying number on the shoulder of each mare. The shaved number is usually easily visible throughout the breeding season. Some ranches prefer that the mare wear an identification tag around her neck fastened to a snug collar that will easily break if it becomes caught. Often very large ranches brand their own broodmares with permanent identification numbers. For example, a ranch may brand the year of birth on the buttocks on one side of the tail and the mare's ranch number on the other. For the convenience of the ranch personnel, to eliminate errors in records and to avoid the chance of accidentally breeding a mare to the wrong stallion, an identifying system is necessary where a large number of mares are to be bred during the season.

A complete set of records should be kept for each mare on the ranch. Such records will be covered in chapter 14.

7.7 Ranch Mares. Ranch mares should be separated and divided into classes of barren, maiden and foaling mares. It is better to further divide these classes into aggressive and nonaggressive groups so that the feed will be shared more evenly and the chance of injury will be lessened. Aggressive mares are less troublesome when with their own kind.

Barren mares, maiden mares, foaling mares that are to be bred on the foal heat and sutured mares each require specific handling. Because of the variation in nutritional requirements, observation and breeding procedures, it is practical to have mares of each category (with the exception of sutured mares) grouped together.

When a new horse or a visiting mare is added to the herd, a day or two should be allowed for the horses to become acquainted through a fence. A horse should never be turned into an unfamiliar pasture at night, for injury is very apt to occur from running into objects or from being kicked.

Another method of introducing a new mare to the herd is to place

Numbers shaved on the mare's shoulder provide a quick, easy means of identification.

her and the boss mare together in a large separate pen (preferably round so the mare can't be cornered) for a few hours to establish a relationship, then turn them out together with the rest of the mares. This method usually reduces the excitement, fighting and chance of injury.

a) *Barren Mares.* No matter how expertly the breeding procedures are performed, a number of mares will be barren each year. Barrenness may be the result of malfunction of the reproductive tract, thyroid deficiency or other hormone imbalance, cystic or fibrous tissue in the uterus, natural body rejection, improper nutrition, infection, accident or incorrect ovulation determination. Injections of cortisone drugs often used to prevent soreness in race horses may make mares temporarily sterile. Regardless of what is suspected to be the cause, the barren mares should be examined for breeding soundness by the ranch veterinarian. Most ranches conduct the examinations in the fall so that several months will be allowed for the correction of disorders prior to the next breeding season. Such an examination should follow an established routine that includes the following:

1. The general health of the mare: Nutritional deficiencies usually are quite easily detected and corrected. Age and condition of the teeth

may be factors contributing to failure to conceive. Not much can be done about old age but if the teeth do not properly mesh for grinding the feed or sharp edges have developed, much corrective work can be done and a softer diet may help the mare get the nutrients she needs. Pain of long duration, such as from chronic lameness, can be a deterrent to breeding health. Some painful conditions can be relieved through surgery. Unhappy or psychologically disturbed mares often will not settle.

2. Obesity is often accompanied with unpredictable heat periods. Settling such mares can be difficult. Vigorous exercise and lighter feeding may be helpful. A thyroid condition may be suspected if a properly fed and exercised mare remains in an overly fat condition, usually with a heavily crested neck and lethargic or sluggish disposition (such mares are often predisposed to founder).

3. Mares should be inspected for signs of parasite infestation that might impair breeding health. The most serious internal parasite is the strongyle or "bloodworm."

4. A blood sample from a mare in poor health should be taken. It may indicate anemia or one or more hormone imbalances.

5. The veterinarian will examine the uterus and the ovaries by reaching into the rectum with hand and arm encased in sterile rubber or plastic glove and locating the organs through the rectal wall. Abnormalities found should be entered on the mare's chart and corrective measures taken if necessary. Causes of failure to conceive or early abortion that may be noted might include the fact that the uterus is filled with cysts or tumors so that the embryo cannot implant securely in the uterine wall, or that the fallopian tubes or the cervix are closed, making conception impossible, or that infection is present.

6. Examination of the genital tract is accomplished by insertion of a sterile, well-lubricated vaginal speculum through the vulva, expanding it and illuminating the interior with an examination lamp or a flashlight. The insertion must be made in an upward curving motion to follow the contour of the vagina. Any indication of infection, such as abnormal discharge or color, should be noted. Acute infection is usually detected but mild chronic infection may be difficult to detect and can cause alteration of the heat cycle, prevent conception or cause early abortion. If a mare has normal heat periods and her ovaries and other reproductive organs appear normal but she does not settle or aborts early in pregnancy (2nd to 4th month), she may have a low-grade infection. Some infected mares will carry to full term only to produce a dead or diseased foal. Infection is most easily detected during estrus when the cervix is dilated. If the wall of the vagina shows a bright red color and a white or yellow pus is visible on the floor of the vagina, an infection is indicated and a culture of the fluid should be made and studied for bacterial growth. To detect low-grade infections, cervical cultures should be taken at the height of estrus and incubated for twenty-four to forty-eight hours in a suitable medium. If an infection is found to be present, the bacterial

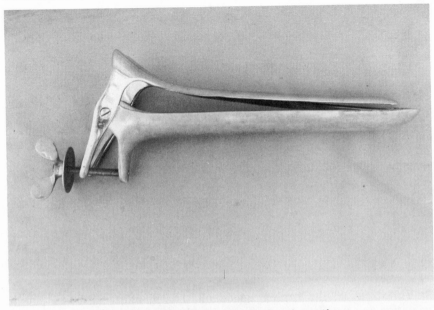

1. Vaginal speculum closed for insertion.

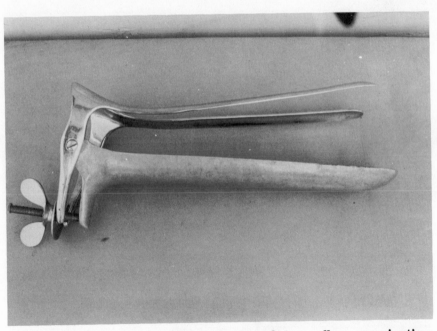

2. Upon insertion the speculum is expanded to allow examination.

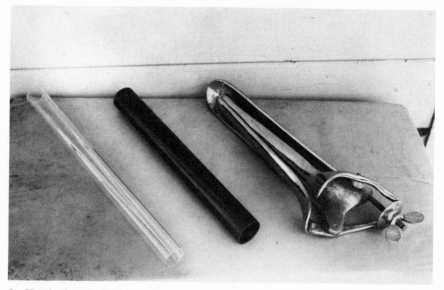

3. Variations of the vaginal speculum. The smaller glass and plastic varieties are excellent for routine culturing and insemination, for they allow less air to enter the vagina. The glass is easily sterilized and both glass and plastic are inexpensive, thus disposable.

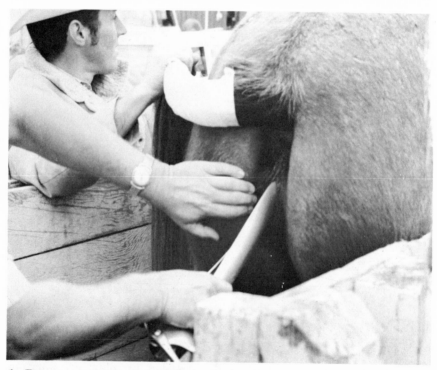

1. Proper method of inserting the vaginal speculum. The speculum is inserted sideways in an upward curving motion.

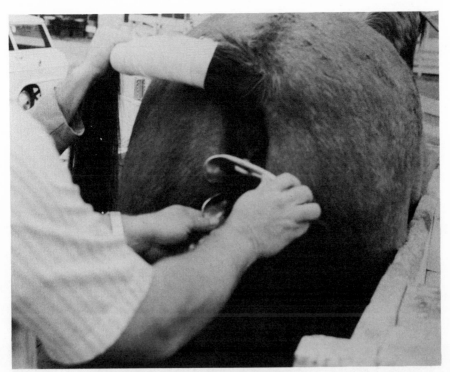

2. Once inserted to its full length, it is turned.

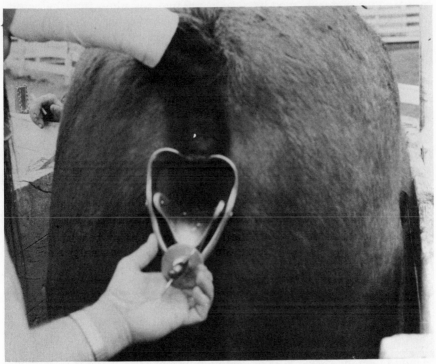

3. When turned to the proper position, the speculum is expanded so that the vagina and cervix can be observed.

culture should be sensitivity tested and treated with various antibiotics to determine which will be the most effective in combating the infective organisms. Infections should be treated as soon as they are discovered. If allowed to become deep-seated they will be extremely difficult to clear up and can result in permanent sterility.

The genital-tract examination may also reveal scars or adhesions that may impede breeding or conception.

b) *Maiden Mares.* Maiden mares are those that have never had a foal. They arrive at the ranch in various physical conditions. Those that have been in training or have been used heavily are usually more inclined toward nervousness; an adequate let-down period should be provided for them.

Although breeders often dispense with complete examinations to determine the fitness of the maiden mare for breeding, they do closely inspect the vaginal area to be certain that a physical condition does not exist that might lead to injury. The conditions of particular concern in this regard are imperforate or persistent hymen, obstructing vaginal septum (membrane partition), and underdeveloped genital tract (infantile genitalia).

Some farms follow the practice of "opening" the maiden mares by removing the resisting hymen, to lessen the chance of infection during breeding. Opening should be done under the most sanitary conditions by an experienced handler or veterinarian. The area around the vulva should be thoroughly cleansed and the tail wrapped. The procedure is to introduce a sterile, well-lubricated vaginal speculum into the vagina and gently expand it to tear the hymen. Usually very little bleeding follows and no damage is done to the surrounding tissue. The same thing can be accomplished by the hand encased in a sterile rubber or plastic glove; however, this method is more difficult and should not be attempted except by an experienced person under the direction of the veterinarian. Such opening should precede breeding by two or three weeks so that the tenderness will be gone and the tissue will be healed.

The practice of opening a maiden mare should not be confused with the outmoded method of "opening mares." Many ranches used to open mares by inserting a dilator into the cervix, the theory being that more sperm could then pass through the cervix. While this may have been true, more germs also entered the uterus and a high percentage of uterine infections resulted without any increase in conception rate.

The obstructing vaginal septum or membrane partition, while not particularly a detriment to conception, can result in infection if torn during breeding. The veterinarian usually prefers to eliminate the obstruction during examination procedures.

Underdevelopment of the reproductive organs (infantile genitalia) usually is cause for elimination of the mare from the broodmare band. Some such conditions, if treated with hormones, can be corrected, but unless the mare is exceptional, the extensive treatment and the doubtful chance of correction do not warrant the time and expense. A young mare is more likely to show improvement, but under no circumstances should hormone therapy be undertaken except by a veterinarian. There are several hormones involved in the estrus cycle and the delicate balance can be easily upset by unprofessional tampering.

In early spring a maiden mare will often act "in heat" almost daily for weeks, particularly if the weather is warm. This prolonged heat usually is not a true estrus and it is pointless to breed the mare. Eventually such a mare will go out of heat; when she comes back in, she will probably be exhibiting true estrus.

Ranches with very valuable stallions often practice "jumping" maiden mares and mares never before bred by the ranch. A gentle teaser is allowed to mount the mare, but the penis is diverted to one side to prevent actual breeding. This is recommended for the maiden mare's introduction to the breeding procedure. The mare is usually not so apprehensive of a teasing horse (if she has seen him frequently) as she would be of a strange stallion. Jumping also gives the handler an indication of how the mare will behave when actually bred. Anxious mares can be "jumped" immediately prior to actual breeding to help settle them down.

c) *Foaling Mares* (breeding on foal heat, or post-parturient estrus). The many special problems peculiar to foaling mares are discussed in some detail in Chapter 10. This section is focused upon the advisability of breeding the mare that has just foaled on the first heat after parturition.

This first heat cycle after foaling is commonly referred to as "foal heat." It generally occurs between four and fourteen days after parturition; the average is about 9 days, consequently it is often referred to as the "9-day-heat." The advisability of breeding at this time is a matter of much controversy among highly qualified horsemen, breeders and veterinarians; but whichever position they take, all agree that breeding on the foal heat should not be done under circumstances where:

1. Foaling was very difficult for the mare.
2. Lacerations accompanied foaling.
3. Afterbirth was not expelled within three hours.
4. Afterbirth appeared infected or abnormal.
5. Afterbirth weighed over 14 pounds, indicating it was infected.
6. Vaginal discharge evidences infection or abnormality.
7. Vagina or uterus does not appear to be returning to normal.
8. The foal was unhealthy or stillborn.
9. Forced breeding is required.
10. The maiden mare has given birth to her first foal.

In spite of favorable circumstances, breeding on the 9-day heat is far from satisfactory as a general practice. The conception rate is well below normal expectations, 25% to 35% being about average. This is very unproductive use of a stallion, unless he would otherwise be idle. Almost all mares pick up an infection during foaling. As the foal is expelled a negative air pressure is created inside of the uterus and germ- and dust-laden air rushes in. Most healthy mares are able to overcome such infections within one or two weeks of foaling. Because of the abnormally high number of microorganisms in the uterus during the 9-day-heat breeding, the abortion rate is about three times the normal and stillbirths may exceed six times that of foals conceived by breeding on later heat periods. These high numbers of failures are costly, because they remove the mares from production during the periods involved as well as subjecting them to excessive strain. These disadvantages, added to the high risk of uterine infection during the foal heat period, discourages the practice of breeding at this time. If a mare is to be bred on the foal heat, a cervical or vaginal culture should be made prior to breeding to insure that no bacteria are present that could infect the stallion or cause the mare to abort.

Maiden mares after having their first foals should not be bred on the post-parturient estrus (9-day heat), since foaling usually results in more tissue damage in them than in mares that have foaled more than once before.

The mare that conceives with the foal-heat breeding will foal about three weeks earlier the following year. Breeding at this time provides one additional opportunity for conception and may be the only opportunity to breed mares who have a history of showing no subsequent heat during the remaining season.

If a mare is not likely to breed later during the season, if the additional opportunity for conception is thought to be necessary, or if an earlier foal is highly desirable, the decision to breed on the foal heat should be made well in advance of foaling. The mare's feed should be very carefully balanced and well fortified for two months prior to and three months after foaling. Such feeding will enhance and quicken her foaling recovery and speed up the repair of tissues strained, distorted and damaged during foaling. Special attention must be given to genital hygiene, because danger of infection is much greater at this time.

d) *Sutured Mares.* Many mares that have sutured vulvas must be opened prior to breeding. Vulva suturing is done to prevent vaginal windsucking (Pneumovagina) or inspiration of air, manure, dirt and other foreign matter into the vagina, that is damaging to vaginal tissue and can result in infection. The condition is due to faulty conformation of the vulva in that it fails to constrict and close properly. This leaves the entire genital tract unprotected and exposed so that it is easily infected by foreign matter and germs carried inward. The admittance of air into the tract produces an unfavorable chemical environment that shortens sperm life. The principal causes of vaginal windsucking are diseases of the membranes, overstretching, injury to the vulva during foaling and a conformation that allows the upper region of the vulva and anus to sink forward in advanced age. Under these circumstances the vulva becomes relaxed and gapes open. The condition is found more often in Thoroughbreds and Standardbreds than in other light horse breeds. Veterinarians correct windsucking of the vulva by Caslick's operation, which consists of closing the upper half of the vulva by suturing the lips so that they will grow together. A strip of membrane lining is removed so that the raw edge of each lip is grafted to the other. Many veterinarians leave a small opening at the extreme top of the vulva, which will allow for easier slitting of the suture line and tends to reduce the danger of tearing if the mare should foal unexpectedly.

Prior to breeding the suture must be examined to determine if there is sufficient room for easy access by the stallion. If not, the mare must be opened in time to permit healing so the breeding process will not be uncomfortable and infection will be less likely (also see Section 7.10 (a)).

7.8 Breeding Time. When a mare is to be rebred after foaling (excluding the 9-day or foal-heat), teasing should begin when the foal is 21 days of age. The mare will usually come in heat between the 25th and 31st day after foaling. If the mare is not bred, subsequent heats will occur on the regular 18-to-21-day estrus cycle.

The female egg (ovum) is viable for a very short period of time, three to six hours, after ovulation. If it does not meet the male sperm within its short period of life it will die. The stallion sperm, after ejaculation into the vagina, survive longer than the egg but they too live a comparatively short time. Sperm are usually viable for only one to one a half days after ejaculation. Under ideal conditions the sperm life might be lengthened to as long as three or four days. Some of the time (4 to 6 hours) is consumed while the sperm is traveling to the upper end of the fallopian tube, where it must be in order to meet and fertilize the ovum. If it fails to rendezvous with the viable egg during its lifetime, it too is lost. The breeder's problem is getting the two together in the proper place while both are alive and can unite to start a new life.

The normal healthy mare will continue to produce viable eggs at regular intervals for her entire life, although some older mares may produce a foal only every other year. One egg is released during each estrus cycle. The normal cycle takes from 18 to 21 days to complete, but some mares vary from 14 to 32 days and are not considered to be abnormal. When the mare approaches the time the egg will be ovulated, released from the ovary by rupture of the follicle, hormones are secreted that bring about "heat." Due to the short life of the egg and the fact that another will not be ready until the next cycle, the mare is receptive to the stallion for only a few days, generally for from four to eight. At this time she is said to be "in heat." It is during this heat period that the short-lived egg can be fertilized. Mares usually ovulate a day or two before the heat period is over. Of course if the mare received the stallion sperm every day of the heat period, it would be nearly impossible for the egg and sperm to escape meeting, all things being normal; but the breeder cannot afford to use up the stallion on the few mares that one stallion could handle on that basis. The breeder must use the stallion so that maximum coverage of the critical ovulation time is assured and yet the stallion is spared for other mares. The best use of the stallion is attained by breeding on the second and fourth days of heat and again only if the mare is in heat three days after the last covering. The first two days of heat are missed by this schedule but it is highly unlikely that the ovum will be in the reproductive tract that early. The remainder of the heat period when the egg is most likely to be present will be covered by the presence of viable sperm. Any more breeding than this will not appreciably increase the rate of conception, but will give additional opportunity for infection. If the breeder wishes to spare or conserve the stallion, the most likely schedule is breeding on the third day of heat and not again unless the mare is still in heat after three more days have elapsed.

Before the breeder can calculate the chances of successful coverage he must know the true estrus cycle of the mare. Cycles are usually quite consistent for each mare throughout the season. This heat pattern can be established by regular teasing. The indications of heat are: relaxed vulva, spasmodic winking (opening and closing) of the vulva lips, slightly raised tail, desire for company, mucous discharge often seen on the tail and rear legs and frequent urination of a slightly discolored and odorous nature. Teasing is more fully covered in Section 7.9.

Maiden or barren mares, older mares and mares in poor condition often exhibit abnormal estrus cycles. Long periods of heat often occur early in the season and the mare may remain in heat for 30 days or longer. It is a waste of a good stallion and time to breed mares in this condition unless rectal palpation of the ovaries and vaginal examination are used to determine the exact time of ovulation. Some mares will ovulate and can conceive though they show no signs of heat. Through examination at very close intervals, the proper time to breed these mares can be determined. Irregular heat periods are a form of abnormal estrus cycles. Such mares may often have a divided estrus cycle; they may be "in heat" for one or two days, then out for two or three days. If the right time is determined by examination, they usually can be settled. The vaginal speculum is useful in these instances in determining the proper time for breeding. At times a mare may not indicate estrus at all. This may be caused by low estrogen production at the time of ovulation. Upon careful examination the time of ovulation can be determined, and proper hormone treatment by the veterinarian may bring on a definite estrus. A mare being bred when not strongly in heat must be well restrained so she cannot kick the stallion.

A mare that is always in heat is often suffering from a cystic ovary. She may be helped by manual rectal massage of the ovary or by an injection of Luteinizing hormone (L.H.). Treatment of cystic ovaries should be done by the veterinarian.

False pregnancies occasionally occur and are due to hormone imbalance. The mare, although she has not been bred, will show all signs of pregnancy, even to ceasing estrus, developing a milk bag and becoming large as if in foal. Early detection and treatment will avoid loss of production for the season.

Mares sometimes do not show estrus while nursing foals. As with false pregnancies, this condition is physiological and may be overcome with proper veterinarian treatment.

There are many reasons for abnormal estrus cycles. While hormone therapy can be quite helpful, it should be administered only by an equine practitioner familiar with hormone use. The mare's hormone balance is very delicate and involves several different hormones, the most im-

portant of which are FSH (follicle stimulating hormone), LH (Luteinizing hormone) and progesteron. Depending upon the situation, the veterinarian may use one of these hormones, HCG (human chorionic
gonadotrophin) or a synthetic hormone. Improper use of hormones can
cause irreparable damage to the reproductive organs.

Teasing is the most practical method of determining the heat cycle
in mares, but the vaginal speculum should be used to examine the
cervix and to establish the correct breeding time. Through use of the
speculum the farm manager can identify mares that show in heat before
they are at the peak of their estrus period and ovulate after going
out of heat.

The reproductive tract changes in appearance during the estrus
cycle. When the mare is not in heat the vaginal mucous membrane is
pale pink in color and is relatively dry; the area is small and crowded,
the cervix is tight and raised from the floor of the vagina. During
estrus the amount of mucous secretion increases and loses its viscosity
until it has approximately the consistency of egg white, and the color
is clear. The area becomes enlarged and takes on a slight red appearance. The cervix is dilated and its labial folds hang downward toward
the floor of the vagina. It is light red in color and has a glossy appearance. These phases of estrus are discernible with speculum examination.

Air entering the reproductive tract through the speculum causes some
irritation; because of this the speculum should not be used more than
twice a week. The expandable speculum stretches the vaginal walls and
allows air to enter, often resulting in a ballooned vagina. Some stallions
refuse to cover a mare with this condition. Therefore, use of the
speculum on the day the mare is to be bred should be avoided. If, however, there is any question concerning the condition of the reproductive
tract on the breeding day, the speculum should be used. Disposable
speculums have been developed that are much smaller than the stainless
steel expendable type. With the smaller speculum there is less air entering the vagina, resulting in less irritation and less ballooning of the
vagina.

With proper use of the speculum the farm manager can increase
the percentage of pregnancies, avoid unnecessary use of the stallion
and save time consumed in unproductive breedings.

7.9 Teasers and Teasing.

a) Teasers. The teaser is one of the most valuable assets of the breeding farm. Whether a gelding or another stallion, he must be well-dis-
ciplined and completely manageable. The teaser must show consistent
masculine interest and must demonstrate a courting attitude with soft
nickering or "sweet talk." A loud, rough or ill-mannered teaser is un-
acceptable. Some breeders tease with the stallion that is to be bred
to the mares. Such a practice is not advisable because teasing may cause
the stallion to become ill-tempered or over-anxious, or he may become
frustrated and lose interest in the breeding process. Young stallions may
be an exception; used occasionally as teasers they can be disciplined and

will learn that actual breeding is determined by the studman rather than whenever a mare in heat is encountered.

Several types of teasers may be used; an "entire" stallion, a horse that is "proud cut," a stallion that has had the vas deferens severed so that it is no longer fertile, a gelding that will tease mares, or a pony stallion. A small pony stallion is safe to maintain on the farm, for even if he should get loose he would be unable to breed a mare. However, some mares will not show heat to a pony.

A teaser should be allowed to breed a barren or grade (nonregistered) mare when he begins to lack interest.

b) *Teasing.* There are many teasing techniques practiced. A mare may be teased over a solid board partition built for the purpose or over a stall door. Teasing bars or teasing chutes are satisfactory. If the mare has a foal by her side, a pen for the foal should be placed near the mare's head so that the mare will not be upset by separation. Some mares, however, will not show to a stallion when they have a foal at side; they must be watched closely in pasture for signs of heat—they may show to other mares. For them regular speculum examinations are required. This trait should be noted on the mare's record.

Riding a stallion past the mare's stall or paddock is often a very simple way to determine whether or not the mare is in heat. Pastured

A teasing chute is invaluable to a breeding farm.

This mare is being teased immediately prior to breeding and is showing strong signs of heat (estrus).

mares may have to be brought to the stud and formally teased.

The most common method of teasing a group of mares is done by turning them loose in a corral next to a fenced alley where the teasing horse is placed. The mares in heat will come to the fence and flirt. The heat cycles can be determined much more accurately when the mare is allowed to approach the teaser rather than the other way around. Whatever set-up is used in teasing, if it successfully and safely establishes the estrus cycle of the mare, it is satisfactory.

It is advisable to begin teasing maiden and barren mares in January so that the estrus cycle may be determined well in advance of the breeding season. A teasing record indicating the most favorable time to breed should be maintained for each mare. After the mare is serviced, daily teasing should be continued for the duration of the heat period. Rebreeding is in order if the mare remains in heat for three days following the first attempt. If breeding on the foal heat is anticipated, the foaling mare should be teased every day starting with the fourth day after parturition, then every other day following the first heat. Mares should be kept on a teasing list throughout the breeding season to assure that none will be overlooked. The majority of mares that go 33 days from the last breeding without showing heat when teased regularly are probably in foal.

As was mentioned above, it is best to know the exact cycle of the mare so that the number of services per mare can be kept to a minimum. By doing this the risk of infection can be reduced and overtaxing the stallion can be avoided, yet there will be maximum opportunity for conception.

7.10 Covering Procedure.

a) *Preparing the Mare and Stallion.* A mare with a sutured vulva, not required to be opened for breeding (see Section 7.7 (d)), should nevertheless be equipped with a cross-stitch "breeding stitch" at the lower end of the suture seam to avoid tearing. A mare with a slight windsucking problem may have a mild infection in the vagina that does not affect the uterus. Such a mare may be successfully bred by artificial insemination (A.I.), thus bypassing the point of infection.

It is an excellent precaution to tease the mare just prior to breeding her to reaffirm heat, psychologically prepare her for breeding and to encourage her to urinate and defecate. After she has been teased the upper portion of her tail should be wrapped with a roll-type clean sterile bandage to prevent the tail hairs from cutting the stallion's penis. A tail bandage made from an innertube cut into a strip three inches wide is an exceptionally good wrap that can easily be sterilized. Disposable tail wraps are sanitary and inexpensive and do away with the need of sterilization. Sterile cloth or gauze tapes will do as well. Some breeders, however, believe that gauze tail wraps are rough and may injure the stallion's penis. They prefer to use soft cotton leg wraps for the tail. Such wraps may be washed, rinsed in disinfectant, dried and stored for future use in covered jars. The anus and area around the vulva should be scrubbed thoroughly with water and mild soap (castile soap is effective) and followed by a clean, warm-water rinsing. Paper towels are often used to clean the mare, since the rough texture seems to remove dirt from hair better than cotton. Cotton is recommended for cleaning the stallion's penis. The penis should be washed with mild soap and water and thoroughly rinsed before covering the mare.

If a mare has a foal at her side, the foal should never be allowed to run loose during breeding. The mare will become nervous while trying to watch the foal, which is likely to be in the way or get hurt. The foal should be placed in a foal cage or a paddock nearby.

The restraints, if any are to be used, should be placed upon the mare immediately before she is to be covered. The most popular types of restraints are the twitch, breeding hobbles, front leg tied up or hind leg tied up (sidelining). Sidelines and the older type of breeding hobbles present the hazard of the stallion's front legs becoming entangled, which can result in harm to both animals. Quick-release breeding hobbles are available and are much safer than the older type, which present difficulty when the animals must be freed in an emergency. A mare is much more likely to fall when a front leg is tied up. Many breeders feel that the less the mare is restrained during breeding the better the chances are that she will settle. It is felt that anxiety will interfere with ovulation.

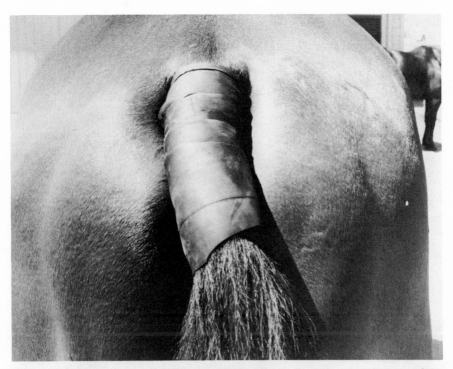

An inner tube cut into a strip four inches wide and four feet long makes a tail-wrap that is easily disinfected.

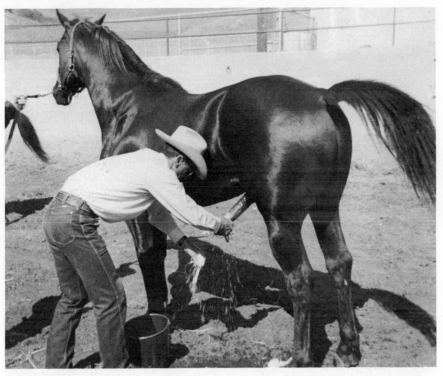

The stallion should be washed before and after servicing the mare.

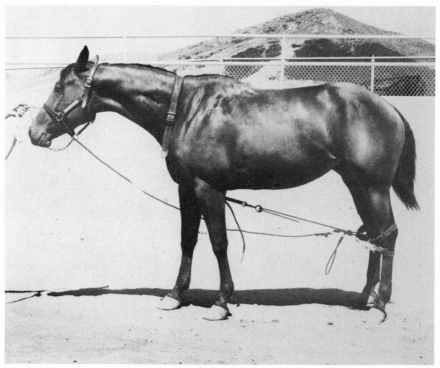

Breeding hobbles will prevent the mare from kicking the stallion. These hobbles are tied so that they can be released quickly in case of entanglement.

A breeding roll is often used to prevent injury to the maiden mare or any mare having small genitals. This device is a padded roll measuring 4 to 6 inches in diameter and approximately a foot and a half long. The roll can be quickly placed crosswise between the mare and the stallion just above the penis as he mounts and prepares to enter the mare, thus acting as a buffer to reduce penetration.

Since shoes can hurt or injure a mare during breeding, the stallion is usually left unshod or the shoes are removed from the front feet during the breeding season. If the shoes must remain on, padding should cover the hooves during breeding to protect the mare. The back shoes of the mare should also be removed for the protection of the stallion.

If the stallion tends to bite excessively during the breeding procedure, he should be muzzled or a heavy protective cover should be strapped in place over the mare's neck and withers.

A very tall or short mare, one that will not match the stallion's reach, may be bred on a slight incline or in a depression, provided the ground will give good balance and sure footing.

b) *Covering.* During the breeding procedure the mare's handler should be at the right side of her head, holding the mare and tending the twitch if one is used. After the stallion is mounted the mare handler should cross in front to the mare's left side, where he should remain.

The stallion should be held on his handler's right side and allowed to approach the mare's left side but should be made to wait until fully erect before being permitted to mount. When he does mount, he rises onto the mare's left side then works his way backward and aligns with the mare until in proper position. At this point the stallion's handler may assist by placing the breeding roll, if one is to be used, and by directing the penis into the vulva of the mare.

The stallionman should insist that the stallion mind his manners and refrain from rushing or lunging toward the mare. Some stallions do not ejaculate during the first mounting and must be made to remount. Ejaculation is always signified by the pumping action (flagging) of the stallion's tail and should be carefully watched for. If the tail does not pump the stallion has not covered the mare.

The stallion should be permitted to dismount in his own time. The mare's handler should be watchful so that as the stallion dismounts he can turn the mare to her left toward the stallion in order to prevent her from kicking the stallion. This is the point at which kicking is most likely to occur. Turning the mare may discourage her from kicking; if she does kick, the blow is not likely to strike the stallion.

1. The stallion is always made to mount on the left side of the mare to avoid being kicked.

2. He is then allowed to work his way back . . .

3. into breeding position. Note that no restraints are used on this mare. When a mare is very strongly in heat and shows no inclination to kick the teaser, many studmen prefer to dispense with breeding hobbles, since they feel that the danger of entanglement is greater than that of kicking.

Breeding concluded, the stallion's penis should be thoroughly washed with soap and water, rinsed, then swabbed with a mild antiseptic solution. The chest, legs and underline may also be rinsed with disinfectant solution after breeding to further lessen the chance of infection.

Stallions will not usually transmit an infection from one mare to another if there is a 24-hour period between servicing the infected mare and servicing the healthy mare.

It is the practice on many farms to walk the mare around calmly for half an hour after breeding, the theory being that straining is more quickly reduced so more semen is retained, thus enhancing conception. However, the benefits of this procedure are very doubtful and do not seem to be borne out by any reliable statistics.

7.11 Determining Pregnancy.

a) It is desirable and quite essential that the mare be pronounced in foal or not in foal as soon as possible so that those not settled can be rebred at the earliest opportunity. The mare should be teased every other day through her next two scheduled estrus cycles to see if she comes back in heat, then at weekly intervals until determined to be in foal by a pregnancy test. There is sometimes a slight risk of abortion if indeed the mare is in foal but shows false heat and is bred again. The largest percentage of mares (80% or more) that do not indicate estrus within 33 days after breeding are probably in foal. A more positive diagnosis may be obtained by several different methods; these methods require much more effort but are also more certain.

Clinical examination, biological assay and chemical assay are the three procedures most commonly employed in testing for pregnancy. If the knowledge is required very soon after breeding, the manual clinical examination is the only one available. While it is quite reliable for a positive finding where the indications are pronounced, it may not be for either a positive or negative finding if the signs are very faint. The biological or chemical assay results are quite accurate; however, they are not useful for early pregnancy testing.

b) *Clinical Examination.* A highly experienced veterinarian may be able to detect pregnancy as early as four or five weeks after the last breeding, but can be better assured in his determination if the examination is made nearer to sixty or ninety days. The veterinarian, having his hand and forearm encased in a rubber or plastic sterile glove, manually enters the rectum and feels the uterus through the rectal wall. The fetus is very small and difficult to detect early in pregnancy but the uterine arteries will have begun to slowly enlarge (by the fifth month the arteries will be as large as five-eighths or three-fourths of an inch in diameter). The expert may be able to determine that a fetus is present. The veterinarian may also insert a sterile speculum into the vagina and by flashlight or examination lamp will be able to study the cervix and vagina. If a fetus is present, the mucous membranes surrounding the vulva and vagina will be dry and pale; the mouth of the cervix will be small and tightly closed; the neck of it will tend to

The veterinarian can detect pregnancy by entering the rectum and examining the uterus through the rectal wall. This method of examination is also used to determine estrus by palpating the ovaries, or to determine abnormalities that may be interfering with estrus or conception. (Note sanitary tail-wrap.)

slantingly protrude into the vagina; and the cervix mouth will be plugged and covered with a protective coating of very sticky mucus.

The advantage of clinical examination is that it can be performed at the ranch, does not require a laboratory, does not require time for biological or chemical analysis, and, if negative, rebreeding can start as soon as the mare shows readiness. Of course the disadvantage is inaccuracy if attempted very early.

c) *Biological Assay.* The biological or blood-serum test is performed by laboratory technicians usually at some place other than the ranch. Two or more ounces of the mare's blood drawn between 50 and 120 days after the last breeding are sent to the laboratory. The maximum concentration of gonadotrophin (sex hormone) is present in the blood at 70 days and disappears about the 120th day. Blood drawn on the 70th day is optimum.

In the laboratory, the blood is allowed to divide and the serum is injected under the skin of an immature female rat from 18 to 24 days old. After 96 hours following the injection the rat is killed and examined for effects. If the ovaries and uterus are enlarged and the tissues are oversupplied with blood, the activity indicates the presence of sex-stimulating hormones that can be attributed only to the pregnancy of the mare that produced the blood. This test is very reliable. The Friedman test for pregnancy is similar to injecting female rats but virgin rabbits are used instead.

d) *Chemical Assay.* The unfortunate female rat must in this case be mature and sprayed with urine from the mare taken 120 days after the last breeding. If the urine contains estrogen, the rat will show symptoms of heat. The test is either positive or negative, but if negative the test should be rerun with urine taken some time after 160 days from last breeding. If this last test proves negative, it is almost certain that the mare is not pregnant.

8
STALLION CARE

(SYNOPSIS)

This chapter considers many special matters pertaining to the stallion and his role as a sire. The following synopsis by section numbers is provided as an aid for quick reference to the specific subjects.

8.1 *General Statement.* Whether the farm maintains one or several stallions, the stud is far more important to the farm breeding program than is the individual mare. The stallion will sire up to 40 or perhaps as many as 50 offspring each season, while a mare should not be expected to produce over four foals in five years and some poor producers never have over one or two foals in a lifetime. The stallion's offspring are prime factors in the success of the ranch. The youngsters produced by the farm stallion will have a great bearing upon the reputation of the farm. A sizable portion of the farm's income will be realized from the stallion services to outside mares. The stallion's prepotency for siring quality offspring will be his best advertising and greatest reputation booster.

The stallion's pedigree, prepotency and conformation have much to do with his desirability, but unless breeding temperament, performance, health and fertility are developed and maintained, the reputation of both stallion and farm will suffer. Careful attention to these qualities through training, exercise, nutrition and health procedures is required if the stallion's full potential is to be realized. Discipline can never be relaxed even though the stud's natural disposition and temperament seem to be accommodating.

The reputation of the animal will be greatly impaired unless the breeding program is carefully planned and conducted to assure properly spaced covers and to avoid excessive breedings. A stallion used too often will ejaculate immature sperm that is incapable of fertilizing the egg and the conception rate will suffer.

Campaigning the stallion is important, because there is not much demand for the offspring or services of an unknown sire. The fact that a horse has not been campaigned will detract from his appeal and without an outstanding performance record the stallion will be difficult to promote. It takes many years for the worth of a sire to be proven through his offspring, and without an enviable campaigning record the stallion will not draw outstanding outside mares. Without servicing top mares, his chances of producing high-quality offspring that will enhance his reputation are not good. In most cases an unknown stallion is too costly to promote over the long period it takes for him to develop a reputation as a good sire. A possible exception may be the stallion that comes from a very strong line of well-known, outstanding horses; even without campaigning he may draw attention to some extent, but his offspring and services will not bring the price a championship record can demand. The farm should strive for as many Grand Championship awards from important shows as the stallion can acquire during his show career.

The stallion must be highly publicized in the breed news media. His name and picture should appear whenever possible along with his campaign record and the winning of his successful get. The farm should make him available for the viewing public whenever possible. His stall door and the farm office should be well posted with proof of his outstanding successes. It is costly to promote the stallion, but if he is unknown, potential income will never be realized.

A young stud prospect should be shown early. He should earn his halter points as soon as possible and become known on the show circuit. His future should be directed to becoming a breed Grand Champion as soon as his age will allow. There are many area breed associations that are offshoots of the national parent organization. For example, the AQHA (American Quarter Horse Association) is the parent organization of all area Quarter Horse organizations. Before a Quarter Horse is two years old the points he earns apply only to the area Quarter Horse association ratings. After he is two years old the points apply toward national ratings. The younger horse should earn a reputation under many different judges at area shows in halter classes. The older horse should compete in working classes as well as conformation classes and earn points toward an AQHA Champion award.

The young stallion that has not learned about mares is more manageable; therefore, campaigning before standing an animal at stud simplifies handling problems in unfamiliar surroundings. Judges will not tolerate an unruly entry; if he is not quickly straightened out, most judges will disqualify the horse. Such a disqualification is a blemish against his record and frequent occurrences will seriously detract from his reputation.

The young, unproven stallion should be bred primarily to outstanding registered mares so that the quality of his get can be maintained at the highest level; standing free to such mares is justified for the start of his career. He should be bred to a limited number of mares the first year at stud to determine the type and quality of foals he will sire before the decision for the farm to promote and publicize him is made. A young stallion should be bred to mares of different bloodlines to determine the type he will best cross with.

When establishing the breeding fee for a young stallion, it is better for the animal's public image to at first ask a low fee and raise it later than to start with a high fee that must be reduced. Should the latter be done, the public will feel that the animal was not worthy of the fee asked previously if the price had to be lowered to attract mares. It will then be difficult to raise the fee as the stud's offspring performance record improves.

The older stallion, having earned his reputation through many outstanding offspring, can occasionally be bred to mares of somewhat lower quality, but under no circumstances should he be bred to obviously inferior registered mares; the offspring of such mares can severely damage the stallion's reputation.

Proper handling and training of a young stallion is essential for developing a good disposition and will greatly influence his reputation and consequently his value to the farm as a future sire. Manners must be taught from the start. Early training will form a basis for his future behavior while near or covering mares. Most stallions are not a problem to handle if they are properly trained when young, receive plenty of exercise, are taken out of the stall for activity other than breeding and can associate with other horses through a safe barrier. A gelding placed in a run or corral next to the stallion will provide companionship without antagonizing the stud.

Vigor, stamina and appearance are all beneficially affected by exercise and sunlight. Vitamin D, synthesized from the ultraviolet light of the sun rays, bears directly upon the health of all horses. The stallion's quarters should have a very large paddock or a pasture of several acres adjacent to his stall and the stall door should be left open unless the soil is so wet that the pasture will be damaged by deep hoof prints. The paddock or pasture fencing must be high and strong. The smaller the quarters, the more daily attention to grooming, exercise and sanitation is needed. When occupied 24 hours a day a small area becomes unsanitary and depressing.

8.2 Stallion Control and Discipline. Stallion management requires extensive knowledge and appreciation of horse psychology by a studman who genuinely likes and respects stallions. A well-bred stallion is very alert and energetic. In general he is more intelligent and responsive to his surroundings than are mares or geldings.

Studmen rarely have to discipline the well-trained stallion, but when disciplined the stallion usually knows he deserves the reprimand and does not easily forget the punishment he receives or the reason for it. The studman must demand obedience and any insubordination such as charging, rearing, striking, biting or other form of misbehavior must be stopped immediately while the horse is first experimenting with the limits of his privileges. As soon as the animal has been punished, the handler should proceed with what he was doing and not show any continued anger over the earlier incident. The sooner the relationship again becomes friendly, the less resentment the stallion will hold.

A studman must never lose his temper and beat the animal in anger; instead, the response to the punishment should be observed and when the animal behaves the reprimand should be stopped immediately. Once the administration of punishment has begun, it should never be left half finished; if so, the stallion will be more difficult to control in the next similar incident. It is better to let a stud go unpunished than to begin a fight the stallion will win. However, if the animal goes unpunished, the handler must be prepared to meet the next similar incident with force. Experienced stallion men never let the situation arise where they may have to back down from the stallion.

The studman should never beat the stallion to demonstrate his superiority over the animal; this is often the case with men who are afraid of stallions—abuse is intolerable. The reason for punishment should always be apparent to the horse. A stallion should never be provoked or goaded unnecessarily or baited into a confrontation. If a critical situation arises it should not be avoided in a way that is obvious to the stallion. For example, if a mare in heat comes upon the scene while the stallion is being exercised and the stallion becomes unruly, the handler must confront the situation and discipline the stallion then and there and then proceed with the program. If a problem can be anticipated in advance the situation should be avoided. The handler should not ride an unruly stallion in an area where he is likely to meet a mare in heat unless he is fully prepared to cope with any situation that may develop. A careful studman recognizes that a stallion completely manageable and docile at home may not be the same in unfamiliar surroundings and that he must have complete control at all times to avoid catastrophes. The responsibility for the stallion's actions rests exclusively with the handler, although the owner as well may be legally liable for any injury and damage the horse may cause if he is known to be unruly. The same rule of law applies to all horses—mares, fillies, colts, geldings —but of course the stallion poses the greatest danger.

Stallions are quick to detect any nervousness or fear on the part of the handler and will often become aggressive and dangerous in the hands of a person unsure of himself. While a studman should not be afraid of the horse, he must take precautions for his own safety and not become lax in discipline or careless in routine stable safety. At times and under unusual circumstances all animals are unpredictable. A loose stallion cannot be controlled by even a competent studman. An uncontrolled stallion is quick to seek out another stallion or gelding to fight, and if he has been bred frequently, he is likely to try to mount a mare or gelding that is being ridden. Unfortunately such incidents occur all too frequently, with serious injury resulting.

Too much petting and affection not only make discipline more difficult but can lead to injury for the handler. Sometimes a stallion will become sexually aroused and attempt to mount his handler. This may happen when the person is grooming or washing the stallion, but most instances occur when the stallion is allowed to follow on a long lead rope. He should be led by a short lead rope, and a chain if necessary, with his head even with the handler's shoulders so that his actions can be watched closely. (See Section 4.11 (c)).

The relationship between the studman and the stallion should be one of mutual admiration and respect. The stallion should be praised for the work he does well. A word of praise and an appreciative pat will encourage the horse to strive to please the handler while rough, inconsiderate and hurried handling will cause him to become nervous and irritable. The stallion should be able to trust his handler, although the handler should never completely trust the stallion. Fairness and consideration on the handler's part will result in a relationship conducive to cooperation and more efficient performance from the stallion. When training a stallion the handler should use only as much force as is necessary. Training stress for a long period can cause irritability and may often lead to outright orneriness. It is better to have frequent short training sessions—for any horse.

Many studmen do not use whips on their stallions at any time; they believe that whipping merely smarts and serves to irritate the stallion. This is especially true of cold-blooded stallions. Most studmen will either strike or jerk the animal with something that really hurts, or will not punish him at all. All horses resent having their noses handled, stallions in particular. If a stallion must be struck, an area other than the nose should be chosen; unnecessary hard feelings will result from punishment in that area.

Some studmen prefer to use a breeding bridle in place of the stallion chain or metal nose band during the actual breeding procedure. The

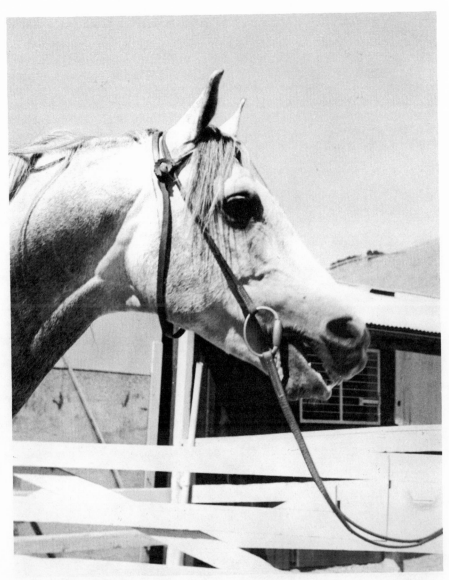

Some farms use snaffle bits to control stallions. These are also used as breeding bridles.

breeding bridle, consisting of a headstall and a breeding bit (similar to a snaffle) is placed on the stallion exclusively for breeding purposes, so that he associates the breeding bridle with the breeding performance; when otherwise haltered or bridled he will know that he is not to breed. This method, if never deviated from, proves extremely helpful for the campaigning stallion. He is more likely to mind his manners when he knows he will not be used for breeding.

The stallion chain or metal nose band is quite harsh and should be used only when the stallion cannot be controlled otherwise; however, safety cannot be sacrificed. Many stallionmen prefer to use a strong halter made of leather or webbed nylon on stallions properly trained and disciplined from colthood. The halter should never be used in place of the chain or metal nose band unless the studman is absolutely certain that the stallion can be controlled under whatever circumstance may arise.

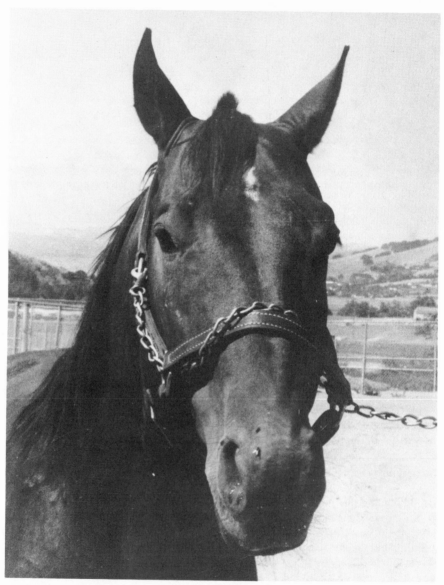

Proper attachment of the chain to the halter for control of the stallion.

If the young stallion has been disciplined, as he should have been, for showing undue attention to mares while being shown, his training may carry over and retard his enthusiasm for his first breeding. A shy breeder may be turned loose in a corral with a gentle mare in heat until he overcomes his timidness. (Pasture and corral breeding are discussed in Sections 7.2 and 7.3.) A mare strongly in heat and one that is unlikely to kick should be selected for a stallion's introduction to breeding. Some young stallions have no respect for a mare; such a stallion could be pasture bred to a band of older, experienced mares that will teach him manners. This method of course presents the obvious danger of injury to the stallion.

The young stallion that is too aggressive and difficult to handle around other horses can create havoc and is not an asset to the farm. His attitude will change if pastured with and outnumbered by a few domineering geldings. A full-grown stallion should never be turned out with geldings, for fighting and serious injury are likely to result.

8.3 Stallion Separation. When the farm has more than one stallion, they must never be allowed to get together; an atomic explosion could not be worse. Their natural instinct is to fight. If two stallions should get together, an effective method of separating them is by the use of strong water pressure from a fire hose. If this is not possible they may be roped, thrown or choked down if enough help can be summoned. In the fury of a fight there is great danger to the handlers as well as other animals in the area. It can never be assumed that a fighting stallion will see anything except the animal he is at odds with.

8.4 Stallion Feeding. Aside from the general health of the stallion, nutrition has a direct bearing upon the quality and amount of semen produced. The formation of spermatozoa by the stallion is absolutely necessary if he is to settle the mares he covers. The stallion must be fed a diet well fortified with protein, vitamins and minerals. During the breeding season the stallion's nutritional requirements approximate those of the lactating mare. He should receive about 13 percent protein and 60 percent total digestible nutrients, supplemented with adequate vitamins and minerals.

Vitamin A is especially important to fertility. An important function of Vitamin A is to improve and maintain the health of the germinal epithelium, which is the tissue that produces spermatozoa. Vitamin deficiencies are best avoided by feeding a commercial vitamin-mineral supplement.

Good green pasture contains a large amount of Vitamin A as well as other vitamins and minerals essential to horse health, and also encourages exercise. If good pasture is not available, high-quality alfalfa is an essential component of the daily ration. Sections 6.3 (a), (b) and (e) should be read in connection with vitamin requirements of the stallion. Volume of semen will be lacking if the stallion must accept dry feeds exclusively. Inferior semen usually cannot be detected by the stallion's willingness or reluctance to cover, but is demonstrated either in the failure of the mares to conceive or through a semen examination.

Calcium and phosphorus are the minerals of greatest concern to sperm production. Steamed bone meal furnished free choice usually supplies satisfactory amounts of these minerals in their proper ratio. Section 6.4 (c) contains a further discussion on calcium and phosphorus.

8.5 Stallion Diets. Following are two diets for a light horse stallion, one to be used during and the other prior to and after the breeding season. Variation and additions according to the needs of the individual stallion should be made where necessary:

    During Breeding Season
        Twice daily: 4 lbs. grain mix
                       12 lbs. combination of alfalfa and oat hay plus
                       pasture if available, or vitamin and mineral sup-
                       plements
    Before and After Breeding Season
        Twice daily: 2 lbs. grain mix
                       12 lbs. combination of alfalfa and oat hay plus
                       pasture if available, or vitamin and mineral sup-
                       plements.

When the ration is altered, the type of feed should be changed gradually to prevent digestive disturbances and to assure that the stallion will remain on feed. Pasture or green feed is highly recommended during the breeding season because it provides vitamins not always contained in dry feeds and is mildly laxative. Some horsemen feel that green feed should not make up over half the roughage diet.

The horse's digestive system can handle small amounts of feed more efficiently than large amounts. Fed at frequent intervals the horse will not be so hungry and will be less apt to bolt his food, which can result in digestive complications as well as loss of food value. It is better to feed small quantities often to a confined horse than to give a full day's ration at one feeding.

Clean water must be available to the stallion at all times. An automatic water basin with no standing water is the most sanitary for the stall, but even this type should be cleaned daily. It is important to the well-being of the horse that it consume generous amounts of water. Dirty or stagnant water is not appealing and the animal will not drink enough to stay in good health. See Section 6.2 (d) for further discussion on water.

Trace mineral salt should be available free choice or added to the feed to meet individual requirements. Salt and other minerals are more fully discussed in Section 6.4.

8.6 Stallion Exercise. Adequate daily exercise is essential during the breeding season. Regulated exercise tones the muscles and keeps the stallion in a virile condition. Insufficient exercise will have an adverse effect on the breeding process and will be evident in the manner the mares are served and the decrease in consistent successful coverings.

The methods of exercise used will depend on the facilities of the individual farm. Riding is usually the easiest, most effective method; walking and jogging alternately from three to five miles or more a day is excellent for conditioning the breeding stallion. Longeing on a 30- to 40-foot line for fifteen minutes twice a day is sufficient. Many stallions do not exercise themselves adequately when simply turned loose alone, and must be longed, ridden or put on a mechanical hot walker to be conditioned.

Some stallions will require a great deal of exercise to remain in proper breeding condition, while others may be maintained with a minimum amount of activity. The extent of the exercise and its effect must be carefully watched; each individual must be regulated to assure top performance. The stallion's strength should not be overtaxed by work or exercise, since fatigue will greatly reduce his sex drive. Too little exercise can result in obesity and poor health.

8.7 Hormonal Imbalance. Hormonal imbalance may affect fertility in the stallion and cause disinterest in breeding. If such a condition is suspected, a veterinarian should be consulted. The imbalance can usually be corrected by injection of the hormone lacking; however, most cases require careful study to determine the extent of the problem. The farm manager should not, under any circumstances, administer corrective hormones without the advice of a competent veterinarian.

8.8 Genetic Inheritance. Genetic inheritance of fertility is an important factor to be considered before purchasing a stallion or when fertility appears to be on the low side. Sometimes low fertility is a family trait about which nothing can be done. It is common knowledge that Thoroughbreds tend to be lower in fertility than most other breeds. Service can still be obtained from a stallion whose low fertility is of a genetic origin, provided he receives proper nutritional requirements and ample exercise. The number of mares he can settle is increased by reducing the number of mares he is required to service in a single breeding season, thus conserving and concentrating the smaller quantity of sperm by fewer services but at the same time increasing the conception rate of the mares bred.

8.9 Parasites. Fertility of the stallion can be greatly reduced by heavy infestation of parasites common to all horses. Treatment for him is the same as for other horses. During the off-breeding season the stallion should be scheduled for worming like any other horse, but the procedures should be timed so that worming is not required during the peak of the breeding season unless absolutely necessary.

8.10 Masturbation. Masturbation is frequently the direct cause of lowered fertility in stallions; much semen and stamina can be lost in the practice. Stallions that masturbate often lose interest in breeding mares. It occurs more often in stallions that are stabled than in those in pasture, since confinement leads to boredom and nervousness while pasturing allows for exercise as well as many activities of interest. An unclean or irritating sheath is sometimes the cause of the initial practice. When an attendant approaches the stall where the stallion is housed the horse's attention is diverted from himself, which probably accounts for the fact that stallions are seldom observed in the act of masturbation. Sure evidence of masturbation is the presence of semen on the abdomen, the hind legs above the hocks or on the back of the front legs. The practice is often retarded by allowing the stallion to be in sight of other horses. If he still persists, a "stallion ring" may be obtained through a veterinary supply outlet. The ring should fit the individual stallion when slipped over the penis so that it will not slide up or off and will not impair circulation. The ring is in a very delicate area and must be kept clean and free from irritating dirt to avoid rawness and soreness. A forgotten ring will mean infection. The ring must be removed during the breeding process, for it prevents the dilation of the penis necessary for ejaculation.

8.11 Savaging, Biting. Savaging or biting is a vice peculiar to stallions. Some will savage themselves; they are more likely to do this after servicing a mare. If possible, savaging must be prevented. Washing the stallion after breeding, changing his environment so that he can see other horses or allowing more room for exercise will sometimes alleviate the problem. A companion such as a goat may help draw his interest from himself. If all else fails, the stallion should be muzzled for the time during which he has the inclination.

8.12 Stall Weaving. Stall weaving is a debilitating vice often practiced by stallions. Weaving back and forth in his stall usually starts out of boredom and may develop into a bad habit that can cause reduced fertility. A change of environment or companionship may stop this irritating habit and certainly will curtail it somewhat.

8.13 Plan for Stallion Use. Over-use of the stallion can result in infertility or cause the stallion to be inefficient in settling a large percentage of the mares he services. Approximately 21 days are required to produce a group of sperm cells. Therefore, the sperm present at the time of ejaculation were developed during the previous 21 days. When a stallion is over-used, the sperm will show reduced quality within a week or two, and when examined under the microscope, the sperm cells will be immature, lacking in volume, motility and morphology and will be unable to fertilize the egg. Over-use of the stallion can greatly reduce production of healthy cells and can cause serious disturbance of the reproductive tract.

Although some stallions are not adversely affected by over-use, most will have lowered fertility. In addition to lowered fertility, over-use will limit the number of years that a stallion will be useful as a productive sire.

Attention should be given to the number of services a stallion is allowed to perform each week. At least one day of rest weekly is advisable. Some authorities recommend a limit of one breeding per day or six services per week. Others, however, believe it is safe to "double" the stallion once or twice a week providing it is at 12-hour intervals. This would make a maximum of eight services per week possible if the animal is allowed one day of rest. The March through May breeding season is 13 weeks long and would permit 104 services per season by a mature light horse in good breeding health. As an example, if 30 mares are booked for the season and an average of 2.5 services is required to settle each mare, a total of 75 services will be required for the stallion within the 92 day season, allowing a minimum of 17 days rest if no "double" service is performed on any day. This is four days more rest than one day per week and, if properly spaced, the program is well within the level of safe limits for the healthy, mature light horse stallion.

Without the assistance of artificial insemination, and assuming it takes 2.5 services to settle a mare, light horse stallions cannot be expected to successfully settle more mares than are listed for his age group on the table below:

Recommended Number of Mares That Should Be Bred to the
Average Stallion per Breeding Season

Stallion Age	Number of Mares	Service Notes
2 years old	up to 12	2 to 3 per week maximum
3 years old	up to 18	spread, with rest
4 years old	up to 28	seldom "double," with rest
mature	up to 45	2 "doubles" a week possible
18 to 25 years old	up to 18	spread, with rest

Although sperm is generally present in stallions 12 to 15 months old, at this age the horse is decidedly immature and should not be allowed to breed. A stallion 24 months old usually is quite developed sexually. Sperm is present in sufficient volume and potency to allow limited breeding use of the horse, but most stallions are not truly mature until nearly five years old.

8.14 Artificial Insemination. The use of artificial insemination can greatly increase the number of mares bred to a stallion. If artificial insemination is a factor, mares settled must have been inseminated on the same ranch where the stallion is located and insemination must have been immediately after the semen was collected from the stallion or they will not be recognized by most breed associations as being in foal to that stallion. The foal cannot be registered if the requirements are not satisfactorily established. Before establishing a program based upon artificial insemination the breed association's current requirements must be studied (see Section 7.5).

8.15 Stallion Examinations. If a deficiency is not recognized until after the breeding season has begun, it may be too late for fast corrective measures to be taken in time to realize the full value of the farm stallion. There is a sad collapse of the breeding plan when three or four weeks have passed before it becomes known that Mr. Sire has an infectious disease or is operating below the required level. This surprise does not arise if the manager has the veterinarian fully examine the stallion for fertility well ahead of the season. In addition to the complete physical examination, the veterinarian will collect an honest sample of the stallion's semen in a sterile artificial vagina, culture it and study it under the microscope for bacteria, motility, sperm concentration and all other characteristics that have a bearing upon the quality of semen. A urine sample should also be studied for any sign of infection. The veterinarian can diagnose an existing infection or probable deficiency and can usually prescribe treatment, diet or procedure that will correct any impairment of fertility. The veterinarian should also be consulted in regard to the plan of use proposed for the stallion during the ensuing season.

In spite of a fine showing in the pre-season semen evaluation, the stallion may require further attention if he is not obtaining a high conception rate in the mares he is covering during the first part of the breeding season. Routine semen checks should be made to detect problems early and at a time when they are easily corrected.

In addition to the semen evaluation, the penis should be examined for any abnormalities or injuries. First-season stallions should be checked for adequate testicle development.

8.16 Diseases Affecting Fertility. The most troublesome bacterial infection usually encountered is Pseudomonas a. If a stallion breeds a mare infected with this organism, he will become infected and transmit the organism to the other mares he breeds. Semen from an infected stallion may be off-colored and contain no viable sperm, and the stallion's testicles may be swollen. Often the first sign of this infection is low fertility followed by a large number of mares showing signs of infection. If the disease goes long undiscovered, an entire breeding season will be lost. The best protection against this organism is to culture all mares and check the stallion's semen periodically throughout the breeding season.

Stallions that have recovered from influenza (viral arteritis) may carry the virus in the semen for a few weeks to several years after recovery. It is wise to have the semen checked for the virus prior to using the animal for breeding again. If the virus is present and the stallion is bred, the disease will be transmitted by the semen and a mare that conceives will abort.

The equine venereal disease, dourine (*Trypanosomiasis*, equine syphilis), has been eradicated from the United States, the last case being reported in 1953. However, there is always the danger that this deadly, debilitating disease could be introduced from foreign sources. Even though incoming horses are bloodtested and checked for the disease, newly imported horses should be inspected carefully to insure freedom from dourine.

The early symptoms of dourine are quite mild, usually consisting of painless swelling of the sheath and redness of the mucosa membrane of the urethra in the stallion and swelling of the vulva, abnormal discharge and redness of the vaginal lining of the mare. Infected animals may run a low temperature. Circular, flat lumps will appear on the body in about four to six weeks. These lumps may remain for one or two days or may last over a week. The animal is definitely anemic and listless as the disease progresses, and usually drags its hind legs. In most cases death results from paralysis. Because of the seriousness of the disease any stallion or mare manifesting the mild, early symptoms must not be bred until it is definitely known that the swelling, discharge or redness is not a symptom of dourine.

A venereal disease that is sometimes similar to dourine in the severe form is coital exanthema. This disease is probably caused by a virus and is spread by direct contact (copulation) and indirect contact (dirty hands, grooming tools, etc.). Coital exanthema is characterized by painful swelling of the genitals accompanied by small vesicles that develop into pustules and then into ulcers. In severe cases the scrotum, sheath, abdomen and thighs of the stallion may be affected, or the udder and dock area of the mare, accompanied by edema (swelling) along the underline. Because the disease is highly contagious, animals must be isolated immediately and a veterinarian contacted. Under veterinary treatment recovery is usually complete and the animal's fertility is not impaired.

9
MARE CARE

(SYNOPSIS)

This chapter considers many special matters pertaining to broodmares. The following synopsis by section numbers is provided as an aid for quick reference to the specific subjects.

9.1 Mare Pre-Breeding Nutrition. General horse care, nutrition, health and sanitation as covered throughout this book all apply to the broodmare. However, because she is producing a foal, her body requirements are greater and special health problems are encountered. This chapter is concerned with the special requirements and problems of broodmares.

If a mare is to conceive, produce and nurse a strong, healthy offspring, she must be in excellent health. A mare may be free from disease, parasites and other physical impairments, as discussed in Section 7.7, and yet may be suffering from undetected nutritional deficiencies so that she does not conceive, or if she does, she may not carry the foal full term. Failure to settle can often be traced to poor nutrition if a mare is otherwise in good breeding health. This condition is rarely encountered in the foaling mare that has been properly fed during pregnancy but is often found in barren mares and maiden mares. If the mare is free from illness and the reproductive tract appears normal but the mare fails to conceive, the problem could very well be a nutritional deficiency.

Prior to breeding, the mature nonlactating mare should be fed the same ration required for maintenance, plus allowance for the amount of work she does. For feeding purposes the mare being ridden more than an hour a day is considered to be in a working category; two or three hours per day is light work, three to five hours per day is medium work and five to eight hours per day is heavy work. The following table shows the basic minimum daily nutrients required for the maintenance of a 1,000-pound adult horse and the work it does.

Work	TDN in lbs.			Dig. Prot. in lbs.	Ca gm	P gm	Carotene (Vita. A) mg
	Maint.	Work	Total				
Idle	7.0	0	7.0	.70	15	15	50
Light (1-3 hours)	7.0	3.5	10.5	.70-1.0	15	15	50
Medium (3-5 hrs.)	7.0	5.0	12.0	1. -1.2	15	15	50
Heavy (5-8 hours)	7.0	9.0	16.0	1.2 -1.4	15	15	50

The single most important nutritional reason mares, especially barren mares, fail to conceive is obesity. These mares are overfed and underexercised and as a result the ovaries become imbedded in a layer of fat, which interferes with the organ's normal functioning and hampers ovulation, thus leading to low fertility. The majority of overweight mares will conceive readily once they have been trimmed down to proper weight. It is better for a mare to be 150 pounds underweight and in good health than to be 150 pounds overweight. In addition to being healthier, the thin mare can be "flushed" or fed heavily prior to breeding, while such feeding will only aggravate the problem of the fat mare. Generally speaking, as breeding season approaches, mares to be bred should be fed rations high in protein and low in fat.

Most breeding farms classify mares into four broad categories as maiden mares, barren mares, mares in foal and lactating mares. These four categories differ enough in their general needs to make it impractical for the farm to feed them as a single group. In comparing the general requirements of the four categories it should be noted that:

1) Maiden mares are usually still growing and need extra rations until they are at least five years old. Many farms breed mares as three-year-olds. Mere maintenance will not suffice for such young mares. Both their development and work requirements must be considered and care must be taken that the calcium-phosphorus ratio does not drop below 1:1; preferably it should be 1.1:1. (See also Section 6.4 (c)).

2) Barren mares are mares that failed to produce a foal that year either because they were not bred, did not conceive, or aborted. Such mares usually benefit from the practice of "letting down" and "flushing," as discussed later in this chapter.

3) Mares in foal are discussed in Section 9.2 (a) as pregnant mares.

4) Lactating mares are those nursing foals and are discussed under Section 9.2 (b).

Maiden mares and barren mares being retired from performance and put into the breeding program present a special problem. Racing weight is ordinarily from 175 to 225 pounds under ideal breeding weight. Time is required, perhaps as long as five to eight months, for such mares to put on enough nonfat flesh to be in the best breeding condition. Weight, however, is not the only consideration; mares that have been conditioned for such strenuous activity as racing or cutting will almost always be tense and nervous and should be given time to completely relax before being turned out with other mares and teased for breeding. Usually, however, the time it takes to "let down," "cool off" or "unwind" does not exceed the time required to regain full flesh.

When horses are running free on open range, their weight varies with the amount and quality of feed available. During the winter months the animals lose weight and are usually quite thin by the time spring comes. As the breeding season approaches the green grass comes on strong and the animals begin to gain weight rapidly. At this time conception rates are usually high. Many progressive breeders take advantage of this natural cycle by letting their maiden and barren mares down in the winter and then "flushing" them as the breeding season approaches. When letting the mares down, care must be taken that they receive properly balanced rations and supplements so that they will not develop nutritional deficiencies. Individuals must be watched closely, because some mares will stay fat on very little while others will become too thin. Many ranches let maiden and barren mares down by leaving them on pasture alone during the summer and fall and adding hay as the pasture diminishes, plus small amounts of grain high in vitamins and protein as winter approaches. A gradual increase of hay, up to 15 to 18 pounds, and a few pounds of grain per day brings the mare along in slender condition so that as the breeding season nears an increased ration will cause her to be gaining without being overweight at the start of the breeding season. This "flushing" process has been quite successful in increasing conception rates.

Occasionally barrenness may be due to a deficiency of Vitamin E, as discussed in Section 6.3 (e). Although this possibility is unlikely, it is advisable to consult a veterinarian for his recommendation well in advance of the breeding season so that Vitamin E can be supplied at least one month before breeding if found to be lacking. Many farms feed some wheat germ for its Vitamin E content as insurance against such a deficiency.

Mares in good health are capable of producing healthy foals throughout their entire lives, but will need soft feeds as their teeth deteriorate. Alfalfa meal and molasses, steam-rolled oats and similar feeds are good substitutes for harder feeds for the more aged mares.

9.2 Pregnant and Lactating Mare Nutrition.

a) *Pregnant Mare.* The nutritional requirements for the pregnant mare differ greatly between the first two-thirds of pregnancy and the last one-third, due primarily to the rate of fetal growth. For the first two-thirds of pregnancy (about 7½ months) the mature mare needs little more than maintenance and working energy replacement. The young mare needs a growing ration to meet her own developing body needs. For this period feeding as discussed in Section 9.1 applies, adjusted of course to meet the needs of each individual animal. Fetus requirements need not be considered during the first two-thirds of gestation as the fetal development is extremely slow and makes very little demand upon the mare. In that period the fetus develops to only about one-fifth of the size of the foal at birth. The mare's health is the foremost concern at this time. The quantity of feed is far less important than the quality. During this period the pregnant, nonworking adult mare can be maintained on excellent green pasture alone, but if not on pasture she will need from three-fourths of a pound to one-and-one-fourth pounds of grain per day for each 100 pounds of body weight, and as much hay as she will eat. If legume hay is fed, it should be limited to about one-and-a-quarter pound per day for each 100 pounds of body weight. The nutrient table shown in Section 9.1 applies to the mare in early pregnancy; however, more than medium work should not be required of her. The manager must adjust the quantities to compensate for the young growing mare and the work the mare is doing.

Whatever the required amounts, the nutritional content of the ration should be proportioned as suggested below:

Suggested Component Values for Mare's Ration
(first 2/3rds of pregnancy)

Total Digestible Nutrients (TDN)	50%
Protein	10%
Calcium (Ca)	0.5%
Phosphorus (P)	0.45%
Trace Minerals	0.5%
Vitamin A (Carotene)	50 mg.—1,000-lb. mare
Iodized salt	free choice

The above component values do not suggest the types of feed or quantities of the feeds. The ration will of course include both grain and hay, the type depending upon cost, availability and choice of the manager.

A simple grain mixture prepared by some farms and used successfully is as follows:

Suggested Components of Grain Mix

Rolled Oats	60%
Rolled Barley	10%
Cracked Corn	10%
Wheat Bran	10%
Molasses	10%

This grain mixture when fed with good alfalfa hay and proper vitamin and mineral supplements is adequate in all respects and ordinarily is not extravagant from a cost point of view.

Vitamins and minerals are discussed in Section 9.3 (a) and (b).

During the last one-third of gestation the fetal development is greatly accelerated. It grows the remaining four-fifths of its size and weight during the last three-and-a-half months of gestation and makes greater demands upon the mare. The mare must have about 7.5 pounds of TDN ($\frac{1}{2}$ pound over maintenance requirements) and 0.95 pounds of Dig. Prot., which is considerably more than for maintenance. She should be furnished 15.4 grams each of calcium and phosphorus and 60 mgs. of carotene per day. If she is performing light work she will need an additional 3.5 pounds of TDN and if performing medium work, an additional 5 pounds of TDN.

All through the pregnancy period care should be taken to feed so that the mare remains thrifty and without excess fat. A fat mare will have more difficulty foaling. The foal has first access to the nutrients the mare's system absorbs, so even a thin mare will produce a strong, healthy foal. There is a tendency to overfeed pregnant mares, due mainly to the thought that a foal that will weigh from 70 to 100 pounds at birth draws heavily upon the mare. However, about three-fourths of the foal's weight is water, not nutrients, and most of the other 25% is acquired over the last period of three-and-a-half months.

The mare should be fed a mildly laxative ration a day or two before foaling and for the first feeding after foaling. This helps the mare clear her bowels and makes her feel better generally.

b) *Lactating Mare.* The lactating mare producing between 25 and 35 pounds of milk per day is working continuously and expending a great amount of energy that must be replaced, and she must furnish large amounts of protein and other nutrients needed for milk production. Above her own maintenance she needs 0.25 pounds of TDN and 0.04 pounds of protein for each pound of milk produced. The nutrient requirements for the 1,000-pound lactating mare producing 30 pounds of milk is near that shown in the following table:

TDN in lbs.			Dig. Prot. in lbs.			Ca gm.	P gm.	Carotene mg.
Maint. + Lactation		Total	Maint. + Lactation		Total			
7.0	7.5	14.5	.70	1.20	1.90	24.4	22.2	60

The "Suggested Component Values for Mare's Ration" and the grain mix mentioned in Section 9.2 (a) are as suitable for the lactating mare as for the pregnant mare; however, the quantity fed must be adjusted to supply the additional needs.

Mares that are lactating will generally need a pound of grain for each 100 pounds of body weight daily (10 pounds for a 1,000-pound horse) and good green pasture, but if pasture is poor, a pound to a pound-and-a-half of good-quality hay will be needed for each 100 pounds of body weight. Additional vitamins and iodized salt should also be made available.

9.3 Vitamins and Minerals.

a) *Vitamins.* Vitamins B and D, while important to all horses, are not required in additional amounts by the pregnant mare or fetus and are adequately discussed in Section 6.3 (a), (c) and (d). Vitamin A (Carotene), also considered in Section 6.3 (b), is vital to the mare, especially the young, growing mare, the fetus and the nursing foal.

If Vitamin A is deficient in the mare's ration it must be drawn from the reserve supply stored in the body. Depletion of the reserve has been proven to greatly reduce fertility. If a mare conceives while she has a Vitamin A deficiency, and the deficiency progresses into pregnancy or if the deficiency develops during pregnancy, the result may be an abortion, a stillborn foal or a foal so weak it cannot survive. Vitamin A serves to maintain the surface (epithelium) tissues of the mucous membranes so that they can ward off bacterial infection. A deficiency of Vitamin A may result in the hardening of the mucous membrane tissues and the loss of their protective functions, causing adverse effects on the reproductive tract as well as general susceptibility to infection. It is therefore imperative that the mare receive adequate amounts of those feeds which contain carotene, from which the body synthesizes Vitamin A, or a suitable Vitamin A supplement, prior to conception, throughout the gestation period and during lactation.

The fetus must draw upon the mare for vitamins. At birth the foal has very little Vitamin A stored within its body. For a few weeks after birth the foal derives its entire daily supply from the mare's milk and is dependent upon this source until eating enough solid feeds containing carotene to meet its Vitamin A requirements.

Natural sources of Vitamin A (carotene) are green pasture, leafy legume hay and corn. Animals should receive some alfalfa hay if not maintained on green pasture to assure that there is no Vitamin A deficiency. Synthetic Vitamin A is inexpensive and readily available. Most farms include it in the mare's grain mix to avoid any chance of a deficiency.

b) *Minerals.* A well-balanced ration under normal circumstances provides all trace minerals required by the pregnant or lactating mare. However, she requires larger amounts of iodine, salt, calcium and phosphorus than the average horse, discussed in Section 6.4.

Mares being fed iodine-deficient diets tend to produce weak foals and appear to have a high stillbirth rate. The iodine content of the feed is dependent upon the iodine contained in the soil where the feed is grown. Most soils contain an adequate amount of iodine, which is absorbed by the growing plant and is passed on to the horse, but if the iodine content is not known it is wise to supplement for it by providing stabilized iodized salt free choice. If the iodine is not stabilized it will sublime and disappear.

Salt must be provided free choice at all times, either in block or crystal form, plain, iodized or treated with other trace minerals. Lack of salt will result in decreased milk production, retarded growth, poor appetite, weight loss and rough hair coat. Deprived animals will crave salt and will often lick fences and rocks, drink muddy water and chew wood, but excessive amounts of salt should not be forced on the horse.

Calcium and phosphorus requirements for gestation and lactation should be maintained at a balance close to 1:1. These two elements are very important for proper fetal development, for maintaining healthy bone structure of the mare and for continued skeletal growth of the young growing mare. The daily requirements of a 1,000-pound pregnant mare are approximately 15.4 grams of calcium and 15.4 grams of phosphorus, and are much greater when the mare is nursing a foal.

9.4 Pasturing Pregnant and Nursing Mares. The ideal environment for pregnant mares and mares with foals is pasture. Pastures provide plenty of exercise for the mares and room for foals to play, they are cleaner than stalls and enhance the general well-being of the animals. However, even good green pasture alone will not supply the nutrients required by lactating mares or mares in the final third of pregnancy. They must be fed grain and, if the pasture is not excellent, hay may also be required. Proper adjustments will be needed to fulfill additional requirements and vitamin and mineral supplements must be provided.

Pasturing during the last third of pregnancy should be avoided where footing is slippery, since a hard fall may result in abortion.

Pregnant mares and mares with foals at side are much more content when pastured with other horses, but they should be pastured only with other foaling mares or mares with foals at side. Barren mares and young, playful horses create too much excitement and can be ornery toward a foal. Geldings are apt to tease a mare and cause her to abort, especially if the mare is in the very early stage of pregnancy. Some geldings and older mares will steal a foal and prevent it from nursing the mare (Section 10.9).

Pregnant mares and mares with foals at side are more content when pastured with companion horses of the same status.

Section 10.3 considers the desirability of permitting the pregnant mare to foal in pasture and should be read in connection with this section. The advisability of pasturing the mare after foaling is discussed in Section 10.9.

9.5 Exercise for Pregnant Mares and Nursing Mares.
   Pregnant Mares. One of the main reasons for foaling difficulties is that mares often do not receive enough exercise. Exercise stimulates

the circulation and aids the well-being of the mare as well as the fetus. Lack of exercise can cause foaling complications due to weak muscles, poor muscle tone, slowed circulation, nervousness, depression or obesity. Most mares loose in a large pasture exercise themselves adequately, but if a mare is stabled or kept in a small pen she must be walked for at least an hour twice a day if other exercise is not provided. It is perfectly safe and desirable to ride a pregnant mare for an hour or so a day with due consideration for her condition. Fatigue, strain or over-exertion can be a cause of abortion (see Section 9.7 (b) (3) and (7)). Toward the end of gestation the exercise should be very moderate and quick starts, stops, turns and other jarring maneuvers should be avoided. An older mare should not be exercised so extensively as a younger, more vigorous mare.

Nursing Mares. Most large farms do not ride their broodmares; however, a small breeder may wish to ride a lactating mare occasionally. When the foal is a month old the mare may be taken away for very short periods of moderate exercise. The foal should be shut in a safe stall or a sturdy paddock (preferably in the company of other foals) where it cannot hurt itself by struggling to follow the mare. For the comfort of both animals, the foal should nurse before the mare leaves. Three or four minutes of separation is sufficient for the first day. The time may be increased gradually until the mare is absent for a half hour by the time the foal is four or five months old. With this method of separation and exercise for the mare there will be less stress when the foal is weaned. As the length of separation progresses, the mare's udder may become uncomfortable, especially if she is being exercised extensively. Aspirin may give relief if the mare seems to be having discomfort.

It is never safe to let the foal tag along while the mare is being ridden. Until the foal is halter broke and trained to lead, it cannot be controlled, and the mare will be difficult to handle because her attention and anxiety will be directed to the foal. Many accidents have occurred because a foal was allowed to follow the mare through a pasture with other horses or along roadsides.

9.6 Drying-Up Milk Bag. Starting a day or two before and continuing after the mare's foal is weaned, the mare should be placed on a maintenance ration to help decrease milk production. If the mare is not on pasture, she can be fed high-quality hay alone. If she is on very good pasture, she should be removed to an area where the feed is less plentiful or dried up.

When nursing stops, the udder will become full and tight. Depending upon the method of weaning (see Section 11.7), the udder generally should not be milked out except perhaps once lightly the first night of separation, for milking will only stimulate more milk production. The uncomfortable pressure is needed to deactivate the milk-producing cells. In about a week the udder will become soft, indicating that milk production has ceased. If at that time any milk remains, it can be milked out.

Applications of cold towels to the udder may help to relieve the mare's discomfort when the udder is tight. Wiping the bag gently with baby oil or hand lotion may serve to soften the skin and lessen the tension. Ten mashed aspirin tablets in one-half pound of grain may be given twice daily. Heavy milking mares that are not in foal at the time their foals are weaned may benefit from hormone injections to help dry up their milk.

9.7 Abortion

a) *General Discussion.* Abortion is the expulsion of the fetus when it is so underdeveloped that it cannot survive. When the fetus is dead or the uterus or fetus are diseased or badly damaged, the mare's body regards the fetus as foreign matter, which it rejects and expels.

Usually the fetus is expelled very soon after its death but occasionally, if no infection is present, the dead fetus remains in the sterile uterus without decomposing, where its tissues liquify and are absorbed by the circulatory system of the mare. Later the bones and skin are expelled. Because of their appearance, such aborted fetuses are referred to as mummified fetuses.

If a mare appears to be on the verge of aborting (straining while in a prone position), it is usually best not to try to prevent it, for the mare's body probably has a good reason for rejecting the fetus. Most prevented abortions result in abnormal foals unless the abortion was attributed to physiological causes (see Section 9.7 (b) (7)).

When a mare begins to make a bag far in advance of the parturition date, it is an indication that she is about to abort.

Prefoaling colic a month or so prior to foaling is often mistaken for abortion symptoms. The mare shows cramp-like symptoms and appears to be straining. Warm bran mash with 50 grams of aspirin or antispasmodic and analgesic medication may give relief.

When a mare that is fairly far along in pregnancy aborts due to traumatic injury, poisoning or other rapid cause of death of the foal, she is likely to have a very difficult time because her body did not have time to condition the muscles and ligaments for easy foaling. Such a mare is likely to need help and should be watched closely.

Many factors may contribute to the death of or injury to the fetus. The causes of abortion can be divided into two broad classes, "Noninfectious Causes" and "Infectious Causes"; the types of abortion the farm is most likely to encounter are discussed in the following paragraphs.

b) *Noninfectious Causes of Abortion.*

1) *Feed and Water.* Although malnutrition is rarely a cause of abortion, primarily because the fetus has first access to the nutrients ingested by the mare, moldy or spoiled feed or contaminated water almost always creates physical conditions that can cause abortion. A sudden change of diet may sufficiently upset the system to terminate pregnancy. Vitamin or mineral deficiencies often result in a weak and sickly foal and may cause rejection of the fetus.

2) *Poisons or Toxins.* Poisons or toxins may bring about abortion. Consumption of water containing excessive amounts of copper sulfate or other chemicals used in water troughs or the ingestion of poisonous plants (Section 2.20) or frostbitten Sudangrass of the types that produce excessive prussic acid (Section 2.17) are likely to cause miscarriage. Rodent poisons and insecticides used near pregnant mares create the hazard of abortion. Ergot, a parasitic fungus of grains and meadow grasses, will cause abortion if consumed.

3) *Fatigue.* Fatigue, overexertion and excessive excitement are not uncommon causes of abortion. Mild exercise is necessary for the pregnant mare (Section 9.5) but should not exceed medium work during the first two-thirds of gestation or light work after that. Jumping, straining, quick starts, stops and turns and fast gallops should not be permitted during the latter two-thirds of gestation. Care should be taken to avoid situations where the mare is apt to become extremely excited. The pregnant mare should be separated from geldings and playful, barren mares (Section 9.4).

4) *Injury.* The fetus is well cushioned by the shock-absorbing water in which it is suspended, nevertheless traumatic injuries can cause abortion. A rupture of the membranes surrounding the fetus is almost sure to cause the mare to abort. A kick, a bump, a severe squeeze or a jolt from a jump or fall may be enough to cause expulsion of the fetus.

A twisted navel cord may result when the fetus rotates within the uterus. This cuts off the fetal blood supply and results in the death of the fetus, which is rejected almost immediately.

5) *Heredity.* Occasionally a genetically defective fetus is conceived. For some yet unknown reason the mare's body usually rejects a severely malformed fetus or a fetus whose organs will be nonfunctional early in development. The horse does have a few true lethal genes, genes that cause death before or shortly after birth, which the breeder must contend with. A few such lethals are lack of an anus, no connection between two sections of the large intestine, hydrocephalus, hemolytic disease, stiff forelegs, absence of eye orbits, convulsive syndrome, white lethal and a few others. Fortunately, with the exception of white lethal, these lethal genes are quite rare. The white lethal gene is the gene that would produce an albino horse if two of these recessive genes were present. However, when the white lethal gene is not masked by another dominant gene, the albino fetus dies and is aborted. This is why there are no true albino horses (white horses with pink skin and red eyes). The closest to a true albino horse is a white horse with pink skin and blue pigmented eyes. Apparently the gene that causes the small amount of blue pigment in the eye is enough to offset the lethal white gene. The Albino registry, which registers and promotes white horses, and Albino breeders have long

been aware of the lethal white gene, because of low colt crops
and lack of true albino foals.

6) *Twinning.* A very small percentage (probably less than 1%)
of the twins conceived survive. Few twins are conceived by
mares, but those that are have very little chance of surviving
the crowded conditions of the uterus. The twins usually live
until the last half of gestation, but when their rapid growth
commences, the pressure from overcrowding usually kills one
of them and both will then be aborted. Since the abortion
occurs quite late in the pregnancy, it is not uncommon for the
mare to "make bag" and drip milk prior to aborting twins.
Ordinarily the mare that "makes bag" will dry up a week or
so prior to aborting.

7) *Physiological Causes.* There are two periods during pregnancy
when critical physiological changes occur quite abruptly; if
either change is not successfully completed the fetus cannot
survive.

The first change occurs during the second or third month
(between the 50th and 70th day) of gestation when the center
of control over hormones necessary for fetal development shifts
from the ovary to the pituitary gland in the brain. The major
hormone of concern, progesterone, is still produced by the
ovary; all that is transferred is the control center that sends
the signals demanding hormone production. If the change is
not satisfactory, the necessary control is lost and the mare
will abort.

Occasionally an early clinical diagnosis indicates pregnancy
and yet the mare will come into heat during the third or fourth
month. Often the mare's heat is improperly explained as an
error in the pregnancy diagnosis; however, it is quite possible
that the mare inconspicuously aborted or completely absorbed
the small, boneless fetus after the examination. The cause may
have been a faulty shift of the hormonal control from the ovary
to the pituitary gland.

The second physiological change occurs during the fifth
month. Up to this time the corpus luteum (a small ductless
gland in the ovary) produces the hormone progesterone.
Progesterone is necessary to maintain the pregnancy by sup-
pressing the development of follicles in the ovary. In the fifth
month the production of progesterone shifts to the placenta;
if that shift is not complete, the lower level of progesterone
will allow the pregnancy to terminate.

Abortion due to a hormone failure does not mean that the
mare should be removed from the broodmare band. Such a
condition may not be present in future pregnancies. If a mare
has lost more than one foal due to faulty shifting of hormone
production, she can be given hormone injections under veterinary supervision during the critical change-over period.

(c) *Infectious Causes of Abortion.* An infection of the reproductive
tract may cause abortion by injuring or partially destroying the pla-

cental tissue or by affecting the fetus itself. If the mare is bred while infection is present, she may not conceive, and if she does, she will probably abort or the foal will be born dead.

Harmful amounts of bacteria may gain access to the reproductive tract if the mare's vulva and buttocks are not washed thoroughly before breeding or examination, if an unsterilized vaginal speculum or artificial insemination equipment is used, if the stallion is not cleansed, or if the stallion himself is infected. The cervix of the mare becomes relaxed and slightly open throughout estrus, thus inviting infection of the uterus unless examination and breeding are done under sanitary conditions.

In years past many horsemen "opened up" all mares to be bred on the farm. This procedure entailed inserting a dilator into the mare's cervix in hopes of making passage easier for the sperm. While the rare individual mare possessing a deformed cervix may benefit from being opened, the procedure does not increase the conception rates for normal mares and has been responsible for many serious uterine infections.

Vaginal windsucking is a common source of bacterial infection. Mares with this condition should have the upper portion of the vulva sutured, an operation known as Caslick's operation, to prevent aspiration of air and fecal material into the vagina. Such mares must all be reopened for foaling and many require reopening for breeding also. A more detailed discussion of vaginal windsucking is contained in Sections 7.7 (d) and 7.10.

The bacteria and virus usually associated with abortion in the mare are *Streptococcus genitalium, Salmonella abortivo-equinus* and Equine Virus Abortion, which are discussed in the following subparagraphs. Other somewhat less common bacteria causing abortion are *Staphylococci, Bacterium coli, Leptospira sp.* (rare) and *Pseudomonas aeruginosa*.

1) *Streptococcus genitalium* is without doubt the most prevalent of the bacteria causing genital tract infection of horses. This organism is found in every part of the country and causes about one-sixth of equine abortions and 25% of the foal deaths. This infection and other bacterial uterine infections can cause such scarring that the glands in the mucous coat of the uterus are replaced by scar tissue and the fallopian tubes can be blocked, resulting in permanent sterility. The bacteria grow and thrive in the female reproductive tract. They gain access to the mare's reproductive tract through unsanitary breeding conditions and unsterilized vaginal speculums, from infected stallions, from vaginal windsucking or by foaling on contaminated bedding. The infection may be indicated by inflammation of the cervix or of the uterus, or by abortion. Definite diagnosis of the infection is possible through examination and culturing of the genital tract, aborted fetus or fetal afterbirth. There is no reliable vaccine available that will protect against these bacteria. Prevention through adequate sanitation is the best protection. If treated early the infection responds well to antibiotics.

2) *Salmonella abortivo-equinus*, once a dreaded cause of abortion, is now well controlled or prevented by vaccination. This contagious digestive-tract infection is usually confined to horses and causes the condition that is commonly known as "Equine Contagious Abortion." These bacteria are carried from one horse to another through ingestion of contaminated forage or water. An infected mare usually has a brown, foul-smelling discharge from the vagina, and diarrhea. Diagnosis is by means of an examination of the blood serum of the infected animal and by microscopic identification of the organism in the aborted material or in the uterine discharge. "Equine Contagious Abortion" can be and has been greatly controlled by annual blood testing of the breeding animals and the elimination of any positive reactors from the herd. Vaccination with a specific bacterin during the fourth or fifth month of pregnancy has been very successful in reducing abortion from this cause and is highly recommended for the pregnant mare.

Whenever a mare aborts, the fetus should be removed immediately and the area should be thoroughly disinfected to prevent the further spread of disease in the event the mare is infected (see Section 9.8). The aborted material should always be sent to a laboratory for analysis.

3) *Equine Virus Abortion* (Rhinopneumonitis or epizootic abortion). Abortion caused by this virus infection of the respiratory tract usually occurs in the last third of pregnancy. If a mare aborts after the seventh month without another cause being known, this disease should be suspected. Little is known about the disease, which is very difficult to diagnose and extremely hazardous to the breeding farm. The mucous membranes of the eyes, nose and mouth of the infected fetus may appear slightly yellow in color; a considerable amount of liquid may be found in the chest and abdominal cavities; some of the internal organs may show indications of bleeding; and inclusion bodies may be found in parts of the liver or lungs.

The rhinopneumonitis virus also produces cold symptoms in young horses, while it does not seem to affect older animals other than causing pregnant mares to abort. If the farm's weanlings and yearlings appear to have colds at the same time that late abortions occur, the chances are very good that the virus is present. As a precaution, many horsemen follow the practice of separating the broodmares from the more susceptible young stock.

The breeder's primary defense against an epidemic of the disease seems to be isolation and quarantine of infected animals and the use of current vaccines. A modified live virus vaccine that provides temporary immunity is available. For further discussion see section 5.18. It is believed that the virus is transmitted by direct contact with infected animals or by ingestion of contaminated feed. Laboratory diagnosis of the aborted fetus and sanitary treatment of the area it has contacted is essential. The infected mare should be isolated for a month or more and her blood should be tested at the time of abortion and again before she is allowed to rejoin the herd. If an outbreak is suspected, the stallions as well as all mares should be blood tested. Once the mare has contracted the disease, she develops a certain amount of immunity to the virus and won't abort for this reason during the next several succeeding pregnancies.

In cases where isolation is required, the sanitation schedule must be thorough and complete. Feed and water containers must be disinfected with a strong lysol or similar solution and kept strictly at the isolation unit. Tack and other equipment must be burned or buried with lime and must never be discarded where the infectious material may be picked up by other animals or carried by the shoes of the handler to other parts of the farm. If isolated horses are pastured, the same sanitary methods must be maintained. The isolation pastures should be located far enough away from other pastures so that the wind cannot carry the disease to them.

4) *Brucella abortus,* which causes Bang's disease in cattle and Brucellosis or undulant fever in people, can cause abortion in mares. The bacteria should be suspected if abortion is associated with fistulous withers or poll evil or if the mare has been in close contact with goats, cattle or pigs. It is a form of "contagious abortion."

5) *Mycotic placentitis* is a fungal infection that can cause abortion. The fungi gain access to the uterus through unsanitary breeding or examination procedures, where they attack the placenta near the cervix and gradually invade the entire placenta, destroying the villi and cutting off the foal's supply of oxygen and nutrients. If the fetus is not aborted, the foal is usually stunted, weak and underdeveloped.

9.8 Abortion Prevention Measures. It hardly needs to be said that prevention is the most efficient defense against abortion. The practice of preventative measures will greatly reduce the hazard of abortion and will correspondingly reduce costs and losses. Some of the preventative measures every breeding farm should practice are:

1) Never breed a mare to an infected stallion.
2) Never breed a mare in poor breeding health.
3) Breed only under the most sanitary conditions.
4) Vaccinate all mares against abortion as recommended by a veterinarian; the veterinarian should determine the extent of vaccination for the ranch.
5) Feed clean, nutritious feeds without abruptly changing the diet and supply clean, uncontaminated water free choice.
6) Provide proper exercise for pregnant mares but avoid fatigue, strain and excessive excitement.

7) Avoid conditions that may cause falls and other injuries.
8) Have all barren mares examined by a veterinarian to determine cause of barrenness.
9) Find all aborted material as soon as possible and submit it for laboratory analysis.
10) Disinfect the area where aborted material has lain with a strong disinfectant solution such as lysol, followed with a sprinkling of lime and a layer of dirt. Burn or bury bedding. Disinfect the stall, the feet of the mare and the shoes and clothing of the attendant. Disinfect tack or equipment that has been in contact with infected material. (The sun will kill many germs over a period of time but flies and other insects can transmit the infection from the aborted material.)
11) Isolate and quarantine horses having contagious or infectious diseases and those horses that have been exposed. A veterinarian should determine the extent of quarantine and whether the entire band in which the disease was discovered should be quarantined in place.

9.9 Udder Problems. Although mares are less prone to udder problems than are cattle, sheep or goats, udders should be checked regularly to correct difficulties before they become serious.

a) *Abnormal Teats.* Some mares have inverted teats, small nipples or even extra teats. Extra teats are no cause for concern; however, a mare that has inverted teats or small nipples is difficult for the foal to nurse, especially if the udder is swollen with edema. Such mares should be watched closely and if the foal is having difficulty, it should be bottle fed and the mare milked by hand until the udder edema goes down and the foal can nurse easily.

b) *Agalactia (Lack of Milk).* Occasionally a mare will not produce milk after foaling. The foal's attempts to nurse will usually stimulate milk production but in the meantime the foal should be bottle fed. High concentrate rations, udder massage and perhaps hormone therapy should speed milk production. Hormones should be administered only by a veterinarian (see Section 10.6 for further discussion).

c) *Udder Edema.* This condition is often encountered in mares that are far along in pregnancy or in heavily milking mares whose foals have just been weaned. Udder edema is characterized by excessive fluid in the tissues, which produces a warm, painful swelling and may involve the lymph vessels and extend along the mare's underline. Severe cases can become complicated with bacterial infection, although bacteria are rarely the initial cause of the edema. Adequate exercise is usually sufficient to clear up mild cases; however, severe cases present the risk of bacterial infection and cause a great deal of discomfort to the mare. Such cases should be treated by a veterinarian, who will probably administer a diuretic, which will bring immediate relief.

d) *Mastitis.* The mammary tissue with its rich supply of milk provides an ideal situation for bacterial growth. Udders should be watched closely for mastitis, a bacterial infection. Unless treated immediately mastitis can advance to the point where surgical drainage or even amputation of infected quarters will be necessary. Mastitis is characterized by hard, hot, painful swelling of the affected quarter. The milk will probably be lumpy or stringy and may contain pus or blood. A veterinarian should be consulted. Systemic antibiotics or antibiotics injected directly into the infected quarter usually clear up the infection quickly.

e) *Abscesses of the Udder*. Abscesses may form as a complication of mastitis or can be caused by a separate bacterial infection. In California the bacterium *Corynebacterium pseudotuberculosis* (pigeon fever) is responsible for many udder abscesses. This same bacterium causes abscesses of the chest muscles and sheath of stallions and geldings. A veterinarian should be consulted and antibiotics administered.

f) *Cystic Udder*. This condition is sometimes seen in nonpregnant mares. One or more of the udder quarters fill with clear fluid or milk. Such mares should not be milked out unless advised by a veterinarian. A veterinarian should be consulted when this condition is suspected.

g) *Caked Udder* (Borderline infection of the Udder). In this condition the teats are usually plugged up and milk appears slightly abnormal. There may be some heat, swelling and pain in the affected quarter. Warm compresses often provide some relief. Antibiotics are advisable and a veterinarian should be consulted if an infection is suspected.

h) *Cleaning Udders*. Udders should be cleaned regularly to prevent dirt, oil and sweat from accumulating around the udder and between the teats. Such accumulations cause a great deal of discomfort and mares will rub their tails as if they have pinworms.

Pregnant maiden mares should become used to having their udders touched so that they will accept their nursing foals more readily. Massaging the udders and teats during the last six weeks of gestation will be very beneficial.

10
FOALING

(SYNOPSIS)

This chapter considers the gestation period, the birth process, areas for foaling, indications of impending birth, assistance, special health problems at birth, care of the foal and mare, and exercise of the foal and mare.

The following synopsis by section numbers is provided as an aid for quick reference to specific subjects.

10.1 Gestation. Statistics show that the normal horse gestation period is between 315 to 350 days, but the time may vary because of the sex of the foal, the genetic make-up of the mare or the mare's environment.

As a general rule, the foaling date can be expected about four weeks less than a year (337 days) after conception; however, no exact date can be determined. Many unknown factors may bear upon the length of a particular pregnancy. Some horsemen believe that the gestation period for a colt is sometimes a few days longer than for a filly. Long daylight hours and warm climatic conditions tend to shorten pregnancy. Mares in excellent health and well nourished usually foal a few days early. Arabian mares often foal a week or so later than other breeds. The general condition of the foal and its weight or size may cause some difference in the duration of the mare's pregnancy.

Often the only sign of impending parturition up to six weeks or so before actual birth is the progressive increase in the size of the mare's abdominal area. As the foal drops lower the abdomen begins to appear flattened on the bottom. However, the appearance of the mare may be deceiving, for some mares are naturally heavy without being in foal while others carry their foals high and do not appear pregnant.

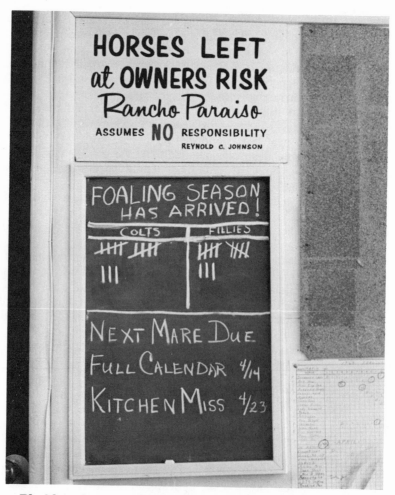

Blackboards are a convenience during the foaling season.

10.2 Parturition Stages and Theories. There are three distinct stages of labor: preparatory stage, expulsion of the fetus, and expulsion of the placenta. Each is a definite and necessary phase of giving birth. There is also involution of the uterus, which might be called a fourth or recovery stage.

Precisely how these stages are brought about is not fully understood. Each may be the result of many other undetected stages. It is theorized that several things occur at the same time or in close sequence, each triggering another physiological change to prepare the mare for parturition. It has been established that near the end of pregnancy the ovaries release greater quantities of estrogen into the blood. The posterior pituitary gland and the placenta begin releasing oxytocin, which stimulates the mammary cells and begins milk production. A little later the pituitary gland begins releasing the hormone relaxin, which causes the muscles and ligaments around the vulva and across the back and hindquarters to relax, thus permitting these areas to stretch more easily for passage of the foal. The fetus does most of its growing during the last quarter of pregnancy. By the time of birth the uterus and cervix are under extreme pressure due to the size and weight of the fetus and fluids. Something eventually must give way. The nerve centers may become so sensitive to the pressures and hormones that contractions are involuntarily set in motion in an effort to reject any further burden.

After parturition nature immediately sets to work repairing the damaged and strained tissues and muscles. Complete involution of the uterus requires two or three weeks.

10.3 Foaling Areas. In warm climates many breeders prefer to allow mares to foal in clean, grassy pastures or paddocks. A pasture is a more natural place for a mare to foal, especially if she has not been stabled. A mare used to being outdoors will be less nervous if allowed to foal in a paddock. In addition, there is not the danger of injury to the foal from stall walls. Unless a regular, large sanitary foaling stall is available, a clean pasture is much to be preferred over a smaller, less sanitary stall.

The foremost drawback to pasture foaling is that it makes observation of the mare much more difficult. If facilities are available the foaling mares may be grouped according to foaling dates, so that mares due to foal soon may be pastured in an easily observed area. If a paddock is used, it should be situated so that all areas can be seen easily, and it should be equipped with floodlights for night observation. Though the sun's rays will kill most germs present in the paddock, manure should be picked up several times throughout the day and the pasture should not be used for animals other than mares expected to foal within the next few hours. Care should be taken that the paddock fence is low enough to prevent the foal from rolling under it during birth or getting a leg caught underneath it. The pasture must be free from dangers such as deep gullies, streams or sharp objects.

Whenever the weather is cold or damp, or the mare is expected to have difficulty, a foaling stall should be used. Most large breeding farms are equipped with several specially constructed foaling stalls. Such stalls are at least 16′ x 16′ and constructed of material that is easily disinfected. Ventilation, lighting and heat, if necessary, must be provided. If the stalls are enclosed, an observation window will be required so that the attendant can check on the mare without being seen, since most mares will attempt to hold back the foal until left alone.

Straw is the preferred bedding for foaling mares, because there is less trouble with particles clinging to exposed membranes, which can result in infection. Solid rubber matting under the bedding is desirable because of the ease with which it can be disinfected. Needless to say, the bedding must be kept clean at all times.

10.4 Foaling Indications. Most mares foal without difficulty and the foal usually enters life sound and in good health. However, the difference between life and death for the mare as well as the foal is sometimes small. Occasionally a slight difficulty may develop into a very serious or fatal condition unless human assistance is given at the crucial moment. Unneeded or incorrect assistance or interference with a normal birth can cause a great deal of unnecessary injury.

As the foaling date nears, the mare that has been sutured should be opened to prevent tearing when the foal is born.

Before the foal is born the hormone oxytocin has prepared the udder for milk production. The mare usually "makes a bag" (begins to produce milk) a few weeks before parturition. Stabled mares "bag up" more before foaling than do pastured mares. While the gestation period varies greatly from mare to mare, individual mares are usually consistent for length of gestation and signs of impending parturition. The length of gestation as well as the time the mare makes a bag and the way she acts should all be recorded on her record for future reference.

Several days before birth the coupling muscles on both sides of the tail, the hips and the abdomen begin to sag and stretch in preparation for the passage of the foal. A day or so prior to foaling, "waxing" of the teats occurs. A light yellowish-brown waxy substance clings to the end of the teats, where it remains for about 15 to 30 hours, then drops or strings off. At this point milk will seep from the teats and parturition is usually no more than a day away. Occasionally a mare will foal without having "made a bag" or waxing. Usually the stimulation of the foal attempting to nurse is sufficient to begin milk production; however, such animals must be watched closely to see that the foal receives adequate nourishment.

Parturition is near at hand when the vulva appears enlarged and the muscles and ligaments in and around the pelvic area relax and become loose and flabby. Some mares discharge mucus and walk with a straddling gait. As birth approaches, the mare may be very irritable and even threaten to bite or kick at her attendant or familiar horses. She wants to be alone and may be suffering from slight cramps or labor pains. She may sweat around the flanks and neck, frequently pass urine, vigorously switch her tail, paw the ground, look at her flanks or make half-hearted kicking motions toward her abdomen. These are only a few of the many ways she may demonstrate her irritation and discomfort.

Just before foaling the mare will lie down and get up several times and will appear to be seeking a comfortable position. When foaling appears imminent, most attendants wrap the mare's tail to keep it out of the way and prevent it from becoming soaked with blood and fluids, and wash her udder, buttocks and vulva with soap and water and thoroughly rinse with a mild disinfectant.

10.5 Parturition and Aid. The first indication of delivery is the rupture of the water bag and the expulsion of considerable amount of fluid (3 to 6 gallons). Should the water bag appear from the vulva and not rupture within five to ten minutes, the attendant should rupture it with his fingers. The mare, temporarily relieved from the pressure, will usually lie down for half an hour before commencing the powerful contractions required for parturition. The muscles of the abdomen and the uterus contract together as labor proceeds. Foaling is usually accomplished within 15 to 30 minutes, but may take as long as an hour or more for completion.

The foaling procedure should never be rushed, nor should the foal be forcefully pulled from the mare. Since the muscles are not relaxed sufficiently to allow ease of passage, such pulling will probably result in severe bruising or tearing. Should the mare want to get up and walk around, she should not be restrained.

One reason that a qualified attendant should be on hand is that as the powerful muscles contract to expel the foal, there is the danger of the mare's being torn. If the foal is in an incorrect position, a hoof may damage the walls of the mare's vagina and rectum, causing a recto-vaginal tear. To prevent this, the foal's position must be corrected.

In the normal presentation both forelegs and the muzzle of the foal are visible; first the two front feet, then the nose and head appear. The correct and natural position of the foal is muzzle between the forelegs, facing downward toward the mare's hind legs. When the head has cleared the vulva the foal is usually expelled rapidly and with force.

Normally one foreleg will extend slightly ahead of the other prior to the time the shoulders have passed through the pelvis. The shoulders do not pass precisely together; one is forced backward a little until the leading one clears. The importance of this knowledge is obvious if assistance is required before the shoulders have passed the pelvis. Pull must be applied only to the leading leg. Often when the two shoulders are pulled through together they will compress and somewhat crush the foal's thorax, which can result in rib fractures and torn cartilage. The attendant should be very slow to volunteer assistance at this point; very few stoppages occur that will prevent normal passage of the shoulders if the mare is given plenty of time.

Birth of a foal sequence

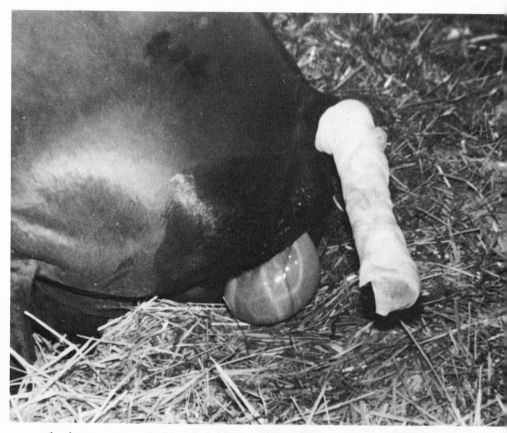

1. Appearance of the water bag. Note that the tail has been wrapped in sterile gauze to keep hair out of the way and prevent contamination by germs present on tail hairs.

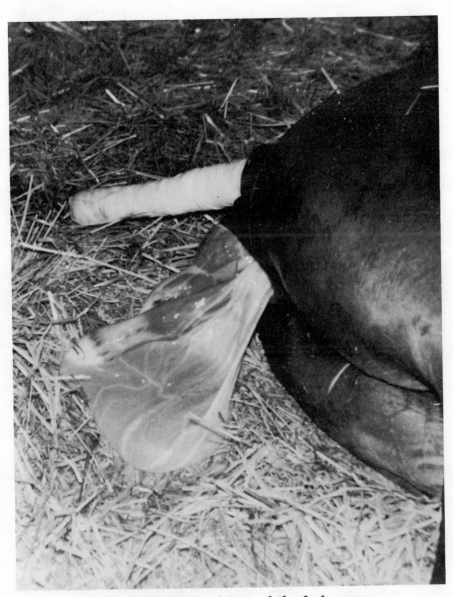

2. The front feet and nose of the foal emerge.

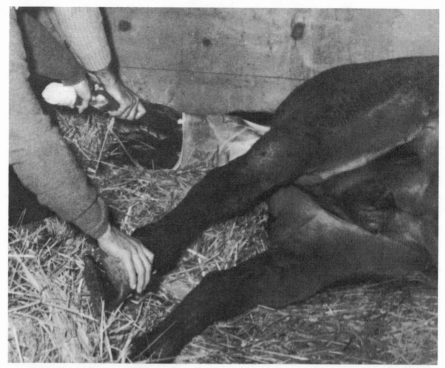

3. Attendant is guiding the foal out and helping it slide through the straw. No pull is being exerted upon the foal.

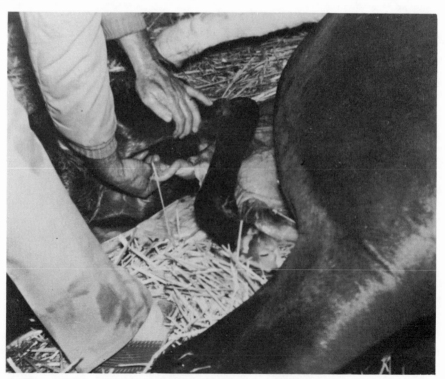

4. The umbilical cord is still attached to the foal and the placenta. It will break as the mare stands.

5. The mare inspects her new foal.

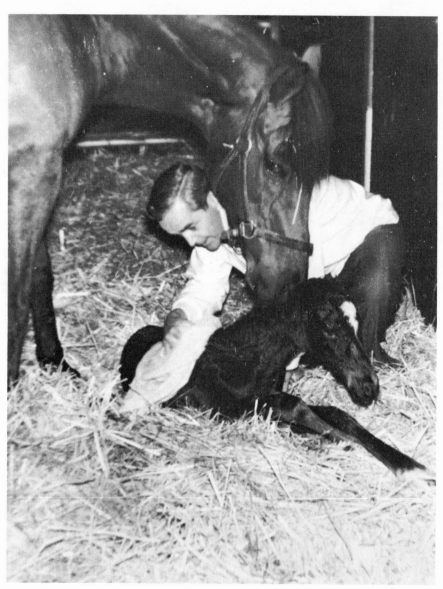

6. Attendant is rubbing the foal dry and stimulating blood circulation. Note that the head and neck are left wet so that the mare will recognize the foal more easily.

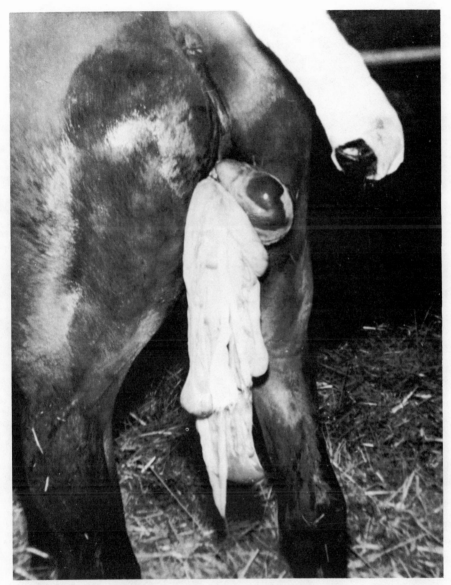

7. The placenta (afterbirth) will hang from the vulva until expelled.

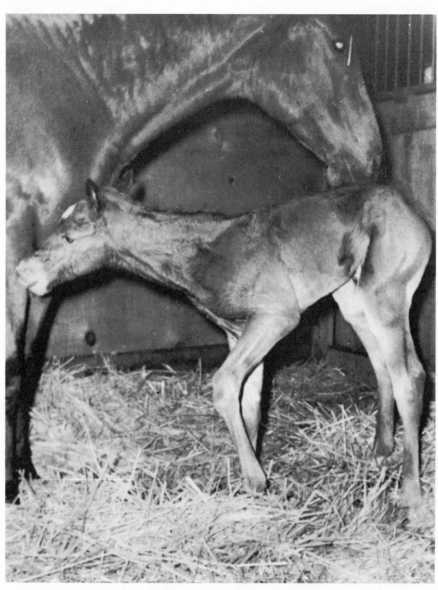

8. The newborn foal will usually stand within an hour and a half.

If the foal appears to be in an abnormal delivery position or other observations indicate a difficult birth, the veterinarian should be called at once and the mare should be gotten up and walked to avoid straining and injuring either herself or the foal. Often when birth is other than normal it is due to infection or a sick or dead foal. An equine veterinarian should be present when such a case is suspected. The veterinarian must be summoned immediately if upon examination after an undue length of labor any of the following conditions is found: nothing can be felt in the vagina, indicating the back of the foal is to the pelvis; one or both forelegs but not the head is presented; the muzzle is detectable but the forelegs are not present; or more than two feet are presented, indicating either twins are vying for priority or one hind foot is improperly forward.

Sometimes a serious condition will occur that cannot wait for the veterinarian to arrive. In these cases the attendant, with clean hands, may have to enter the mare for manual examination and manipulation.

If progress of birth appears to be stopped with the head and shoulders of the foal exposed, the foal's hips are probably caught in the mare's pelvis. The foal can usually be freed by crossing its legs to turn it right and left while pulling downward. Any pull on the foal should be exerted in a downward direction following the natural birth canal in the mare and should be applied at the same time the mare is having a contraction.

When the foal is being presented upside down, it will usually turn to the correct position if the mare can be made to rise and walk around for a minute or two. The muscular activity and weight-shifting ordinarily enable the foal to roll over. If the situation does not correct itself after walking the mare for a short period several times, the attendant should make sure that the head is through the cervix, then grasp the forelegs, cross them and rotate them, causing the foal to turn over. Turning should be attempted only between contractions.

In instances where the birth is posterior (breech birth—hind legs first, hooves pointing toward the mare's tail) the foal must be delivered quickly to prevent suffocation. Often the umbilical cord is squeezed and circulation is cut off or the cord breaks before the head is free. To prevent loss of the foal, assistance must come quickly if the mare appears to have difficulty. If a breech birth is apparent, the mare should not be allowed to foal right away, for more time is needed for full relaxation of the muscles and ligaments to allow quick passage of the large hindquarters of the foal. She should be walked for half an hour before being allowed to lie down. As labor begins, the attendant should pull the hind legs in the same manner discussed earlier concerning hips that are wedged. Occasionally a foal will have one or both front feet extended backward into the uterus or it may have its head back. In this case the foal must be pushed back so that room may be provided for the feet and head to gain the correct position. Sometimes one foreleg will be deflected upward so that as the foal is expelled, the leg is forced through the vaginal wall and into the rectum. This position will tear the structures outward as the foal is born; early corrective action is imperative. If a tear has not yet occurred, the pastern of the raised leg can be grasped and pulled forward into normal position. If a tear has already been made, the foal must be forced back into the mare until the leg position can be corrected.

The mare should never be permitted to deliver in close proximity to the wall, manger or other obstruction that might impede delivery or injure the foal. Mares that deliver while on their feet have been known to brush or bump the half-born foal against an obstacle. They should be tied so the hindquarters cannot make contact with another object. The attendant should aid the foal so that it will not be injured by dropping from the mare.

The umbilical cord (20 to 40 inches long) is attached to the placenta, which usually remains in the mare for several minutes to several hours after birth. Within fifteen to thirty minutes after the foal is born, the mare will get up; this motion normally breaks the umbilical cord between the foal and the placenta. The mare will lick the foal free of the membranes covering it. Her licking and nuzzling will stimulate the circulation and bring the foal to "life."

The placenta (afterbirth) contains blood that is an aid to the foal and should be allowed to drain into the foal. It should not be severed by the attendant unless it is expelled with the foal and does not break by the activity of the mare or the foal. If this should occur, the umbilical cord should be pulled (not cut) apart at the constriction about two inches from the foal's body. The umbilical cord should never be tied. If an infection is present, tying will force the infection into the abdominal cavity, which often proves fatal. If the navel appears abnormal a veterinarian should be consulted. When the placenta is normal and expelled complete, it should be removed from the area and buried or otherwise sanitarily disposed of.

Some mares are overly protective of their new foals and must be watched to prevent injury to the attendant or others who might enter the stall.

10.6 Foal Post-Foaling Care. The placental membranes, which made up the water bag, sometimes cover the nose and mouth and can smother the foal. When this happens the mouth and nose must be cleared, the mucus removed from the nostrils and the tongue pulled forward if it is back. Often lifting the foal by the hind legs will be sufficient to allow gravity to clear the air passages. Occasionally stimulation is necessary to promote breathing and may be accomplished by brisk massaging with a cloth or whatever suitable material is available. The head and neck should remain wet so that the mare will have no difficulty identifying the foal. The front legs may be worked back and forth briskly to encourage breathing if the foal does not appear to breathe. If the foal is diseased, professional care is absolutely essential. If a foal is apparently dead, artificial respiration will sometimes revive it. Artificial respiration should be administered from 15 to 20 minutes before efforts to revive the foal are considered futile. Some large ranches have mechanical respirators on hand for this purpose. Pumping the rib cage with the heels of both hands or with one knee to simulate normal breathing will sometimes suffice. While the foal is lying on its side the foreleg may be drawn forward and slightly upward to give access to the lung area. The foal may be lifted in the middle and dropped gently to encourage breathing. One nostril may be pinched off and air blown into the other. The lungs of a new-born foal are very delicate and care must be taken that a ruptured lung does not result from blowing too hard into the lungs. In rare instances the heart might be stimulated by quick, rather firm jolts from a knee or the heel of the hand.

After parturition the navel stump should be immediately treated with 7% tincture of iodine or other strong antiseptic by either inserting the stump into a bottle of iodine or by swabbing. Any foreign matter imbedded in the stump should be gently squeezed out. Usually iodine swabbing or dusting with antiseptic powder is continued for three days or until the stump appears dry, but it must be remembered that iodine is very severe and should not be used repeatedly for very long periods.

It is a wise procedure to administer antibiotics such as penicillin and streptomycin to provide the foal with protection against infections for the first few days until the natural immunity furnished to the foal by the colostrum (mare's first milk) has adequate time to become effective. Routine preventative treatment for foals has greatly reduced the occurrence of septicemias (blood poisoning). The veterinarian should determine if further injections should be administered. Most veterinarians recommend and administer tetanus antitoxin to both mare and foal promptly after parturition as a routine part of the birth care.

If the mare is slow to dry the foal, it may be wiped with a soft towel. As mentioned above, care should be taken to leave the head and neck wet so that the mare will be able to identify the foal as her own.

The average foal will weigh approximately 10% of the mare's body weight. A healthy foal will stand within an hour and a half. It will be wobbly and will fall as it toddles after the mare but as its legs "unfold" and become stronger, its balance will rapidly improve. The longer the foal's legs are and the rangier it is, the longer it will take to stand. The foal should not be steadied on its feet; the use of its legs, although awkward, aids circulation and helps the foal gain control of its muscles. New-born foals' legs often seem quite crooked and are sometimes down at the pastern. The vast majority of foal's legs straighten up within a few days. Occasionally a foal will be born with severely contracted flexor tendons and may have to be fitted with splints or casts. The eyes of the new-born foal should be guarded against very bright light for the first few hours until the eyelids react voluntarily.

The baby should be let alone until it nurses of its own accord. Instinct will tell it where the milk is to be found and how it is served. It is safe practice to wash the mare's udder with warm, soapy water or suitable mild disinfectant and rinse well before the foal nurses for the first time. If the foal does not attempt to nurse after four or five hours or appears too weak to stand, it must be assisted or veterinary help summoned. Some mares will not allow the foal to nurse until the placenta has been expelled, relieving her muscular constrictions.

The mare's first milk (colostrum) contains a certain amount of antibodies that provide the foal with temporary immunity to many illnesses. It is of the utmost importance for stimulating the foal's unused digestive tract. The colostrum contains all of the necessary substances for putting the foal's body in working order and therefore is its most important meal. Through the colostrum the mare provides the foal with antibodies she has naturally developed against diseases in the area. The same digestive mechanism that allows the foal to absorb the antibodies in the mare's milk the first few days of life also allows germs to enter the young foal's body through the digestive tract. For this reason colostrum and sanitary surroundings are of the utmost importance to the foal as it begins life. (See Sections 11.2 and 11.4 (e).)

Occasionally a mare refuses to accept the foal. If this occurs, she should be haltered and held or tied until she will allow the foal to nurse. However, if the mare's udder is sore to the touch, she may only appear not to accept the foal. She should be calmed and restrained until the foal nurses. If the udder is extremely tender, careful hand milking to relieve the pressure is advisable. If the colostrum is being milked out by hand, it should be saved in a sterile bottle and fed to the foal. (See Section 11.2 for bottle feeding.) The mare will quite likely accept the foal when the pressure has been reduced. If the udder has become badly swollen, heat applications several times a day will usually soften the udder, and careful hand milking will relieve the pressure enough so that the mare will allow the foal to nurse.

If the mare persists in refusal to accept the foal and appears to want to hurt it, she may be ill, the foal may be unhealthy, she may have insufficient milk (if a bubbling noise is apparent when the foal nurses it is an indication that milk is scanty), or if it is her first foal the mare may be afraid of it and in her fright may trample or kick the foal if in a stall or abandon it if in the pasture. A separate stall should be provided for the mare next to where the foal is bedded, or a corner of the stall can be separated to keep the foal from reaching the mare. The foal should be kept away from the mare except at intervals when an attendant is present to hold or twitch her while the foal nurses. To prevent the mare from kicking, it may be necessary to hold or tie up one of her forelegs so that she cannot lift the hind leg on the same side. Care should be taken that the mare does not fall and injure herself, the foal or the attendant. If the mare tries to kick while her leg is held up, another method of restraint must be tried. If the mare still refuses the foal the veterinarian may decide to tranquilize her. The handler should hold the mare by the halter and turn her toward the foal if she appears to want to kick. She may also lay her ears back and bite the foal. An assistant handler should guide the foal so that it does not get injured by the mare while trying to nurse.

If the foal must be taught to nurse, the attendant may dampen his finger with the mare's milk and coax the foal to the udder by getting it to nurse his finger. The finger is held next to the teat and milk is squeezed from the udder so that the foal gets the taste and will gradually transfer from the finger to the teat. The procedure should be repeated every 45 minutes or so until the foal learns to nurse by itself. The foal should not be left with the mare until it is apparent that she has accepted it.

Sometimes the mare is slow to produce enough milk for the foal but usually the nursing action will stimulate the mare's system enough to overcome the deficiency; if not, and the foal seems to be nursing constantly, the veterinarian may induce milk production through hormone injections.

Only the foal health problems requiring immediate attention at birth have been discussed in this section; Section 11.4 covers many other matters of concern in the early life of the foal.

10.7 Mare Post-Foaling Care. One of the main reasons for foaling difficulties is that mares often do not get enough exercise. Broodmares should be exercised daily and may be ridden until foaling, provided care is taken to avoid exhaustion, unnecessary strain and jarring during the last month of pregnancy. A hilly pasture in which the mare must walk a good deal will probably provide adequate movement. Mares that do not exercise themselves should be made to move about either by riding them, longeing them or exercising them on a mechanical hot-walker.

The most common complication after foaling is "foal colic." The pain in the abdomen is caused by the stretched muscles of the uterus cramping as they rapidly contract. Although not related to indigestion, the symptoms are the same. The mare's attention will be directed toward her abdomen and flanks. The irritation may cause her to paw, sweat or roll. The symptoms may appear within a half hour after foaling and usually last from 30 minutes to an hour. A warm bran mash with ten mashed aspirin tablets may give relief. If the condition persists for over an hour or if very severe, a large dose of colic medicine should be given. If the mare does not respond to the medicine within a reasonable length of time, a veterinarian should be summoned.

To avoid colic associated with bruising of the colon during foaling, the mare should be fed a laxative diet such as bran mash with mineral oil added to it. The mare should be watched closely for dry manure passed in small quantities, which indicates pending impaction.

A very serious condition called "foal founder" can result in permanent lameness or death of the mare. It is caused by the retention of even a small portion of the placenta. For this reason the placenta should be spread out for examination to make sure that all of it was expelled. Should part or all of the placenta be retained for over six hours the veterinarian should be summoned to remove it.

Sometimes the placenta will be only partially expelled. If this happens a knot can be tied in the upper portion or a light weight attached to it to cause a gentle pull as the mare walks, or it may be tied to the mare's tail to keep it from flopping about if it makes her nervous. Young maiden mares often become terrified at the afterbirth hanging behind and kick at it or run away, injuring the foal, themselves or those around them. Under no circumstances should the attendant try to pull

The placenta (afterbirth) should be spread out and examined to be sure it is complete and none has remained in the mare's uterus.

the placenta from the mare, since a portion is likely to tear off and remain in the uterus.

The retention of some part of the placenta in the uterus will result in uterine inflammation or possibly foal founder or laminitis. If a portion of the placenta is retained, the mare will show signs of colic or become lame about a day or a day and a half following foaling. Her pulse and breathing will step up and she will run a fairly high fever. If the mare shows these signs at this time the veterinarian will have to examine her and remove any parts of the placenta that were left in the uterus. A delay in discovery or treatment of this condition can have serious consequences. Early treatment may avoid foal founder (laminitis) and recovery can be expected. If laminitis has become involved, treatment for the condition is imperative. Laminitis can result in permanent lameness and is so painful that the mare may go off feed and fail to supply milk for her foal.

The temperature of the mare that has foaled should be checked for the following two days. If she has a fever something is wrong and the veterinarian should be notified.

The sutured mare not reopened prior to foaling is almost certain to suffer severe tearing of the vulva. Such an injured mare should not be bred back on the foal heat. All vaginal windsucking mares should be resutured as soon as the afterbirth has passed. If a mare is to be bred on the foal heat and is not resutured within 48 hours after foaling, she should be left as is until after breeding.

10.8 Internal Hemorrhage. In the final stages of pregnancy the size and weight of the unborn foal place a very heavy burden upon the mare's uterus and the surrounding tissues, muscles and ligaments. The two main blood vessels within the uterus may be as large as a person's thumb and are subject to the same pressures as those exerted on the uterus by labor and parturition. In the vast majority of cases the vessels withstand the strain well. However, when the foal's hoof or leg is temporarily wedged so that the labor constrictions unduly mash or stretch the uterus, one of the blood vessels may be ruptured. If the tissue has not been pierced, which is the usual case, the blood escaping from the vessel will form a pocket that will grow in size until the opening seals or the pocket gives way; in the latter case death will soon occur. While the pocket is unbroken, the bleeding is somewhat restricted due to back pressure from the bubble, but when it breaks there is nothing to impede the flow.

If such a hemorrhage is very slight the blood may coagulate and seal the leak, in which case the body will eventually absorb the wasted blood, but if the rupture is quite extensive it must be repaired quickly. If the condition is detected in the early stage a blood coagulant may be effective.

Unfortunately, early detection of a serious internal hemorrhage is extremely difficult. The mare that foals quickly and easily is just as apt to hemorrhage as the mare that had difficulty. The damage can come about in one quick instance without any indication of what has happened. As with all abdominal aches and pains, the horse will exhibit symptoms of colic when the pressure of the blood pocket builds up. This is not likely to occur so early as "foal colic." Familiar signs of colic are suspect if they appear an hour or so after "foal colic" has subsided or an hour or so after foaling where "foal colic" has not been in evidence. Any signs of colic up to 15 hours after foaling, if "foal colic" can be ruled out, should warn the attendant that the veterinarian is needed immediately. As blood is lost the heart must work harder and pulsates more rapidly; the blood pressure is reduced, resulting in a weak pulse, and the body temperature will decrease as the fluid loss becomes critical. An inadequate supply of blood will cause a paling of the eyes and gums. At this stage the mare is very weak and may shake and perhaps stagger; she is not in control and is dangerous to be near because of the likelihood that she will topple over. The foal should be removed so that the mare will not accidentally injure it.

As noted before, a minor hemorrhage may seal itself. There are no statistics to establish the number or percentage of self-sealing internal hemorrhages but the number, if known, would be quite frightening. It is well known, however, that a sealed hemorrhage may easily be reopened and that a minor hemorrhage may enlarge with exercise.

10.9 Pasturing after Foaling. Instances of post-foaling complications are strong arguments against turning a mare and foal out to pasture soon after parturition. While exercise helps a mare to recover from foaling, vigorous exercise may worsen an unknown condition that otherwise would have corrected itself. Most managers who prefer to have the mare foal in a stall will wait 12 to 24 hours before turning them out

Mares and new foals are not turned out to pasture until health and weather conditions are optimum for the well-being of both.

so that both mare and foal can be watched for complications. In any event they should not be turned out until health and weather conditions are optimum for the well-being of both.

Mares that are able to move about freely will expel any matter left in the uterus. The discharge will be noted around the tail and vulva area. About 5 or 6 days after foaling, a dark, thick discharge may appear. This is normal and the discharge soon stops, leaving the uterus clean and healthy.

Many mares become very protective toward a new-born foal. For this reason persons must be watchful when first approaching the mare with a new foal to avoid being charged, bitten or kicked.

Neither the mare expected to foal soon nor the mare with a new foal should be pastured with geldings or barren mares. These animals sometimes become quite excited at the presence of a new-born foal and may kick or strike the handler or the foal. Aggressive geldings, like aggressive barren mares, have been known to "steal" a foal and prevent it from getting to its mother to nurse. The mare and foal should be pastured apart from other classes of horses.

The mare and foal should never be turned into an unfamiliar pasture after dark, especially with horses the mare does not know.

11
FOAL CARE

(SYNOPSIS)

This chapter considers foal feeding, weaning, worming, special foal diseases and problems, hoof and leg care, castration and early handling.

The following synopsis by section numbers is provided as an aid for quick reference to specific subjects.

11.1 General Statement. Long before birth, the care of the mare takes into consideration the requirements she has for herself and the unborn foal. Foal care begins at conception, continues through parturition and until the foal is weaned.

Special care at birth is so vital that it has been discussed along with

The youngster should be turned out to pasture with its mother as soon as health and weather will permit. Neither animal should wear a halter.

Foaling, chapter 10. Section 10.6 should be read in connection with this chapter even though in some respects the matters discussed overlap. The specific subjects covered in Section 10.6 are artificial respiration and resuscitation, navel treatment, antibiotics, tetanus antitoxin and colostrum (mare's first milk). Section 10.9 has to do with turning the mare and foal out to pasture. The natural habitat of the horse is open space; youngsters should be turned out with their mothers as soon as health of the mare and foal will permit; however, neither horse should be forgotten. Close surveillance for a few weeks is a wise policy. Complications and poor health must be detected as soon as they arise. If for some reason the mare must be confined, the foal should be permitted to run and frolic in the pasture a few hours a day if the pasture is adjacent to the mare's stall and she can see the foal. The company of other foals will encourage play, but a young foal should never be placed with other mares without its mother to protect it. Foals need sunlight, fresh air and all the exercise they desire. This combination builds healthy bodies, strong muscles and firm bone structure.

11.2 Foal Feeding. The new-born foal will usually be able to manage its unused legs well enough to rise and support itself within a half hour to an hour and a half, but some normal foals may need another hour to gain control of their limbs. Soon after the youngster is up and has tried its legs, it will seek the mare's milk. It needs no introduction; it has instinct for its guide. Although the foal should not be forced to take the teat, it should be observed and if it fails to nurse for as long as four or five hours after birth, the mare or foal may have a problem. By this time the foal has begun to need nourishment and must be hungry; the reason for its failure to nurse must be discovered and corrected if possible. Section 10.6 provides further discussion of nursing problems.

All foals need milk to stimulate the unused digestive tract. Very young foals cannot tolerate solids. The colostrum is very important to the new-born foal (see Section 10.6). If the mare has been in the area for two or three months prior to foaling, she will have built up antibodies that resist many types of germs in that area. These antibodies concentrate in her first milk and for a day or two can be effectively absorbed by the foal's system, giving the foal temporary immunity against those germs for a short time, perhaps two or three weeks. Without colostrum the new-born foal is a prime target for disease organisms until it can develop antibodies of its own. If the mare does not have a sufficient supply of milk or if she has been leaking substantial amounts of milk within a few days prior to parturition, her foal may not receive full immunity benefits. This fact should be made known to the veterinarian, who will prescribe antibiotics for the foal in an effort to provide defenses that the foal will need. Other colostrum benefits are its natural laxative effect and its high concentration of protein and vitamins. So vital is colostrum that well-managed breeding farms withdraw it from healthy mares having stillbirths and from pregnant mares leaking badly before parturition and store it under refrigeration for the later benefit of foals that might otherwise be deprived of it.

Whether or not the foal receives colostrum, it must have liquid milk for at least two months and, if possible, until weaning at five or six months. The death, disease or inadequate milk supply of the mare complicates the foal's life and well-being. Many large farms keep nurse-mares on hand in case of such a development. The nurse-mares are usually cold-blooded, have calm, quiet dispositions and will accept orphans. They are bred early in the season so that at foaling time their milk will be available. Some horsemen groups form nurse-mare pools from which a member can draw a mare if needed. "Project Nurse-Mare" at Murietta, California, is such a nurse-mare pool established by southern California Thoroughbred breeders. Some areas that do not have nurse-mare pools do have colostrum banks, which may be located through a local veterinarian.

If the farm loses a broodmare before her foal has received the colostrum, the dying or dead mare (while still warm) should be milked out and the colostrum fed to the foal. Possibly the large ranch will have frozen colostrum on hand or can obtain it from a colostrum bank. If not, enough might be obtained by milking out small quantities from several mares producing it for their own foals. The orphan foal is at a definite disadvantage and needs the colostrum to aid its body in its functions. Mother's colostrum is best, but colostrum of another mare in the same area will afford the same immunity from local diseases and will assist the foal in commencing its body functions.

If needed, the foal should receive a laxative such as milk of magnesia. Vitamin and antibiotic injections are also important for its protection in its somewhat less than normal circumstances.

The best food for an orphan foal is, of course, mare's milk, and every attempt should be made to find a nurse-mare, not only to feed and protect the foal but also to avoid psychological problems resulting from loneliness and to educate the baby in the ways of a horse. While awaiting the arrival of the nurse-mare, the foal should be kept very warm so that it will expend less energy for heat and will require less food. Other mares, if available, should be milked and the mare milk fed to the orphan so that the type of milk will not be switched again in the event a nurse-mare is located. The young foal's digestive tract is very delicate and cannot cope well with changes in milk.

It is often difficult to persuade a nurse-mare to accept an orphan. Every attempt should be made to keep tension at a minimum, because the mare will already be upset about her own foal's absence. If the mare is unfamiliar with the farm, she should not be introduced to the foal for several hours. She must be allowed to get used to the new surroundings and must be relaxed when she meets the foal. It is important that the foal smell proper to the mare when first introduced. Any medication should be thoroughly removed (except for navel applications) and the foal should be rubbed with a cloth saturated with salt water. The nurse-mare's urine may be daubed over the foal, especially on its back, croup and tail. It can then be rubbed with some of the nurse-mare's fresh manure. A little of the nurse-mare's milk rubbed on the foal's head and neck may also be beneficial. The point is to make the foal smell as much like the mare as possible. Some of the nurse-mare's milk may be smeared on the mare's nostrils to make strange odors less detectable.

If the mare seems extremely upset, she may be mildly tranquilized. The two must be introduced very cautiously to prevent the mare from biting or striking at the foal. To avoid injuring the foal the mare should be restrained by a twitch, lip chain, or hobbles, or by holding up a front leg on the side where the foal is nursing. The mare should be allowed to examine the foal carefully and smell it thoroughly. The foal should be quite hungry so that it will want to nurse immediately and the mare's bag should be full and tight so that she is uncomfortable and will welcome nursing to relieve the pressure. Although she may allow the foal to nurse while she is uncomfortable, she may object strongly when the pressure is relieved and the early acceptance may not mean that the mare has accepted the foal. The two should not be left alone and the restraints should remain in place although eased while the mare is calm. If the mare becomes nervous the foal should be made to stop nursing, since the mare's attitude is most important and must be cultivated.

Between feedings the foal should be placed over a partition or in an adjoining stall where the mare cannot hurt it but can see and smell it while becoming accustomed to its presence. Some mares will accept an orphan almost immediately, but others may take a week or more before accepting it. After the foal has nursed for a day or two the fecal material will begin to smell like her own foal's, at which point the mare will usually accept the foal, but the two should not be left alone until the mare is following the foal around and nickering, thus demonstrating that she is content with it.

Sometimes a complacent older mare with foal, if a very good milk producer, can mother an additional foal. A young mare will ordinarily not accept an additional foal not her own.

A milk cow or goat is often maintained as an inexpensive source of milk for orphan foals. Cow's milk is higher in fat and protein and lower in water and sugar than mare's milk. For this reason pure cow's milk should not be fed unaltered to the foal. Milk from Holstein cows (low-fat processed) is generally lower in butterfat than milk from other cows and is therefore preferred. The composition of goat milk is similar to that of mare milk and may be fed to the foal unaltered or the foal may nurse the goat. If a milk cow or goat is on hand but not being used, she will have to be milked out by hand once a day to keep her from drying up. Time and labor will be saved and the foal will benefit from the companionship if a goat is nursed. The nanny should be placed on a bale of hay so that the foal can reach the udder. If nursing the goat is a problem, the milk can be fed by bottle. A new-born foal should not be fed by bottle unless standing or resting on its stomach in an upright position; if fed while lying on its side there is danger that the liquid will go down the wrong tube and into the lungs.

An orphan foal without a foster mother will need companionship to avoid psychological disturbances that will affect its feeding and health; a goat, a pony or even a dog will help in this respect.

If a nurse-mare or goat is not available, the foal will have to be fed a suitable liquid formula. This requires bottle feeding by hand on a day-and-night schedule, beginning within six hours or so of birth, and time intervals and quantities must be closely adhered to, as follows: Feed 5 to 8 ounces of formula (one-half pint), at body temperature

(100°F) each feeding. For two days feed at hourly intervals; the following five days feed at two-hour intervals; then for three weeks feed at four-hour intervals; after one month feed larger amounts four times a day and stop the night feeding.

The needs of each particular foal must be considered and the ranch veterinarian should be consulted in regard to the foal's liquid formula. Unless the circumstances or the veterinarian suggests changes, one of the formulas listed below is a good starting diet.

1. 4 ounces evaporated milk
   4 ounces lime water
   1 teaspoon corn syrup
2. 6 ounces cow's milk (low-fat Holstein is best)
   2 ounces lime water
   2 egg yolks
   2 tablespoons corn syrup
3. Commercial formula (mixed as directed on package)

The average new-born foal will consume about eight or nine pints of formula per day. The consumption will increase in proportion to growth. It is generally considered that for each ten pounds of body weight the foal will need one pint per day. Orphan foals seem to do better if they are underfed rather than overfed. If left a little hungry at the end of a feeding they will continue to eat readily and fewer digestive problems are encountered.

Feeding by bottle is very time-consuming for the busy farm. The sooner the foal will drink from a bucket the better, from a labor point of view. The liquid for the orphan can be stopped entirely at about two months provided it is replaced by adding powdered low-fat milk or a milk replacer to the grain that the foal should be eating by then. The change should be gradual rather than abrupt; decrease the liquid daily over a period of a week or two and supplement powdered milk.

A liquid diet will not suffice long for the growing foal. Milk is not adequate in iron and copper and a foal maintained on milk alone will become anemic. The chapter on nutrition provides a detailed description of the nutritional requirements of foals. Water is covered in Section 6.2 (d); vitamins are discussed in Section 6.3 (a), (b) and (d); minerals, salt, calcium and phosphorus are discussed in Section 6.4 (a), (b) and (c).

All nursing foals, whether orphans or not, over ten days of age need as much legume hay (alfalfa preferred) as they will eat, plus alfalfa meal, milk replacer, light grain mix or colt starter preparation. Nursing mares feel no responsibility to the foal for any food but the milk they furnish; they will take any of the dry feed they can reach. Foals should be fed in creeps placed near areas where the mares congregate, such as close to a water trough, hay rack or stand of shade trees. The foals quite naturally stay near their mothers and will not visit a creep placed where the mares do not go. If the mare and foal are in a stall, other separation methods may be needed to assure that the foal can get all the food it wants. Sometimes a barrier is used, but often a foal feed box is provided and the mare is tied short enough to prevent her interference until the foal is finished eating. Confined foals should be fed two or three times daily to assure maximum assimilation of feed.

Within two or three weeks the foal will be picking at its mother's grain and by a month will probably begin to relish it. The sooner the foal begins eating grain and hay, the easier it can be weaned. The mare's grain ration is probably too heavy for the foal; at first it should be separately fed light grain consisting mostly of rolled oats or rolled barley with some bran and a little syrup or brown sugar. The sweetness will do much to encourage consumption. Powdered low-fat milk or milk replacer can be mixed into the grain if the foal is not getting sufficient milk otherwise. Later a little linseed meal or soy-bean meal supplement can be added to the grain. Ground oats and ground corn can be used after the foal gets a good start, but this must be dampened slightly with syrup and water to avoid dust. For optimum health and development the foal should start at about ½ pound per day for 100 pounds of body weight, soon increased to ¾ pound. By weaning time the ration should be about one pound per month of age per day until it is consuming about 10 pounds per day. At this rate, when it is one year old (12 months), it should have half of its anticipated adult weight and by the time it is two years old (24 months) it should have its full height. If well-fertilized and productive pasture is available in the spring and summer following weaning, the youngster's grain can be cut down considerably (about ½) but gradually increased again as the pasture is grazed down or dries up.

The most important growth period for the young horse is between weaning and 12 months of age. Its feeding should stress vitamins and minerals and its requirements for protein and energy are about 35 percent more than that of an adult horse of equal weight. The animal needs bone growth and muscle development. At this age the skeletal development is more important than muscle, therefore attention to the consumption of calcium, phosphorus and Vitamin D is imperative. These nutrients must be adequate. Fatty feeds should be held to a minimum for the growing horse; excess fat imposes heavy burdens prematurely upon the muscles, tendons, bone structure and joints, and increases the chances of damage to the legs and joints. Excellent pasture is the finest nourishment the developing yearlings can have and at peak seasons will usually suffice for roughage with no additional hay; however, some grain is usually advisable. As pasture feed value changes, so must the grain and hay offered daily. After 12 months of age the horse can be fed much the same as the adult horse, provided adjustments are made to supply about 20 percent more nutrients than required for the maintenance of an adult horse of the same weight and provided the supplies of minerals and vitamins are liberal. Feeding must always take into consideration the needs of the particular animal and the energy it expends daily. Horses being conditioned for showing, selling, working and racing should be fed more generously than those that are idle.

11.3 Worming Foals. As soon as foals are born and sometimes before birth they become subject to parasitic infestations discussed in Section 5.14. Foals are particularly susceptible to ascarids (large roundworms) (*Parascaris equorum*). They should be included in a routine worming schedule very early, even while nursing. The ranch veterinarian will make special provision for them in the farm's worming schedule, which may call for treatment four or five times a year as determined by microscopic examination of fecal samples. The first worming ordinarily is needed near three months of age. Ranch worming schedules are discussed under Section 5.16.

11.4 Foal Diseases. This section emphasizes certain diseases and health conditions that occur predominantly in the new-born foal. As effective as colostrum is, it is not a guarantee against all illnesses. Routine use of antibiotics, antitoxins and vaccines has greatly reduced the incidence of foal diseases. The veterinarian will prescribe medication he feels will benefit the new-born foal. He will be especially concerned with the general condition of the foal; whether or not the foal received adequate colostrum (the veterinarian must be informed if the mare leaked much milk before foaling); the length of time the mare was in the area before foaling to produce antibodies against local germs; the diseases that the mare has or has had during pregnancy; the conditions of sanitation; and the diseases known to occur in the area. With this knowledge he will be able to administer medication that will provide the needed protection.

Although a foal seems to be off to a fine start in life, it must be observed closely for as long as a month. If its health appears to degenerate or if it goes off feed, seems listless or depressed or if its activity appears slowed down, the veterinarian should be consulted. Often the only indication that a foal is sick is the fact that the mare's udder becomes full and tight because the foal is not nursing.

Generally speaking, chapter 5, Equine Health, applies to all classes of horses, including foals. There are numerous foal diseases; this chapter, however, will limit discussion to those diseases the breeding farm is most likely to encounter.

a) *Navel-Ill* (*Omphalophlebitis*), also referred to as joint-ill, is a serious bacterial disease of foals. It is not exclusively a foal disease, although is sometimes thought of as such because it appears so much more often in foals than in older horses. New-born foals are targets for the bacteria because of the ease with which they can enter the body through the unhealed navel stump and because the young animal has not yet developed antibodies against them. The infection may be caused by many different kinds of bacteria. A few known to infect foals are *streptococcus, staphylococcus, pasterella,* and *salmonella.* Severe cases affect the chest cavity, the heart and other organs and advanced cases often cause death. If the mare was infected by *S. abortus equi*, the foal may be born with enlarged joints, but if it has contracted the disease after birth, symptoms will usually not be noticeable until a week or ten days later, at which time the joints enlarge and become very painful and fever is present. All joints may not be affected and sometimes the swelling switches to other joints or may appear to abate but then returns. Any enlarged joint or lameness accompanied by fever, unless definitely attributable to another cause, is highly suspect and the foal needs immediate treatment by an equine veterinarian. Early detection of the disease is essential but prevention is a thousand times better than cure. A foal that has recovered from navel-ill may suffer permanent stiffness or may develop lameness in later life. As long as the navel is not completely healed, danger of infection is great. Treatment of the navel at birth (see Section 10.6) must not be overlooked and the navel should receive daily attention until completely healed. If the navel does not readily seal, remains damp or emits a discharge, a veterinarian should be consulted even if no symptom of disease is evident.

A foal that was deprived of colostrum is highly susceptible to the infection. Such foals must receive antibiotics. When the mare leaks a large amount of milk before parturition, the foal may not receive sufficient colostrum.

New evidence indicates that many cases of navel-ill are contracted through the digestive system during the first few days of life. Occasionally the foal is infected prior to birth through the placenta of an infected mare.

b) *Navel Urine Seepage (Pervious Urachus) and Ruptured Bladder.*

A condition in which urine seeps from the bladder through the navel stump, known as Pervious Urachus, is recognizable by a soft, mushy stump. It is present because the outlet from the bladder through the navel, which carries the urine during the fetal stage, fails to close properly and urine continues to seep through the canal, which retards healing. The condition is usually not serious except that as long as it exists, a pathway for germs to enter is present. The stump should be treated with silver nitrate applied daily until the condition is corrected. Should the condition persist, a veterinarian should be consulted. If no infection is present he will have no difficulty inducing complete closure very quickly. If navel swelling is present, prompt treatment to draw the infection to the outside should be employed in order to avoid the rupture of the abscess into the abdominal cavity, which is almost certain to be fatal.

Rupture of the bladder may occur if the foal's bladder is partially filled before birth. The pressure exerted on the foal during parturition can cause the organ to rupture. If this should happen the foal will usually appear normal for the first day or two and then become dull and lethargic. The mucous membranes of the eyes and mouth will take on a red or yellow color and the foal may strain as if constipated and will usually appear bloated. Early surgery can often save the animal.

c) *Sleeper Foal.* This disease is caused by the bacterium *Actinobasillus (Shigella) equuli*. Foals afflicted with the disease usually die within four days of birth. The bacteria release deadly toxins into the blood stream and cause the formation of abscesses on the kidneys and other internal body organs, including the brain. The mare is probably infected but shows no ill effects and transmits the disease to the unborn foal. The bacteria attack so quickly that antibiotics administered after the disease is detected are too late to be helpful. Death usually follows birth by a day or two and may occur within a few hours. The name "sleeper foal" describes the principal symptom—drowsiness (see note following (d) below). The foal is very weak and sleeps a good deal of the time and although, when awake, it can nurse, it has very little interest in feeding. The disease is unlike jaundice in that there is no discoloration of the outer membranes and listlessness sets in earlier than in jaundice cases. Sleeper foals are often referred to as dummy foals. Although symptoms are similar, more recent studies indicate that the diseases are different.

d) *Barking Foals, Dummy Foals and Wandering Foals.* These three conditions are all thought to be related to one disease. A very violent series of convulsive fits and muscular spasms may occur soon after birth. A series of sounds much like barking is not uncommon. The convulsive foal has to be restrained physically by the handlers until a veterinarian can administer sedatives. Usually the violent condition is temporary but the unfortunate foal is left quite helpless for a long time. It must be fed by hand and nursed for a week or two while it goes through two distinct recovery stages. First it seems to be merely existing in a dumb state. It is then known and commonly referred to as a "dummy foal." It seems to be inert and totally uninterested in its environment. Later it will become active and will roam about the stall. It then is a "wanderer" without vision. It will helplessly and persistently walk, bumping into everything in the stall. It is temporarily without sight or at least unable to associate images with distance and whereabouts.

It is not uncommon for the later two stages to arise without any evidence of the convulsive onset. How this happens is not known but it is believed that the first stage passes before birth or is so mild that it is not noticed. Whatever may be the cause, if carefully fed, nursed and protected, most of these foals seem to fully recover. The foal must be fed the mare's milk through a stomach tube because it will not suck.

Latest research indicates that the condition is probably associated with extreme lack of oxygen just prior to, during or shortly after birth. Suspected causes are premature separation of the placenta from the uterus, interference with the normal foaling process or severing of the umbilical cord too soon, thereby depriving the foal of the oxygen-rich blood still in the afterbirth.

Note: The term sleeper, barking, dummy and wandering foals describe the symptoms, not the causative agents. Other illnesses manifesting similar symptoms are often referred to in these terms.

e) *Jaundiced Foal (hemolytic disease, isohemolytic icterus).* Jaundice is a nonbacterial disease. Ironically, the very antibodies the foal receives from the mare through the colostrum are the cause of hemolytic disease. This condition is very similar to a human disease, erythroblastosis anemia, caused by the Rh negative factor associated with blood types.

The disease will result only if:

1. The sire's blood type is incompatible with the mare's blood type;
2. The unborn foal's blood type is the same as that of the sire;

3. Some of the foal's incompatible blood cells escape into the mare's system.

When these three conditions exist, the mare will develop antibodies that will destroy the foal's red blood cells. These antibodies will be concentrated in the colostrum and, when taken by the foal, the foal's system will absorb the antibodies, which immediately attack its red blood cells and break them up so they can no longer carry oxygen.

Since the blood of the mare and foal do not naturally intermix, some minor injury to or strain on the fetus or protective placenta that causes bleeding must occur before the mare's blood will pick up the incompatible cells and begin producing antibodies against them. If the mare has ever produced a jaundiced foal, a blood compatibility test should be made before the foal is allowed to nurse. If the condition is present the foal must be bottle fed while the colostrum is milked out. (Formulas are discussed in Section 11.2.) The mare should be safe for nursing after a day or two of hourly hand milking, which will dispose of the colostrum and its concentration of antibodies. If left with the mare, the foal must be muzzled until the mare is safe for nursing.

If the condition is present and the foal receives the colostrum, it will soon begin to weaken, become listless and sleepy. The ailment, not being an infection, does not cause a fever. When the very young foal shows signs of lost energy, the epithelium tissue of the eyes and mouth (eyelids and gums) should be examined at close intervals for a dull, whitish color, which always appears in the serious case of jaundice. In the recovery stage the tissue will develop a pale yellowish color. If the condition is present the foal must be muzzled immediately, fed by bottle and remain muzzled until the mare is milked clear. A veterinarian should be summoned immediately if the disease is suspected. Severe cases require blood transfusions. Blood from healthy, cold-blooded horses is usually the safest for this type of transfusion. The sire can often be used as a blood donor; obviously the mare's blood must not be used. A foal cured of the disease will have no adverse after-effects.

Undoubtedly many mares are bred to sires having incompatible blood types, but jaundice is rarely encountered. Most farms regard mares that have had a jaundiced foal as a risk and muzzle the subsequent foals immediately upon birth and milk out the mare in order to avoid the condition. Such foals must receive special antibiotic protection to make up for immunity lost by not receiving the colostrum.

Science has progressed greatly in typing the bloods and in recognizing the possibility of jaundice. However, the disease so seldom occurs that most farms do not resort to blood analysis unless unfavorable circumstances are indicated. If a mare has produced a jaundiced foal, her blood type should be checked with the stallion's before breeding to determine that their blood types are compatible.

f) *Hernias*. Most hernias found in foals result from the failure of a prenatal opening to close properly by the time the animal is born. The tendency toward hernias is considered highly heritable and runs in certain families. The umbilical hernia is the most common. If the opening remains large enough, abdominal tissue or a part of the intestine

Umbilical hernias, while not too serious, occur frequently. Some will disappear with time, others require surgery.

may protrude and press against the outlying skin, forming a very apparent soft lump. The protrusion may not be evident at birth; soon after, however, when the digestive tract is expanded by food intake, the gut movement and added weight may cause the tissue or part of the intestine to escape into the opening and form a lump protected only by the animal's skin. If the protruding intestine is not constricted, time and growth will often repair the hernia. The intestines do not grow so rapidly as the body and thus the gut becomes relatively shorter, causing it to tighten and draw back into the abdominal cavity. If the hernia is large or if the gut is constricted, it should be corrected before the constricted intestine blocks the movement of food through the digestive tract.

Colts are sometimes afflicted with a scrotal hernia, which is the protrusion of part of the gut through the inguinal canal leading from the abdominal cavity to the scrotum. The testicles drop down through this canal into the scrotum shortly before or after birth. The problem arises when the inner opening is large enough to allow a section of the gut to escape through the canal and into the scrotum. In severe cases the scrotum appears enlarged; however, a great many scrotal hernias go undetected until the foal is castrated. Usually the growth of the horse will correct the condition within several months, but it should be carefully watched for enlargement or constriction. Surgery is imperative if the abdominal wall begins to pinch off the protruding intestine. When surgery is performed the colt must always be castrated. Even when the condition appears to have naturally corrected itself, the veterinarian should be advised of the prior condition before he castrates the animal, so that he will be aware of the danger of damaging any gut that may still protrude.

g) *Constipation* resulting from the retention of meconium (waste accumulated by the unborn foal during gestation) is a hazard if the foal does not have a bowel movement within 12 hours after birth. The laxative effect of the colostrum will usually soften the meconium enough so that it is easily passed. However, a few foals will become constipated. Some farms administer an enema to all foals as a matter of routine but most observe the new foal for stooling; if no stool has been passed in 12 hours or so and the foal is unsuccessfully straining with the tail raised, a laxative, such as 12 ounces of milk of magnesia given orally by syringe, is in order and an enema may be helpful if the impaction is not too far inside. Great care must be exercised when an enema is administered to the new-born foal to avoid rupturing the fragile wall of the rectum. Only a soft-end 3/8-inch rubber or plastic tube should be used, and it should not be inserted more than 2 or 2½ inches. Two quarts of warm, mildly soapy water or plain warm water (body temperature), with an added tablespoon or two of mineral oil or glycerin, supplied by gravity from an enema container should be ample. Liquid should not be forced into the rectum. If hard balls of feces are impacted at the rectum opening, they should carefully be removed by hand. The first feces of a new-born foal is very dark or black. The feces of foals on a milk diet will be yellow.

If the colostrum, laxative and enema fail to produce results, the foal will begin to show signs of severe colic and is in dire need of veterinarian assistance. The veterinarian can administer an analgesic to relieve the pain and will determine further treatment needed.

h) *Diarrhea (scours)* occurs frequently in young foals. Any slight irritation of the intestines will cause an excessive flow of intestinal secretions. If the foal is otherwise well, the condition usually passes within two or three days. As long as the irritation exists, the area is more susceptible to infection; therefore the foal should be watched closely for a fever or other indications of some more serious disorder.

The colostrum, being laxative, may cause diarrhea. Scours in foals often accompanies the mare's nine-day heat and is probably caused by a chemical change in the milk. Mares that milk heavily should be milked out occasionally to prevent scours in new foals due to consumption of too much milk. Diarrhea from these causes should not be alarming, but if fever is present, diarrhea may indicate an infection and a veterinarian should be consulted.

Diarrhea may result in dehydration. Symptoms of dehydration are dry skin, sunken eyes and listlessness. A foal can quickly die from this condition and it is therefore imperative that the veterinarian be summoned at once. He will probably administer an electrolyte solution intravenously and give antibiotic injections.

i) *Entrophian.* Occasionally a foal is born with the eyelids rolled inward so that the eyelashes rub the eyeball. If a foal has this problem the affected eye or eyes will water excessively. The condition can be quickly and easily corrected by minor surgery done by the veterinarian. If the condition is not corrected the continued irritation can lead to inflammation of the cornea (keratitis), which can lead to blindness.

11.5 Foal Handling. There are as many different ideas of how a young horse should be handled as there are horsemen. Some prefer to work with the foal every day until it completely overcomes its fear of man. Others advocate no handling at all until the animal is old enough to be ridden. Most farms prefer a compromise between the two. They want the young horse gentle enough to be caught, tied, led and to allow its feet to be trimmed. Unless the foal is to be shown at halter, the training stops there until the horse is old enough to begin intensive training. There are two very good reasons for limiting the handling young animals receive. First, the cost of labor for everyday handling is prohibitive. Second, animals that are still suspicious of people are usually much easier to train than the "pet" horse, because they are worried about what the handler is doing, not about when the next meal will be served.

It takes far less time and trouble to accustom a foal to handling from the day of birth than it does to cope with an unmanageable older horse. The sooner the youngsters are gentled and can be led about, the more easily they can be cared for. The handler should begin to pat and scratch the very young foal as soon as the opportunity is present. Even the most skittish foal may allow some scratching while nursing. Many handlers will start gentling a very timid foal by reaching around the mare and scratching the foal's back or hindquarters. Light brushing and picking up the feet should be started as soon as possible. Halter breaking can begin when the foal is only a few days old. The halter should not be tight but should be snug enough so that a hind foot used in scratching cannot be caught in it. Under no circumstances should it remain on the foal unattended, except perhaps in the stall, and not then if there is anything that it could get caught on.

Foals must often be caught and held. When the foal will not allow itself to be caught and haltered, the mare should be led into a stall or small area such as a corral or alleyway; the foal will tag along. Following a short quieting period, the handler can crowd the foal against the mare or against a wall and catch the youngster under the neck with one hand or crook of the elbow and grasp the tail with the other hand, twisting it upward toward the back. In this position the foal can be held and moved about fairly easily.

Halter training may be accomplished easily if started while the foal is quite young, but the procedure must not be rushed and the foal must not be frightened. A foal that is merely dragged along and is not

The proper method of controlling a colt is by holding it around the neck and grasping the tail.

taught to lead correctly will lug on the lead rope the rest of its life.

The young foal should be fitted with a colt halter that is snug enough so that it cannot slip off; a long lead rope is attached to the halter ring and the foal is encouraged to follow the mare as she is slowly led away. A steady pull should be exerted on the rope until the foal moves forward, then all pressure should be released. If the foal fights the rope and begins to rear, an assistant handler should grasp the foal's tail and twist it upward to encourage the foal to move forward while the handler pulls steadily on the lead rope. The tail aid also discourages the foal from going over backward in an effort to avoid the pull. In the absence of a second man to assist from the rear, a large loop can be made in the end of a soft cotton rope and placed around the hindquarters so that the lowest portion of the loop rests against the back of the gaskins just above the hocks. The end of the rope can be run through the halter ring and held with the lead rope. By the pull on the rump rope, the foal is encouraged to move forward and discouraged from rearing at the same time.

Different methods are required if halter breaking has been postponed until after weaning. A few ranches tie the foal to a gentle burro and let the burro teach it to lead. The burro is fitted with a snug neck strap

The proper method of haltering a colt for the first time.

This foal is being led for the first time. The mare is kept close by while one handler leads it and the other holds its tail to discourage rearing and to encourage forward movement.

to which the foal is tied, allowing 18 inches between the strap and the halter. The two are then turned loose in a corral and the wise old burro soon educates the youngster. A very popular, widely used method is to tie a thick cotton rope in a bowline around the heart girth or neck, run the end through the halter ring and tie the foal short (about 18 inches) to a strong inner tube that is securely anchored. The inner tube will give to a certain extent, but after the foal stretches it as far as it can, the action becomes reversed and the tube will pull it forward. The lowest part of the inner tube should be even with the foal's withers so that there is little danger of dislocating the neck. Ideally the inner tube should be suspended from a stout tree limb so there is nothing nearby for the foal to ram into while fighting; however, serious injury rarely occurs when the foal is tied to a post or a wall, so long as there is no place it can catch its feet. Once the foal realizes that it is tied, it will probably put on a very lively show. Furious fighting is to be expected, and falls will probably occur. Unless severe injury results, the foal should be allowed to struggle until it has convinced itself that fighting is futile. At this point the foal can be sacked out (slapped lightly with a flapping object such as a gunny sack or rain coat) to encourage it

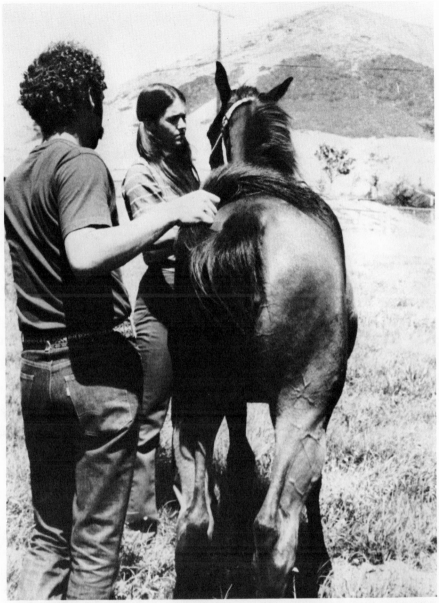

When holding or leading a foal that is not halter broke, one handler turns the tail up over the foal's back to discourage rearing and to encourage forward movement.

to test the rope to the limit. The sacking also helps in getting the animal used to being touched. If a fighting animal gets free either purposely or through equipment breakage, it will be encouraged to fight more vigorously the next time it is tied. Winning the struggle is how chronic pull-backs are made.

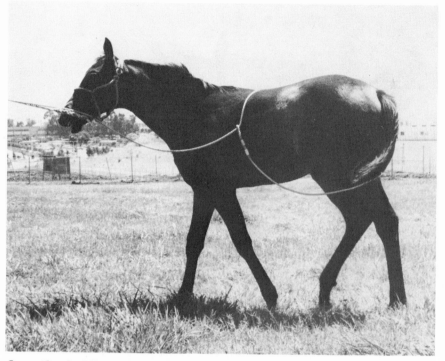

Once the foal has stopped fighting, the rump rope will encourage it to move forward.

Once the foal has decided it cannot win by fighting the rope, it can be taught to lead. The foal may be a little tender where the rope dug as it fought; if so, leading instructions will progress more rapidly. The idea is to let the foal learn to come forward to avoid an uncomfortable steady pull. The handler exerts a firm steady pull on the rope and gradually increases the pressure until the foal is quite uncomfortable. The foal and handler will be in a tug-of-war stance, each hoping the other will give. Eventually the foal will make a slight movement forward to relieve the pressure. The instant it does this, all pressure should be released from the rope. It should be spoken to kindly and if it is not afraid of people, it may be patted a little. Some foals are more stubborn than others and some will still try to rear to get away from the pressure. If a foal refuses to move forward the handler should pull to the side. When the foal turns, it is immediately rewarded by a slack rope. If a foal is exceptionally stubborn, it may be stung on the hindquarters with a whip or its tail may be twisted upward to encourage forward movement. Training to lead takes only a few minutes a day. Foals learn quickly when a firm, steady pull is used to encourage it to move forward and such movement is rewarded by a loose rope. Under no circumstances should a foal be pulled as it moves forward. If a foal

When being led for the first time, most foals resist. A rump rope will help prevent the foal from going over backward.

that is dragged along ever learns to lead at all, it will always lug on the rope.

11.6 Corrective Hoof Trimming. One of the most easily neglected chores on the farm is that of regular hoof care, which has been discussed in detail in Section 5.13. Proper trimming is especially important for horses that are still growing. Because their cartilage and bones are still very soft, young foals must be watched closely for faulty hoof growth. Unevenly worn hooves can produce faulty ways of going very quickly and can change bone structure, resulting in a permanent conformation fault. Poorly conformed feet and legs will only become worse if neglected.

A foal with a faulty way of going should receive corrective trimming as soon as it has enough hoof to allow reshaping. Corrective trimming is much more effective if it is done frequently, perhaps every two weeks. Only a small correction should be made each time so that no undue stress is placed upon bones, tendons and joints. Abrupt, severe alterations can result in permanent damage.

A foal that is so deformed that it cannot move about without effort is best destroyed or donated to a veterinary research laboratory or school where extensive corrective surgery may be tried. A youngster having a severe malformation that does not significantly interfere with performance could be sold without registration papers.

All foals will require trimming by the time they are four months of age, some much earlier. If they have not been handled frequently they may be rather difficult the first few times they are trimmed. Perhaps the best place to trim such foals is in a roomy box stall where they can be caught and held. The presence of the mare held or tied nearby will relieve the foal's anxiety to a great extent. The handler immobilizes the foal by holding it against the wall with his body; foals seem to calm down more quickly and are less afraid when held in this manner. A badly frightened foal will have a tendency to throw itself over backward. Such a youngster can be controlled by two handlers, one at the shoulder and one at the hindquarters, both applying body weight. The foal should be haltered and held short enough so that the handler will have some control should the foal break free. A foal that has never been tied should not be tied during trimming, because it will only become more frightened. However, if the foal is broken to tie, it can be secured to and held against the wall by one handler using body weight. The experienced farrier undoubtedly will suggest the restraining methods he prefers.

11.7 Weaning. Most ranches wean their foals at five or six months of age, with consideration given to their strength and health. Occasionally it may be advisable to wean the foal earlier; if the mare has been rebred and is doing poorly, or if the foal is receiving very little milk from the mare, weaning may be moved ahead, but in most cases it is beneficial to leave the foal with the mare until it is safely eating on its own. Foals taking all of their milk from a bottle or pail can be weaned from the liquid as early as two months, provided they are consuming hay and grain well; however, powdered low-fat milk or milk replacer should be mixed with the grain until the foal is about five months of age.

Weanlings must receive all the water they will drink, because their bodies are used to a liquid diet.

If the foals have been fed separately from the mare or fed in a creep and are consuming a considerable amount of hay and grain, weaning will be accomplished with little difficulty. Weaning at the proper time is more of a psychological problem than one of nourishment. It is usually more feasible to wean all of the foals at once or at least in groups of similar age. The time for weaning a group is usually determined by the age of the youngest foal in the group. If all broodmares and foals are pastured together, some farms prefer to remove one or two mares from the pasture each day, starting with the mare that has the oldest or strongest foal. The foal is left in the pasture in the company of its friends, and because of the security of familiar buddies and surroundings its loneliness will not be acute; it will call for its mother and fret but is unlikely to panic or become depressed. It will not attempt to nurse a mare that it knows is not its mother, so there is little chance of its being kicked.

An entire foal crop weaned at the same time is an easy matter when the foals are transferred to a large sturdy corral or to box stalls (at least two per stall) and the mares are moved out of sight and hearing. The foals will quiet down in a day or two and then can be released a few at a time into a large pasture. The first ones released will run frantically for an hour or so and by doing so will wear off the edge of their fright. Their disinterest in further exertion will have a calming effect on the next few turned out with them. If the ranch has only a few foals it is sometimes beneficial to place a gentle old dry mare with them to console and calm them.

Another method of weaning and a kinder one is by feeding the mares on one side of a smooth wire mesh fence with the foals on the other, to reduce the stress by close proximity for a few days. Ranches with the time and help available often allow the foals to nurse three times the first day, two times the second and once the third, before stopping nursing altogether. This method lessens the chance of the mares' udders becoming caked and places less stress on the foals.

If either the mare or foal becomes extremely upset, tranquilizers are sometimes administered by the veterinarian.

The foal's ration should be increased at weaning to compensate for the absence of milk, and a small amount of bran should be added to avoid constipation. A cup of powdered low-fat milk or milk replacer may also be added. Newly weaned foals are under stress and should be watched closely for colds and digestive disturbances. Water should be available free choice. The mare's ration should be decreased starting a few days before weaning so that the udder will reduce with less discomfort (see Section 9.6 for drying up udder).

The practice on most ranches is to separate the colts and fillies at weaning time. However, some farms that raise only a few foals may put this separation off until they are yearlings. Onset of puberty for both colts and fillies occurs between 10 and 24 months of age (the average being 18 months). Separation should not be delayed long or some very young fillies may become pregnant.

11.8 Castration. Unless a colt is a very exceptional individual, he should be castrated. A mediocre stallion will often make an outstanding gelding. Such a gelding will do far more to enhance the farm's reputation than will the below-average stallion.

Colts that are to be raced should remain stallions. Entire horses usually have more drive and boldness than the average gelding and are not castrated unless they become vicious or do not run well.

Most managers prefer to castrate colts as yearlings, before they become "studdy" and take on stallion appearance and disposition. Castration may be done as soon as both testicles have descended into the scrotum; however, most colts will develop more adequately if the operation is postponed until they are yearlings. Colts that are slow to mature may not be castrated for an additional six months to allow time for more muscle development. Castration becomes more difficult as the stallion becomes older, because the testicle development causes enlargement of the spermatic blood vessel. For this reason, castration should not be postponed too long.

Occasionally one or both testicles will remain within the body cavity. An animal with this condition is referred to as a cryptorchid. If no testicle is present in the scrotum, the animal will be sterile; however, he will have a stallion's disposition and is sometimes even more difficult to handle. Care should be taken to assure that no cryptorchid is accidentally sold as a gelding. Such incidents are not uncommon and have resulted in lawsuits and severely damaged reputations. Castrating a cryptorchid requires abdominal surgery, consequently special arrangements must be made for this animal.

Castration should be done only by a veterinarian, since there can be complications that require skilled handling and a sloppy job can cause excessive bleeding or produce a "proud-cut" gelding. Tranquilizers are unpredictable; a dose that merely quiets one animal can kill another. Occasionally a tranquilizer will have the opposite effect, causing the horse to go berserk. Professional veterinarians are always prepared to cope with such occurrences. Care must be taken to remove the entire testicle. If a portion remains intact, the animal will continue to act like a stallion even though sterile and is said to be "proud-cut." The artery supplying blood to the testicles must be crushed and severed properly or excessive bleeding can result. Any prior indications of a scrotal hernia, discussed in Section 11.4 (f), should be brought to the veterinarian's attention before the operation.

The appropriate time of year for castration is in the spring, when green grass and dampness keep the dust down and flies are not a problem. A clean, grassy paddock is usually the best place, since the area is not dusty, the animal is less likely to hurt himself when he falls or is trying to get up while tranquilized, and he can be left where he is to recuperate. Such pastures or paddocks are by far the preferred recovery area, since they are generally cleaner than stalls and encourage exercise that is needed to reduce swelling and keep the wound draining. If the animal is confined to a stall, he must be made to move about some to help keep the swelling to a minimum.

The sheath will be swollen for several days and serum will drip from the wound. Drainage is very desirable; the cut should never be sutured. Should the incision heal over too soon it must be reopened so that it will drain. Excessive swelling, which produces edema along the underline, or very hard swelling, is cause for concern. A foul-smelling discharge or pus indicates infection. When any of these conditions appears, especially if accompanied by fever, the veterinarian should be consulted immediately.

Castrated colts should be kept separated from fillies and mares until they have recovered (usually four to six weeks), to avoid the rare possibility that enough viable sperm remained in the reproductive tract to settle a filly or mare.

12
THE YOUNG STOCK

12.1 Culling. After the foals have been weaned, the breeder must decide what their future will be. The first thing an effective manager must do is carefully evaluate the merits and failings of each individual animal. It is far better for him to be overly critical than to imagine potential in an animal that has little.

Even the best bloodlines and breeding programs will produce an occasional "cull," an animal far below the quality of the ranch standards. What is done with these animals will have a direct effect upon the reputation and success of the breeding farm. The future of such a horse should be planned so that it will not have a detrimental effect upon the farm's reputation. Under no circumstances should a poor-quality

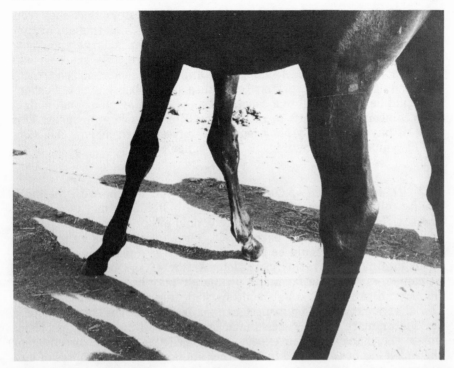

Every ranch will produce an occasional "cull." This very well-bred Thoroughbred colt was born with deformed front legs due to a congenital disease.

animal be put through the farm "production" sale where the breeder's produce receives notoriety. Physical fitness, glossy coat and fine bloom will not replace the lacking qualities. Every effort should be made to keep the poor-quality animals out of the public eye, not as a deceitful measure, but to avoid their becoming recognized as representative of the breeder's horses. Private selling arrangements can always be made for such animals if the price is low enough. Of course they must be honestly represented for what they are. The poor attributes must be pointed out and openly discussed with the buyer.

As protection to the breeder's business, which is in great part based upon the farm stud's reputation, it is best to castrate all colts not of top quality. If the colt happens to be a very poor animal, it is good business to not register it and to sell it as nonregistered, thus destroying its official identification with the farm. Though a loss is taken on that animal, the breeding farm's reputation is protected.

The dilemma of what to do with the very poor or fair-quality filly is a little more complex. Occasionally the quality of her breeding may reappear in her offspring. Before selling at a substantial loss, the breeder may elect to breed such a filly to a carefully chosen stallion just to see if she can pass on the quality of her ancestors that she, from appearances, seems to have been denied. A filly such as this might turn out to be a good broodmare, but unless the farm is large enough to absorb the loss in the event the mare is a failure, this business risk should not be taken. When selling such a filly, she should not be treated as broodmare material, primarily because she probably is not. Perhaps she could be very useful to a casual breeder operating on a small budget or to someone who wants to raise a horse for his own enjoyment. This class of buyer is usually not prepared to invest much money, therefore the filly must be sold within his price range.

12.2 Use and Disposal of Quality Horses. What is done with the high-quality colts and fillies will depend on each individual's merits, pedigree and sex. Those bred and built to race should either be sold as racehorse prospects or raced by the farm itself. Those qualified as halter, show or working horses should be put to those uses as a means of developing or maintaining the farm reputation. A few of the best fillies should be campaigned by the farm and ultimately may be put into the brood-mare band. All of the colts should be sold unless an animal is so superior that a better stud prospect cannot be found. The retention of a stallion or mare of the ranch's own breeding will pose many problems unless the manager is an expert in the genetics of close breeding or the ranch has or can afford to buy other stallions and broodmares of out-

Young stock must have an area large enough for romping and playing in order to develop good bone and strong muscle.

side bloodlines. Only exceptional horses should be held for future breeding stock.

Whatever is to be done with the young horses, competition with animals produced by other farms requires that they receive the best nutrition economically possible. Feeds high in protein plus a little calf manna every day will do much to facilitate growth. Vitamins and minerals, especially calcium and phosphorus, are needed by young animals to attain maximum growth. Green pastures not only provide most of the extra requirements of young horses but also give them room to romp and play so that they can develop strong bone and good muscle. Hilly pastures are beneficial for growing horses; such pastures force them to exercise vigorously and helps develop sure-footedness.

12.3 Selection of the Halter Horse.

a) *Points to Consider.* Competition in all horse-show classes, especially halter classes, is becoming keener every year. Today it takes an exceptional individual to stand out and place well. Good conformation and soundness alone will not guarantee winning. The animal must have good muscle tone and not be too fat or too thin. It must have a glossy, sleek coat, be well groomed and trimmed of long unsightly hairs. Its mane and tail must be thinned, trained to lie smoothly and conform to popular styles. Its feet must be properly trimmed and in healthy condition. The animal must be trained to stand alertly with its feet in the proper position and to move out quickly and quietly when led. It must always mind its manners even when others near it do not. Preparing a horse to show not only requires time and hard work but also takes a certain amount of special skill. Amateur handlers must be especially diligent to compete with professional showmen.

Farms that show halter horses will select three or four new prospects from every colt crop. Now and then the farm may elect to show an exceptionally outstanding weanling; however, because of the special handling and transporting care such young animals require, they are usually not shown until they are yearlings. By having several prospective halter horses, the farm is always prepared to present a good individual in the event one is injured or does not grow to meet the farm's expectations. Every judge has his own personal preference, which usually runs toward one type of horse. In addition, desirable types differ from area to area. Knowledge of the prevailing preference enables the farm with a large show string to select the animal for showing which has the best chance of winning under the circumstances.

The earlier the show string is selected, the more time there will be to condition the animals, correct minor leg faults and accustom them to handling.

The prospective halter horse must be selected with breed characteristics in mind; certain desirable characteristics vary somewhat from breed to breed. The following is a description of what to look for in a Quarter Horse halter prospect; however, most of the basic body structures desirable in the Quarter Horse are also desirable in almost all light horse breeds, with only slight breed variations in degree and emphasis on a point of conformation.

Selection should begin by study of the animals from a distance. The halter prospect should have good balance, being as heavily muscled in the forequarters as in the hindquarters. Carriage should be graceful and lively, demonstrating a great deal of action and freedom as the animal moves. The horse should be alert and carry its head proudly. Once the animal's overall merits have been observed, it should be inspected at a close range.

b) *Viewed from the Side.* Refinement of head is desirable; the horse's eyes should be large and set low, with plenty of width between the eyes and ears; nostrils should be large and set high; the mouth should be moderate to shallow in length. A shallow mouth is usually a lighter mouth, since it is better conformed to take the bit. The head carriage is considered more important than the shape of the head itself. The neck should gracefully curve into the head and provide a roomy free attachment at the throatlatch; it should be very long and graceful, not overly thick and heavy, and should flow smoothly into a long, sloping, well-muscled shoulder. Legs should emerge from the shoulders well forward of the cinch so that sores will not develop. The withers should be prominent; however, young horses will develop withers later in life and should not be discounted for slow wither development when young. If a horse is three or four years old and still appears very "mutton" withered, it will probably remain so. The animal should have a deep heart girth, indicating adequate lung capacity. The back should be relatively short and give the appearance of great strength. More length is tolerated today than was the case a few years ago, provided the back and loin are well muscled; length reduces the danger of overreaching with the hind feet and injuring the heels of the forefeet. Loins should be well muscled and the croup long and powerful. The tail-set should not be too low. The Quarter Horse croup should slope gently; however, a very steep croup, "rainy day croup" or "goose rump," is to be discriminated against. There should be great length from the point of hip to buttocks and also from the point of hip to the hocks; both indicate good driving ability and length of stride. The underline should be long and should slope gently up to the flanks but should not give the appearance that the animal is pinched up at the flanks.

The legs should come straight down from the body and be set squarely under the animal. All joints should be clean, lack puffiness and present an appearance of strength. The bone should be flat, which indicates sufficient separation between the cannon bone and flexor tendons. A large circumference of the cannon midway between the knee and fetlock indicates good "bone." The pasterns should have adequate length and slope to absorb shock but need not be so long and sloping as those of some other breeds. The foot should be well rounded, open at the heel and of sufficient size to support the weight of the animal without appearing clumsy. The forearm of the front legs and the gaskin of the hind legs should be long and well muscled, while the cannon should be short and clean. Muscling should carry well down toward the knee or hock but tie in above the joint. If the muscle ties in too close to the joint, problems with stiffness and thoroughpins are much more likely.

c) *Viewed from the Front.* Great width between the animal's eyes and ears indicate intelligence. The chest should be wide but not muscle-bound and cumbersome. The space between the forelegs should form an "A" shape going into the chest, and the forearm should be heavily muscled inside and out. A horse well conformed in this area is said to "V-up well." Horses that don't V-up well are usually not so agile when turning, especially if they have a wide chest. A line drawn straight down from the point of shoulder should bisect the center of the knee, fetlock joint, and hoof. When moved forward, the animal should track straight and show no indication of paddling or dishing in.

d) *Viewed from the Rear.* The widest point of the hindquarters should be at the stifle, which gives the hindquarters a pear-shaped appearance; the hips should never be the widest point.

The gaskin should exhibit a great deal of muscling with as much muscle present inside as outside. A line drawn from the point of the buttocks should bisect the hocks, cannon bones, fetlocks and heels. The hocks should not turn in or out and the animal should track straight as it moves.

12.4 Conditioning the Halter Horse. After the suitable animals for halter classes have been selected, they must be thoroughly conditioned for showing. The halter prospects should be housed in box stalls. Stabling reduces exposure to sunlight, which dries and bleaches the hair coat, and accustoms the animals to confined living so that the unfamiliar fair ground box stall will be no more upsetting than a new area. The colts should be watered with a bucket and become accustomed to drinking water mildly flavored with wintergreen so that drinking strange water at a show will pose no consumption problem. Two or three drops of wintergreen per bucket of water is palatable to the horse and disguises the taste of water, which varies from area to area.

Top-quality feed is essential. It should be high in protein, vitamins and minerals, with special attention given to the calcium-phosphorus ratio and quantities. Calf manna is an excellent protein supplement. High protein mineral blocks are also available. A great variety of conditioners are marketed that can be fed to give extra luster to the coat. Excess fat is undesirable in today's show horse, and care must be taken to avoid developing an overweight animal. Proper exercise will usually remedy this problem.

A halter horse must receive a great deal of exercise to look and feel its best. The most satisfactory method of exercise is riding; however, if the animal is too young for riding, the best way is to "pony" the animal from another horse for an hour or so a day. If this method is not practical, the halter horse can be placed on the hot-walker for an hour or more a day, longed or turned loose and driven around in a small corral.

The show horse should be blanketed continuously; a stable sheet is

Mechanical hot walkers are used by many farms as an aid in exercising horses.

sufficient in hot weather and should be worn when the horse is being led outside if the sun is strong. The cover will prevent the bleaching of its hair coat, which causes loss of luster.

The feet should be trimmed often, especially if correction is needed. Hoof dressing should be applied daily to aid in developing healthy, good-looking feet.

Grooming should be done at least once a day and should be very thorough. Vigorous brushing stimulates oil production and circulation, and thus aids in developing a healthy, handsome coat. Vacuum cleaners especially designed for horses can be a great aid in cleaning and stimulating the skin. A light application of mineral oil once or twice a week will greatly help hair and skin that tend to be dry. Though there are several excellent animal soaps on the market, most showmen avoid shampooing show horses, since soap removes the natural hair oils and causes the hair to stand up. If a horse must be shampooed, it should be done at least one day before the show so that the coat will have time to return to normal. Rinsing with clean water to remove dirt and sweat is always beneficial.

The mane, tail and forelock should be thinned by pulling the hairs from the underside, and shaped by pulling out the long hairs. Many showmen clip bridle-paths in the mane, especially if the horse appears a little thick where the neck attaches to the head. Long hair around the fetlocks, throatlatch, muzzle, under jaws, eyes · and ears should be closely trimmed to give the legs and head a refined, sharp look.

Animals, especially stallions, that tend to develop thick necks should be fitted with a sweat hood to reduce the neck area and should be fed at ground level to stretch the neck muscles as much as possible.

All halter horses must learn to stand properly for long periods of time and must lead easily. Patience is perhaps the most difficult trait to instill in a healthy, vigorous youngster. Training must begin early. The young animal should be removed from the stall and hobbled every time it is groomed, and should be made to stand longer than is needed for cleaning. Three or four minutes is long enough at the start, but this time should be increased gradually until it will stand quietly for fifteen or twenty minutes. The animal can be trained to position its front feet by a signal from the nudge of a boot toe on the pastern, but care should be taken not to overuse this aid or it may not position itself without the signal. If an animal tends to stand with feet too close together in front, this can often be corrected by the handler's stepping to the side of the offending foot and pulling the animal over one step. If the horse stands wide behind, it can be backed into position; if it stands narrow, it should be led forward for correction. Standing properly is very trying for a young horse and should be practiced often for very short periods. Once the animal stands properly, it should be praised for its achievement and the day's session ended immediately.

The halter horse must be taught to lead freely with its head at the handler's shoulder. It should trot collectedly, stop quickly, back readily and move sideways on command. The animal must also be taught to turn either toward or away from the handler when being led so that it can always be presented between the exhibitor and the judge.

The horse should be taught to load and haul easily; it will be on the road a great deal. Its disposition should be such that it can take traveling in stride. Through proper handling, even a nervous horse can become accustomed to travel and will show well after a trip. The trips should always be planned for early enough arrival so that the horse will be entirely rested for its scheduled class.

The day of the show, the animal's tail should be dampened and wrapped so that the hair will lie smoothly. A coat of petroleum jelly or hoof dressing should be applied to the outer hoof wall and wiped off just before entering the arena. A half hour or an hour prior to its class the animal should be brushed with a stiff brush, followed by a soft brush. After brushing, the coat should be wiped with a soft, clean cloth or sheepskin to remove excess dust. The eyes, lips, nostrils, ears and dock should all be cleaned with a damp rag or sponge. Spray starch, hair spray or commercial preparations can be applied to control unruly mane and tail hairs. The tail wrap must be removed as the animal enters the ring; this can easily be done by pulling it downward off the tail.

12.5 Showing. The exhibitor should be dressed tastefully but very conservatively. Attention must be on the horse, not on the exhibitor. The halter and lead shank should be immaculate, as should the horse and handler. A novice should try to position himself toward the center of the competing horses if possible. He should lead the horse from the left if he is right-handed; however, an animal may be led from the right if the handler is left-handed. The exhibitor should be even with the animal's head and should step out at a good brisk walk. He must be certain to leave a horse's length between himself and the horse ahead of him and must regulate his walk accordingly. There should be approximately a foot of lead shank between the exhibitor's hand and the horse's halter, with a good deal of slack in the lead strap.

As the animals line up, the exhibitor must see that his horse is neither ahead nor behind the majority of other animals in the class and that there is adequate space between animals for the judge to walk as he observes the horse.

The animal should be properly set up immediately. Once it is in proper position, the handler must never step between his horse and the judge. He should watch the animal closely and try to keep it alert. Often the running of a fingernail over the teeth of a comb will awaken the animal; however, this must not be overdone or the animal will become accustomed to it and will not respond at the signal.

The judge usually first approaches the left side of the animal and observes the horse as a whole. At this point the exhibitor should be standing slightly in front of the horse and to its right. The judge will then move to the rear. As the judge moves to the right side, the handler must step to the left. When the judge views the animal from the front, the handler should be at either side. Once the horse has been viewed standing, the judge will usually request that the animal be walked directly away and trotted straight back to him. To do this the exhibitor should move to the left side of the animal if he is right-handed and walk the animal briskly forward, staying even with the horse's head and maintaining a straight line. When the designated point is reached, the animal should be stopped squarely for a moment, then turned so that the animal is between the exhibitor and the judge. Upon completion of the turn, the animal should be broken into a collected trot on a loose lead and trotted straight to and usually past the judge, where he is stopped and stood squarely again. The judge will dismiss the exhibitor, and the horse should be returned by the most inconspicuous route to its original position in line.

The exhibitor is not finished at this point. He must reposition the animal and maintain alertness in anticipation that the judge may reinspect at any time. Prior to receiving the awards, so that the judge may review his selection and the audience can see the animals better, the winning animals are usually pulled forward and aligned in the order of their selection. Whether the horse placed or not, the good showman will continue exhibiting his horse the best that he can until he has completed his exit from the arena.

If the animal receives a first- or second-place award, it will be shown again later in the championship class for its sex. If it wins champion

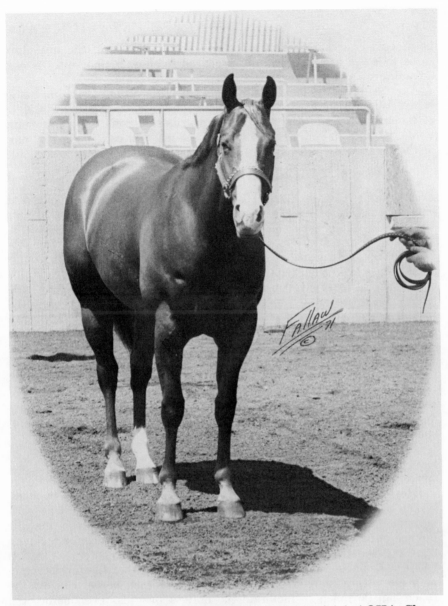

Doc's Sunshine by Doc Bar out of Miss Starnes, AAA-AQHA Champion. Halter horses such as this make the years of planning and months of work conditioning, training and promoting all worthwhile. Courtesy of Double J Ranch, Pacinies, California.

or reserve champion of its sex, it may be shown against the winners of the other sexes for the Grand Champion of breed award.

Procedures vary from judge to judge and from show to show; the exhibitor must be quick to observe the techniques of individual judges. Halter classes prior to the one the animal is to be shown in should be observed so that the exhibitor will be familiar with the procedure to be followed and will be fully prepared to show his horse to its best advantage.

A great deal of work and time goes into producing a winning halter horse. Winning is a source of great satisfaction to the ranch and all of its employees, and the monetary rewards that come from producing winners make the effort well worth while as well as a necessary part of the business.

12.6 Training Young Horses. Intensive training should be delayed as long as possible to allow the young horse to attain sufficient growth and bone development to enable it to withstand the strains put upon it.

Because of the desire to prove horses as quickly as possible for financial reasons, most ranches are forced to start their colts quite young. Race horses are usually started as long yearlings to be sure of gaining a stall at the track. Cutting horse and snaffle bit futurities for three-year-olds are becoming very popular; horses to be entered in those activities are often started as short two-year-olds to be ready for such contests at age three. Animals started young must be watched closely for signs of muscle strain and bone or tendon problems. Most race tracks now require that the knees of young horses be X-rayed before they are raced, as proof that the epiphyseal cartilage has calcified. If it has not, the young animals' bones probably will not withstand the strain of racing. Such an X-ray is recommended for any animal that must undergo a great deal of physical exertion at an early age in order to meet specified show dates.

If possible, heavy training should be avoided until the animal is at least three years old; this is especially so if the horse has light bone and small joints.

Every farm will have some horses to be sent to specialized trainers. Which trainer is selected will depend upon what is to be done with each individual horse, the qualifications of each individual trainer, the difference in fees between trainers, and the farm's preference for trainers. Once the selection of a good trainer, as covered in chapter 1 (Section 1.9 [e]) has been made, it is of great importance that the farm comply with his wishes in regard to the handling of each horse. Usually the ranch can do much to facilitate the horse's progress even before the horse is sent to the trainer. Many trainers, especially working horse trainers, prefer that the young horse not be handled at all so that it can be started in keeping with the trainer's normal program without the confusion created by different handling methods. Other trainers, especially those dealing with race horses, prefer that the animal be gentle and easy to handle when he arrives so that valuable time does not have to be spent gentling the colt. Under most conditions, it is advisable that the colt be broken to lead, tie, haul in a trailer, permit its feet to be picked up, and tolerate brushing; however, the colt must not be a pet. Dog-gentle colts are usually more concerned with people than with schooling and are much more difficult to train; they usually never amount to much as judged by industry standards. Most trainers prefer a wild, snorty colt that isn't halter broke over a "pet" horse, because the snorty horse will try harder and is less likely to injure the trainer. The pet horse is the one most likely to hurt the trainer for the simple reason that it is unafraid and usually does not respect people.

The colt should be conditioned somewhat before being sent to training school. When an animal is in good physical condition, it can be trained much faster. This is especially true when colts are being put into race training. Conditioning a young horse is very tricky, and if done improperly can do irreparable damage to its young legs. Many Thoroughbred farms provide swimming pools as aids for conditioning their young animals. A horse can be built up quite well by swimming, without heavy strain upon the yet-developing legs. The cost of such a pool is usually an expense that most farms prefer to avoid. Some farms provide light-weight boys to ride the colts or have the colts led for a few minutes a day, but this too usually proves to be fairly expensive. A more economical and quicker method is to turn a few colts loose in a small area of loose dirt or sand so that their feet sink in a few inches. In such an arrangement, a man in the center with a whip can keep them moving steadily. They should be jogged 15 minutes a day for the first week and then gradually built up to 20 or 25 minutes a day of trotting and loping. This method has the added advantage of a minimum of handling prior to the time the trainer takes over. No matter what method is used, the conditioning should start at least two or three months ahead of delivery to the trainer, thus assuring fairly good condition at the start of training.

Plans for the training of each individual animal should be determined by the time it is weaned, as assurance that the colt will receive the proper training at the correct time and will be ready and eligible for the events in which it will represent the farm.

The farm may plan to keep a few of the best fillies as replacement mares. To maintain the reputation of the farm's broodmare band, these mares must prove themselves worthy of addition to the herd. Particular care should be taken to assure that these mares will remain sound for a long performance career; they should not be pushed at an early age as other animals might be, in hopes of an early sale. A mare should be campaigned long enough to earn an admirable performance record so that her foals will be in demand. The ranch's top mares should not be sold unless it is felt that equally good broodmare prospects can be obtained reasonably, that the existing broodmare band is adequate and no replacements are needed, or that the addition of such mares will result in too much closebreeding within the herd.

12.7 Promoting Quality Horses. The farm should consider consigning some of the better produce to some popular sales where they can be seen by a great many people. It is especially important that these animals be the best the ranch can produce so that they will bring favorable attention and surpass most of the other animals at the sale. Though

Barrel racing is a fast growing event, prize money is increasing and contestants require top-quality animals. (Photography by Mary Ann Czermak, Cottonwood, California.

they may not bring so much money in the sale ring as they might by sale at private treaty, the publicity and attention these horses will draw to the ranch will attract customers to seek out the farm in hopes of finding comparable horses. In this way any immediate loss will be converted into unlimited potential profit.

There is great demand for Hunt Seat medal class horses for junior riders. Kathy Tohill on Irish Tenor.

Driving classes are gaining in popularity. The horse shown here is Serj, Arabian driven by Murrel Lacey.

Doc's Tassajara, ridden by Charles Ward, demonstrating superb ability to cut cattle. A ranch that consistently produces winners such as this Quarter Horse will never lack customers. Photograph courtesy of Double J Ranch, Pacinies, California.

Thoroughbred racing has always been the most popular of all horse racing. This thoroughbred race held at Santa Anita, California, is being won by Bold Corporal (Stakes winner), owned by Rancho Paraiso, Walnut Creek, California.

" C H A U N D E L L "

BAY MEADOWS, CALIF   6/5/65   POWELL CROSEY UP
PISTOL BAR (2nd)   400 YDS 20;7 MISTER MODECK (3rd)
REYNOLD C JOHNSON   OWNER   LYNDON R SMITH TRAINER

Nothing is more rewarding than joining the horse you own, bred, raised and trained in the winner's circle. Chaundell is shown after winning the Quarter Horse race at Bay Meadow, California. Sold by Rancho Paraiso for $37,000.

Every attempt should be made to place ranch horses with buyers who will show or race them. It is good business to reduce the price a few hundred dollars in order to sell to top trainers or showmen. The winnings these professionals will receive with the ranch's horses will be noted by the public and the demand will be increased. On the other hand a poor rider or trainer can greatly detract from a good horse. The poor results, but not the cause, will be noted by the public and the farm's reputation will suffer. Because of this, good horses should not be sold to men with bad handling reputations unless the price received will offset the loss of a certain amount of prestige. Such values are difficult to measure. It is usually best to avoid sales to inferior horsemen.

When the ranch's young horses are properly nourished and when their future is properly planned, the ranch will never lack for customers seeking to purchase horses or stud services.

13
TRANSPORTATION

(SYNOPSIS)

This chapter covers the various requirements for transporting horses; training in loading, unloading and hauling horses; health of transported animals; road vehicles; trailer qualities and modes of transportation.

The following synopsis by section numbers is provided as an aid for quick reference to specific subjects.

13.1 General Travel Requirements. The breeding farm must transport horses often. Whether buying, selling, showing, racing or sending ranch mares to outside stallions, travel arrangements must be made. Almost all ranches maintain their own road vehicles for transporting horses, such as trucks, trailers or specially made horse vans. Many frequently ship horses by air, rail or boat. No matter what type of travel arrangement has been made, whether transportation involves local shows or overseas exportation, advanced planning, correct handling and proper equipment are essential for maintaining animal health and well-being.

Efficient trip planning begins with obtaining the credentials required to move animals through the various areas to be traveled. Each state and country has its own laws designed to prevent the introduction and spread of disease and to apprehend livestock thieves. The credentials that are often required for entering into a different state or country are listed here; however, since each state and country has different requirements, a check should be made with each individual department of agriculture to insure that all papers are in order and no long delays will be encountered.

Most areas require written proof of ownership and brand certificates, a recent health certificate issued by a qualified veterinarian after examination of the animal and, since the outbreak of Venezuelan Equine Encephalomyelitis (VEE) in the United States, a VEE vaccination certificate. Some states and foreign countries require a certificate showing negative results for an E.I.A. (Equine Infectious Anemia or Swamp Fever) test taken within the preceding six months. Canada began enforcing such a regulation in August of 1971. Any horse entering Canada must possess such an E.I.A. certificate or be quarantined and tested upon entering the country. If an animal is found to react positively to the test it must be destroyed or returned to the country from which it came.

Many countries have rigid incoming quarantine laws. The United States quarantine law for incoming or returning horses is not so strict as in some other countries but nevertheless requires a blood sample to be taken at the time of entry, which is immediately sent to Washington, D.C., for analysis (which usually requires only a few days), during which time the horse is held in quarantine. If the test proves satisfactory, the horse is released.

Exportation of horses should not be undertaken without exact knowledge of the country's incoming requirements. Australian law demonstrates how difficult entry can be. Horses going to Australia must first be transported to England, where they are quarantined for six months. If found to be acceptable they are then shipped by boat to Australia where they are again quarantined for an additional two months before released.

Travel from point to point within the United States usually requires no quarantine, but the agricultural department of each state involved in the travel should be consulted for full up-to-date requirements.

The above notes clearly indicate the need for accurate information in advance trip planning.

Planned departure time must correlate with planned stops enroute and planned arrival at the destination. When road travel is involved, fairgrounds make excellent rest or overnight stopping places. The "Horseman's Travel Guide" put out by R. A. Phillips, 755 Ironton St., Aurora, Colorado, lists a great many overnight horse accommodations available throughout the United States, and is an excellent aid in trip planning.

If the horse is to perform at the destination, the trip should be planned so that arrival will be well in advance of the performance time, thus enabling the animal to recover its best form. At least one day of rest should be allowed if the travel period is long; otherwise several hours should be allowed. If a tranquilizer has been used to calm the horse for hauling, the time for it to wear off must be considered.

Due to the hazards of transportation, travel insurance should always be considered in trip planning. Freight carriers limit their liability for damage or loss to a very small guarantee, such as $100. Live animals are treated as other items of freight in this regard. Obviously such a small guarantee is grossly inadequate; additional insurance is generally advisable. Insurance of valuable animals in transit is available through some, but not many, underwriters. The coverage may be in a blanket trip form to cover all of the owner's horses involved in all transportation or it may be in specific trip form to cover particular horses during a specific trip. Either coverage can be provided by a policy that names the dangers insured against or a mortality policy that includes all losses not expressly excluded. Both types are expensive, the latter being more expensive than the more limited coverage of only named dangers. When transporting by air, rail, boat or commercial truck, the matter of trip insurance, its coverage and cost, should be carefully considered in consultation with an insurance underwriting agent.

Proper trip planning is of the utmost importance. Horses are not common inert baggage. They are living animals that must be well cared for with consideration for daily routine and physical needs.

Confined travel is a hardship for animals. It imposes excessive strain and discomfort and disrupts feeding, watering, resting and elimination of waste matters. Travel for long distances or for periods of several hours is difficult for horses in top physical health; an ailing horse should never be hauled, except to a veterinary hospital, since his condition is almost certain to worsen and a disease is sure to spread to the other horses in the group, whose resistance will be greatly lowered by the stress of travel. Tired horses are very susceptible to illnesses and should not be hauled. Even healthy horses are subject to travel stress and are susceptible to shipping fever and other respiratory ailments, and to colic. Influenza inoculations give some protection against shipping fever, and since the development of long-acting antibacterial drugs, most veterinarians also advise antibiotic injections a day or two before travel and a laxative feed just before and during a long trip. Mild colic responds quite favorably to some commercially prepared colic medicines. Oral and injectable medicines for colic, used only on the recommendation of a veterinarian, should be carried on the trip.

Very nervous horses are sometimes tranquilized for transporting in order to reduce the chances of injury and to minimize stress, which predisposes to disease. There is some risk in the use of tranquilizers; they should never be used except by a veterinarian, or upon his advice. Air flights usually require the use of tranquilizers as a precaution against disaster. California has implemented laws against drugging horses at shows and sales. Other states are sure to follow suit. It should be kept in mind that tranquilizers will show up in a drug test up to 24 hours after administration. Therefore, if tranquilizers are to be given to insure better travel, the animal must arrive in adequate time for the medication to leave the system completely.

A competent attendant should always accompany horses being transported, no matter what type of conveyance is used. Most commercial carriers will not accept a shipment of horses unless an attendant is provided. The attendant must be a person knowledgeable with horses and able to anticipate trouble and cope with emergencies in a calm manner and with a reassuring voice. An excitable attendant will frighten the animals more in the event of an emergency.

In addition to regular riding equipment the attendant should carry credentials, grooming tools, feed, water, hand tools, equipment and supplies that he thinks may be needed for the care and comfort of the horses and to assure a successful trip. By way of suggestion, the following should not be overlooked:

Antibiotics	Hand tools (hammer, pliers, etc.)
Broom	Highway flashers and flares
Buggy whip	Pitch fork
Colic medication	Pocket knife
Cotton rope (30' x ¾")	Rake
Crowbar	Saddle blankets
First-aid kit	Twitch
Flashlights	Wire cutters

13.2 Trailer Training.

a) *Loading.* Young horses should be taught to load by the time they are a year old, but not before learning to lead. The first few experiences will bear greatly upon the animals' reactions to loading for the rest of their lives. Much care should be taken to avoid frightening the young horses during the training. Loading and hauling should be practiced long before traveling is required, so that the animal will be thoroughly accustomed to the vehicle. The horse will then be easier to load and less subject to stress during the trip; trainers have often regretted having rushed trailer lessons. Loading and hauling should be taught in two distinct stages; the first phase, loading and unloading; the second phase, hauling. This is especially true if the animal has a particularly nervous disposition. Hauling adds to the fear of the trailer and therefore should be delayed until the animal loads easily. The best place to teach loading is in an area familiar to the horse, such as a corral where it can be loose. The youngster will feel more secure if its buddies are in close proximity

—just over the fence. The trailer should be put in the corral with the tongue firmly anchored to avoid shifting, the wheels blocked in both directions to avoid movement and corners stabilized with supports from the ground to eliminate jiggling. A ramp trailer is usually preferred for the first few lessons. The ramp should be in loading position and should be baited with grain. Grain should also be placed on the trailer floor and in the manger. During the training period just the one horse to receive the lesson should be in the corral. If left alone it will probably inspect the contraption very carefully, find the grain and, in the search for more food, enter the trailer without urging. The youngster that does not take the bait can be haltered and coaxed but should never be rushed, especially if it is timid. A weanling can be gently shoved into the trailer by two people. Once it is in, the trainer will usually be able to keep the horse there by talking, petting or feeding until it is completely relaxed. The animal should not be shut in or held in on its first experience; if it wants to leave and finds the way barred, it will become frightened.

Animals over a year old that have never been in a trailer will be somewhat more difficult. As with the weanlings, an older horse should be allowed to study the trailer carefully to assure itself that the vehicle is harmless and to find grain on the ramp, on the floor and in the manger. If the horse has never had a bad trailer experience the chances are that it will soon go in for the grain by itself, but if not it should be led in straight, not at an angle. The animal should be coaxed with grain held just beyond its reach and rewarded with a mouthful each time it moves forward. It should be allowed to retreat when it wishes to do so; the horse will become frightened if it senses that it is trapped.

Loading is not taught in one session. The initial lesson ending on a favorable note is a success even though complete loading may not have been accomplished.

An obstinate horse, one that is stubborn but not frightened, requires different tactics. Assuming that the horse was properly taught to lead, it will know that when a steady pull is exerted on its head, relief can be found by moving forward. Such a horse can be led to the rear of the trailer; when it halts, a strong steady pull should be exerted until the horse moves slightly forward. It should be rewarded by a loose rope along with a pat and a little grain. After a moment the process can be repeated. Any forward movement should be rewarded, but if the horse tries to back, it should be stung on the hindlegs or rump with a whip. After each move forward or backward, time should be allowed for the horse to review what has happened and to calm before the next demand. If a horse persists in moving to the side, the trailer can be parked in an alleyway, panels can be arranged to form a chute or two long ropes can be attached to the sides of the trailer and held in the rear to form wings. The ropes should be ¾″ or 1″ cotton to prevent burns. Should additional encouragement be needed, the men handling the two ropes can cross them behind the rump and draw them taut. A wrap can be taken around a post or nearby tree to help hold the animal in the event of a struggle. The key to loading a balky horse is patience and perseverance. If these tactics will not load an obstinate animal, it

Proper use of ropes to create wings leading into the trailer.

One method of loading a difficult horse is to run the lead rope out the front of the trailer and cross two ropes behind, taking up slack as the horse moves into the trailer. The handler on the right is slow in tightening the rope, which could result in the horse moving to the left, making loading more difficult.

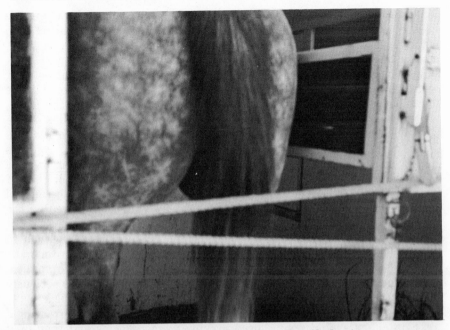

Once the horse is loaded the handlers should pull the two cross-ropes tight across the entrance so that the animal cannot retreat until the butt chain has been fastened in place.

is quite clear that it is either being rushed in its training or has been spoiled by former bad experiences.

The ranch should retrain all animals that are difficult to load. The task of training them will usually be time consuming and aggravating, but almost all animals will have to be hauled at one time or another and should load easily and quickly. These difficult horses should be loaded daily and fed while shut in the trailer. Once the animal loads well, short trips should be taken ending in immediate unloading and return to the pasture or stall. The trailer will eventually become associated with pleasantness, or will at least be tolerable.

The ranch having no trailer-fighting horses is indeed fortunate, but even so there may be times when visiting horses that dislike trailers must be loaded. The trailer-fighter is aggravating, especially when the time schedule is tight. If time permits, the beginner's methods should be tried. Sometimes a different handler, new surroundings and unfamiliar equipment are not associated with past bitter experiences. If possible, the animal should be loaded without fright. When coaxing and other methods for beginners have failed, the easiest solution is loading through a chute. If the ranch is equipped with a chute, the animal can be driven through it into the vehicle. Almost any animal can be loaded into a roomy truck by this method. Because the truck is much more spacious than the narrow chute the animal will usually rush into it, especially when encouraged with a whip.

When there is no chute on hand, one can be improvised by using the side of a building or a fence as one side and some other object such as a van or panel as the other. If neither chute nor substitute is available, three ropes are often used. A one-inch cotton rope tied with a bowline knot around the neck or heart girth, run through the halter ring, extended through the trailer, out the front and wrapped two or three times around an object strong enough to hold the weight of the horse, serves as a lead rope. The others are attached to opposite sides of the trailer, crossed behind the rump and tightened in unison with the lead rope. Each time the animal moves forward, the slack in all ropes is taken up.

Rearing is likely to occur in forced loading; the horse's poll (top of the head) should be padded to prevent serious injury. Should the animal lie back against the rope, a wide, flat, noise-making board used on the rump will usually sting and startle it enough to get results without injury. Even the chronic trailer fighter that has reluctantly given in should be rewarded with grain. While this method is effective for most horses, some animals panic when the pressure and force are applied. If it appears that the animal is fighting so hard that it may be seriously injured, this method should be dispensed with and another tried.

War bridles and "come alongs" are capable of inflicting a great deal of pain. They should not be used except in emergencies and then only by an expert. No horse should be forced to the point of panic. If this seems probable, a veterinarian should be summoned to administer a tranquilizer.

The experienced handler always speaks with a calm, reassuring tone; he does not scream or yell in frustrating situations and never whips, strikes or beats an animal beyond what is needed to focus its attention upon the demand. Loud noises in a frightening situation and excessive punishment will always worsen behavior and may have a lasting effect on an otherwise fine horse.

b) *Hauling.* The first ride in a vehicle should not be undertaken until loading is "old hat" to the horse. When it loads easily and quickly without fear of the trailer, it is ready for its first ride. So far, the trailer has provided the animal solid footing; instability in motion will be a new experience. It must learn to cope with shifting balance while being exposed to the unnatural, loud sounds of the highway. The ride should be of short duration; half an hour in a horse trailer with a calm, older traveling companion is sufficient. The start should be extremely slow, the drive on well-paved roads and the route relatively free from heavy traffic. The progress should be at constant, low speed, the turns with long arcs and the stops with very gradual speed decrease. The novice will be having trouble enough standing up without speed and bumps to cope with.

When home, the animals should remain loaded for awhile to allow the newcomer to calm down. The youngster should be unloaded first; if not, it will become upset when the other horse is removed.

Often a colt is afraid to unload after its first trip. It can be encouraged with a gentle shove or light rap, but should not be rushed.

Sometimes the novice is so unsure of its footing that backing down the ramp or stepping off the trailer is too much for the reluctant animal to undertake. The trailer backed against a gently sloping embankment will help it by making the ramp more level.

One school of thought is that horses do not need hauling experience before traveling is required. True, the horse can be shoved into a trailer and hauled a few hundred miles on its first trip. It will arrive and will have learned to shift its weight for balance. However, high-strung horses cannot take the strain of a long trailer ride the first time or two and will often become exhausted and sick if hauled for several hours. If the first trip is too trying, the horse will probably be difficult to load in the future.

13.3 Travel Comfort and Safety.

a) *Travel Preparation.* Before beginning a trip the vehicle floor should be covered with a substance that provides good footing. Under no circumstances, and regardless of the shortness of a trip, should horses be hauled upon bare floors or hard floors covered with straw. Straw is extremely slippery when placed directly upon a metal or wooden subsurface. Dirt is another cover that should be avoided, because it will become slippery when wet. When animals will be shipped by boat or rail, a sand base covered with straw or wood shavings provides excellent footing. Shavings are commonly used for footing in all types of conveyances; they are, however, dust laden and should be wet down occasionally to lessen respiratory irritation.

Rubber mats provide excellent footing in trailers and horse vans while sand or peat moss are more feasible for use in large trucks.

The use of dividers in vehicles designed for them will prevent injuries, make loading and unloading easier and aid the horse in maintaining balance. Solid dividers should be set far enough apart to permit the horse to spread its legs for balance, otherwise scrambling is likely to result. Partial dividers are preferable in most cases because there is always room below the rail for the horse to shift its feet for balance. All dividers should be well padded to prevent injuries.

If the traveling horses will not need shoes at the destination, or if there will be time to shoe the animals before use, it is best that they travel barefoot. Barefoot horses are much less likely to injure themselves or other horses than are shod horses. Calked shoes are extremely hazardous for horses in transit and should be avoided whenever possible. If such shoes must be worn they should be padded. Smooth shoes are permissible if the horses are separated for travel, but where more than one horse are loose together it is safer to remove all shoes, especially those on the hind feet.

Horses that fight the vehicle are likely to injure themselves; they should be fitted with bell boots and hock guards before being loaded.

Bandages are considered by most horsemen to be crutches. When worn often, the tendons become dependent upon them for support. When the bandages are removed, leg injuries are much more likely to occur. It is not advisable in most circumstances to bandage traveling horses; however, leg puffiness is eliminated by the use of bandages and may be desirable for horses to be shown at halter soon after arriving at the destination. If bandages are used, they should be reset often. While the bandage is off, the leg should be rubbed briskly to stimulate circulation.

Animals traveling in vehicles having low ceilings should be equipped with head bumpers to prevent injuries. Young animals not used to loading and traveling are apt to rear and strike their heads; for them, head bumpers are advisable for travel in all vehicles having solid tops.

Full-grown horses should never be hauled loose (not tied) in any vehicle. Loose horses hauled in a truck tend to crowd to one side, which creates an unbalanced load, and a fallen horse is in danger of being trampled to death. Both hazards are greatly reduced if all horses are tied. A truckload of horses should be tied sidewise in a head-to-tail manner to prevent fighting.

Nylon ropes are best for tying, except for stallions, which should be chained. Ropes are less noisy and can be cut in an emergency. Chains should be secured with safety release snaps that can be unsnapped when under tension. All horses should be tied short enough to keep their heads at body height. Slip knots that can be easily untied should be used when tying an animal (Section 3.2).

b) *Load Composition.* Horses should not be loaded at random. In all vehicles the load must be properly balanced.

When hauling a single horse in a double trailer, or two horses of greatly varying weight, the single animal or the heavier horse should be placed on the left side. All roads are sloped from the center outward to allow water to run off. If the trailer is heavier on the right it will drag to the right and produce more trailer sway, making the ride uncomfortable for the animal and increasing the chance of upset. In addition, should it become necessary to pull off onto the road shoulder for emergency repairs, the heavily weighted left side remains on harder ground, reducing the chance of tipping over and making starting much easier.

When hauling by stock truck, only horses of nearly equal size should be hauled in bulk together. Jostling large horses is hard on smaller ones. The largest and strongest horses of the load should be placed at the rear and other large, strong ones should be positioned at the front. These positions receive most of the force from starting and stopping, which the stronger horses are better able to withstand.

When hauling small loads in large vehicles, swaying will be reduced if the bulk of the weight is positioned toward the front.

If possible, horses acquainted with one another should be hauled together. Their familiarity will lessen tension and fighting during travel. Young, inexperienced horses will travel better if stationed next to quiet old pros. The older horses tend to instill confidence and quiet the youngsters.

If stallions are hauled with other horses, they should be partitioned off and chained. Stallions should not be hauled next to mares or other stallions. If the stallion is excitable, it should not be allowed to see other horses.

The transporting of a mare with its nursing foal presents special problems. The two must be as close as possible to avoid unnecessary stress. The best arrangement is for the foal to be over a partition situated at the mare's head. A shotgun (two stalls in a line) or a four-horse trailer provides excellent arrangements with a place for the foal in front of the mare where she can easily see it at all times.

Sometimes a group of mares with nursing foals must be hauled together. While not conducive to the welfare of either the mares or foals, if such a group must be transported it is best to tie the mares head-to-tail crosswise in the front of the truck with the foals turned loose together behind a rear partition.

The foals, depending upon age, will need to nurse every two or three hours. This necessitates frequent stops. When only one mare and her foal are being transported, a two-horse trailer or small van is suitable. With the divider removed and the mare tied, the foal can be left loose so that it can nurse at will. There is very little danger that the mare will step on the foal, nevertheless, the driver of the vehicle should be particularly careful.

The entire back of the vehicle must be closed or screened when foals are hauled loose, because they are very apt to attempt to jump out.

c) *Weather Conditions.* Mild weather conditions of course are best for animal comfort. Extreme heat is very tiring to horses and causes great discomfort and stress. If travel during hot weather is required,

This type of interior divider in a four-horse stock trailer will allow the mare to see her foal during transit.

A four-horse stock trailer is excellent for transporting mares and foals. The foal is placed in the front half and the mare behind.

the vehicle floor should be wet down often and a 50-pound block of ice placed in the manger. The horse will appreciate the ice and will usually lick a hole in the top, which fills with water it can drink. During movement in dusty areas the vehicle should be as open as possible, preferably without a top. Heavy dust is very irritating to the horse; its concentration in a closed vehicle, even for a very short time, may result in a health problem. Good ventilation without draft is the rule for horses. Frequent rest periods and plenty of water is imperative if the horse is to arrive in good condition.

Movement during inclement weather will suck rain and snow into the open back of the vehicle. The vehicle rear should be covered as much as possible while still allowing adequate ventilation without draft. If the animals may be subjected to wind, rain or snow, they should be covered with waterproof blankets. More than one blanket is needed during extremely cold travel. Wind drastically reduces the animal's body temperature. Weather that may not seem cold while the vehicle is parked will often become bitterly cold while the vehicle is in motion. Cold travel is an invitation to illness.

d) *Stops and Rest Periods.* Horses unloaded, exercised, allowed to urinate and rested frequently during transit not only maintain better mental attitudes but are much less prone to develop leg disorders, digestive disturbances and illnesses associated with urine retention and travel stress. Frequent stops are necessary if the horses are to remain tolerant of travel and physically well. Good horsemen rigidly conform to resting at four- or five-hour intervals, even when hauling horses that are difficult to reload.

Aside from the health aspect of rest, a miserable trip is no incentive for the horses to accept future invitations to travel. A horseman should think about this; the horses do. Traveling eight or ten hours without resting the animals is to be severely condemned. The handler stops when his bladder calls; horses get the same message.

Horses have no difficulty in disposing of fecal material while traveling, but most horses will not urinate while confined in a space too narrow to allow for the usual stance. The average horse-trailer stall is such a place. A horse will have a mental block against urination if it cannot assume the normal position. Retention of urine is unhealthy and can result in serious illness.

Hot weather and crowded conditions indicate that stops should be more frequent than at four-hour intervals. Crowded horses need frequent opportunities to flex their muscles, and hot-weather travel demands that the horses be furnished water often.

If arrangements do not permit a foal to nurse in transit, the frequency of stops must be geared to nursing needs, which generally will be at two- or three-hour intervals, depending on the age of the foal.

A rest period for the traveling horses is a work period for the handler. The horses must be inspected, watered and perhaps fed; the vehicle must be inspected. Adjustments and cleaning may be required.

During rest periods horses wearing bandages require special attention. The bandages should be removed, the legs should be rubbed to stimulate circulation and the bandages should be reset just before reloading.

Horses should be rested between the hours of 2 A.M. and 4 A.M. Studies have shown that horses hauled during these hours tire much more than when hauled at other times.

Stopping, resting, feeding and watering animals in transit is so necessary to their health and welfare that the federal government in 1906 enacted a "humane statute," which can be found in Title 45 of United States Codes Annotated Sections 71 to 74. The reference in the statute to "other animals" applies to horses. The statute is often referred to as the "Twenty-Eight Hour Law" because it forbids continuous confinement of animals in an interstate common carrier for longer than that period except by written consent of the shipper, who can extend the period for an additional eight hours for a total of thirty-six hours. Not even the owner can agree to a longer extension of the transit confinement period. The animals being transported in interstate commerce must be unloaded, watered, fed and rested for five hours before continuing the journey, except that unloading is not required if the car, boat or other vessel contains proper feed and water and furnishes ample space and opportunity for rest, which requires space enough for the animals to lie down if they so desire. The period of continuous confinement applies even though the travel requires the use of two or more connecting carriers. Penalties may be imposed for each violation of the law, whether or not the animals are harmed, and the carrier may be liable to the owner for damage to the animals. Most states have humane transit laws applicable to travel within the state, which are very similar to the federal law.

Whether or not the statute has been obeyed, a carrier may be liable to the owner if he fails to exercise due care that results in harm or death to the animals. Due care is not necessarily measured by the statutory requirements but a violation of the law, not caused by extreme emergency, is almost certainly failure to exercise due care.

The California case of Robinson vs. Schraeder, Vol. 78, California Appellate Report, 2nd Series, page 328 is one among many that demonstrate the seriousness of improper transportation of horses. The owner of Sir Grant, a racehorse, engaged the defendant, a carrier, to truck Sir Grant from the Santa Anita race track near Los Angeles, to the Bay Meadows race track near San Francisco, a distance of a little over 400 miles, which normally requires 15 or 16 hours. The trucker deviated from the direct route to pick up and deliver other horses during the trip, which consumed 45 hours. Sir Grant was not fed, watered or unloaded for rest during the entire period and arrived at Bay Meadows with pneumonia, which caused his death. Veterinarians testified that the length of exposure and lack of proper care in transit was the cause of the fatal illness. The court awarded Sir Grant's owner $25,000 in damages back in 1947.

It should be noted that the penal statutory confinement period of twenty-eight hours is the absolute legal maximum and is much too long for the well-being of horses. That period should never be extended to thirty-six hours. No law prevents the shipper and hired carrier from agreeing to more frequent rest periods.

Travel by motor vehicle is far more strenuous for horses than air, boat or rail transportation and, as stated before, stops for rest and care should usually be made every four or five hours.

e) *Feed and Water in Transit.* The comfort and well-being of the horse in transit are greatly improved if it has familiar feed and water. Enough feed and water from home should be taken to carry the horse through the destination and return, if possible. The animal will be under stress from the transporting and should not have to cope with upsetting changes of feed or water as well. The same type of hay grown in another locality may be different enough to cause severe digestive disturbances to the horse under the stress of travel. Whether home feed can be carried or the feed is purchased along the way, it must be of high quality and free from dust and mold. The feeding of some alfalfa and bran is an aid to digestion that will help the horse greatly during travel; however, the horse should be accustomed to the laxative diet several days prior to departure. The grain ration for the traveler should be reduced somewhat; it does not need so much concentrated feed which, if fed in large quantities during travel, can cause digestive disturbances or founder.

Generous amounts of fresh, clean water must be provided at frequent intervals, especially in hot weather. If the familiar home water carried will be inadequate for the entire round trip, the horse should be accustomed at home to water flavored with wintergreen for at least a week before departing. Thus, a few drops of flavoring added to water acquired in transit will cover the difference in taste and the animal will consume more. If this precaution is not taken, the horse may not drink enough unfamiliar water to stay in good health.

f) *Loading Ramps.* Loading and unloading ramps should be checked for soundness so that there will be no danger of the horse breaking through rotting or weak boards; moreover, the ramp should be inspected for projections that may cause injury.

g) *Loading and Unloading Sites.* Loading and unloading sites should be as free as possible from activity and noise. Horses should never be unloaded onto a slippery surface such as smooth asphalt. After transit, horses are unsteady and are likely to unload rapidly; there is great danger of falling when footing is slick.

h) *Loading and Unloading Procedure.* Loading should be accomplished with the least possible excitement. A horse should never be frightened into loading; if frightened, it will always be difficult to load because of its fear of the vehicle. If level unloading ground is not available, the horse should be unloaded while the vehicle is facing slightly downhill. In this way the animal will back out going uphill and will not have an unexpected, long step down. It is best to train a colt to load before it is a year old. The initial time invested will save a great deal of trouble in the future when time is not available for convincing an unruly horse that it should be locked into a rolling contraption. Horses should be allowed freedom to unload at their own speed. While encouragement is sometimes needed, pushing, pulling or prodding should be avoided; hurried horses are much more likely to injure themselves. People must never stand in the horse's direction of travel. Animals often unload rapidly and may trample those in their paths. The horse must never be tied while the butt chain is not in place. If it starts to back out and finds itself tied, it may panic. When loading, the butt chain must always be fastened before tying; when unloading, the animal must always be untied before the butt chain is released.

i) *Road Vehicle Driving.* The health, welfare and mental attitude of animals in transit are greatly affected by the driver's ability to handle the vehicle. Although most state laws prescribe maximum speed limits of 55 miles per hour for towed vehicles, the driver must drive at speeds conducive to the well-being of the horses. Driving should be done in a manner that will not frighten the animals and will reduce the stress of travel as much as possible. Fear and discomfort increase susceptibility to disease, adversely affect performance and create a reluctance to load in the future.

Sometimes a horse riding in a trailer will literally "climb the wall." This will damage the vehicle and may cause injury to legs. Often these "scramblers" climb on only one side and can be steadied by changing to the other side. The habit of scrambling is usually due to fast turns.

Especially careful, slow driving with particular care on turns will help an animal overcome this problem. Wedging a scrambler tightly into its trailer stall with inflated inner tubes or other soft padding will make it feel more secure and will usually reduce scrambling.

Slow, gentle, gradual starting and equally slow, gentle, gradual stopping is the rule; nothing else is acceptable. Close following of other vehicles inevitably creates instances where rapid speed decreases or quick stops are required to avoid accidents. Twice as much room should be allowed between vehicles when towing a trailer than is usual, because the added weight greatly increases the vehicle's stopping distance. Fast turns, especially on tight curves, are particularly upsetting to horses and often result in such imbalance and fright that injurious scrambling occurs; horses often climb the wall in an effort to regain their balance. The turning radius of the towed vehicle is much smaller than that of the towing vehicle. Poorly calculated turns are apt to result in curbing or ditching a trailer, either of which greatly upsets the animal passengers.

The driver must be sure to allow the trailer ample time to fully complete the turn before accelerating. If acceleration is made too soon, the trailer will whip around the turn and badly frighten or injure the animals inside.

If an official attendant is traveling with the driver, the driver should honor his advice about all things pertaining to the health and welfare of the horses. A driver who is not completely familiar with horse care should not be entrusted with horses unless accompanied by a person who is.

13.4 Road Vehicles. More horse transportation is done by road vehicles than by all other means combined. The great advantage is transportation from the farm to the final destination without switching vehicles. Air, boat and rail transportation require the use of road vehicles to and from the shipping terminals. Because vehicles must be changed, prolonged delays and schedule changes are often encountered.

The most common types of vehicle used for transporting horses on the public roads are towed horse trailers, semi-trucks with large trailer vans, medium trucks equipped with individual stalls and loading ramps, and large stock trucks without partitions for hauling horses in bulk.

Any truck in which animals ride should be equipped with dual wheels and should have a weight capacity of at least one ton. Even a one-ton truck is difficult to hold on the road when the animals shift their weight; any lighter vehicle is very dangerous. A pick-up truck used to pull a trailer should be weighted over the rear axles to provide adequate traction to help prevent jackknifing. A couple bales of hay placed and secured near the tailgate of the pick-up truck will help in this respect.

Horses should never be transported in vehicles that are not completely road safe and free from injurious projections and loud, irritating noises and rattles. Drafty vehicles must be avoided. Drafty conditions added to long travel are almost certain to result in sickness. The shipper

Large ranches may transport animals in modern, comfortable horse vans such as this.

Large, comfortable vans are excellent for hauling a number of horses for long distances.

should carefully inspect the carrying vehicle for suitability and should insist upon correction of faulty conditions.

The transport vehicles over which the ranch has complete control should be maintained in good running order at all times. They require the same care as the family car. Tires, lubrication, bearing packs, brakes, wiring and general condition should be frequently inspected and cared for at regular intervals. Repairs should not be delayed lest a trip date creep up, leaving no time for correction. The tires must be in excellent condition. The remaining tire tread is not a good indication of tire condition; age and direct sun rays alone will deteriorate the side-walls so that a tire may be unsafe even though the tread does not indicate wear. Special attention must be given to the spindle nut to insure that it is adjusted properly. Any welds on the running gear should be checked carefully for cracks or corrosion that will weaken them, especially the weld attaching the spindle to the running gear. Loss of a wheel can result in a serious accident.

Tires should be switched from side to side and tandem tires should be rotated periodically for uniform wear and change of exposure to the sun. Wheels and tires of unequal size or wear should be avoided; mismatched wheels or tires will cause the vehicle to sway. A strong jack and a spare, identical to the size of the wheels in use, should be carried at all times. Tires should not be overly inflated; better traction and smoother travel are provided by air pressure no greater than necessary to carry the load and protect the tires. The tires must have equal air pressure to avoid sway. The trailer-hitch ball should be replaced about every two years. Jerking, strain, and heat created by friction in the socket can cause the steel to crystallize, become brittle and eventually snap off or crumble. Wear on the ball and socket latch can lead to a misfit of the union to such an extent that the trailer may jump off the hitch. The ball should be oiled and covered when not in use to prevent rusting. The trailer hitch should be extremely strong and welded securely to the *frame* of the towing vehicle.

The transportation vehicle should be disinfected regularly, a rented one always before use. Steam cleaning is both effective and time-saving, but soap and water with a disinfectant rinse will serve well. Loose objects that might shift during travel should never be left in with the horses. Nails, splinters and sharp edges should be removed from walls, floors, dividers and ceiling. The structure of the vehicle must be strong enough to hold a fighting horse without splitting or breaking. Rattles, squeaks, clanking and other noises should be eliminated to avoid unnecessary irritation to the horse. Fresh air without drafts must be provided and exhaust fumes must be eliminated from the horse compartment; covered trailers should not be used on dusty roads. If the vehicle is equipped with a loading ramp, it should be faced with cleats of wood or other suitable material to prevent the horses' feet from slipping.

13.5 Rail Transportation. Rail shipping was quite popular in the recent past, but with the better roads of today and the growth of air animal freight, few horses are now transported by train. Most railroads no longer maintain special cars equipped for horses and have no rate provisions for less than carload lots (L.C.L.). A ranch desiring to ship by rail must rent an entire car and build in temporary partitions to suit its special needs. Ordinary-size box cars will easily accommodate a dozen horses in stalls and nearly twice that many horses hauled loose (in bulk). A lined wooden car with a high ceiling and equipped with screens over the doors and vents is much more suitable for horses than the more modern but noisier and hotter metal cars. Unless the weather is very cold, the horses should be shipped without heat; they are less tolerant of heat than of cold. If the car is equipped with heat it can be shut off or regulated by a valve usually located under the individual car. Railroad personnel are not often versed in horse care and comfort; it is the responsibility of the manager to insist upon safeguards. An attendant must accompany the shipment and is usually allowed to travel passage free.

Railroads are very sensitive to the "Twenty-Eight Hour Law" (see Section 13.3 (d)). They must accept the animals from another carrier charged with the time the animals have spent in confined transit since the last rest period and must schedule, and route the animals so that resting areas and facilities will be available at the right time. In spite of the problems presented by the "Twenty-Eight Hour Law," it is not recommended that the shipper give his written consent to the longer period of thirty-six hours. The legal maximum of twenty-eight hours is too long for the horses' welfare and additional time imposes extreme stress upon them.

Horses should never be shipped in baggage cars as crated freight. Baggage cars are routed to the destination of the trains and cannot be switched to another line at a transfer point; the horse and crate must be moved to a baggage car of a different train in order to continue the trip. Such a transfer requires a great deal of time and often necessitates an unplanned layover to await a later train.

If a train trip is planned, the railroad company should be consulted well in advance so that its rules, regulations and schedules can be included in the planning.

Travel comfort and safety for horses moving by train require as much consideration as for horses traveling by other means. Section 13.3 applies in great part to rail shipment and should be read in connection with this type of transportation.

13.6 Steamship Transportation. Large groups of breeding stock are often transported by steamship to countries other than Canada and Mexico. This is the cheapest type of travel for long distances, but does require much longer travel time than air freight. The vessel always supplies water, but the shipper must supply the bedding and feed and pay the cost of the material and labor for the installation of suitable stalls. Usually the animals must be accompanied by an attendant whose passage must be paid. An area and opportunities for exercise during transit can usually be arranged. Quite obviously the "Twenty-Eight Hour Law" (see Section 13.3 (d)) does not apply to open-sea shipments, but the federal "Export Law" (Title 46 of the U.S. Code Annotated, Section 466a) authorizes the Secretary of Agriculture to examine all vessels carrying horses or other animals from the United States to foreign countries and to prescribe rules and regulations and issue orders in regard to accommodations he deems necessary for safe and proper transportation and humane treatment of such animals.

As of August 1971, the approximate cost of water-shipping registered, full-grown, light horses from San Francisco, California, to Europe was near $700 per head, plus $400 passage fare for the attendant and the cost of material and labor for building the stalls. Young colts and small horses travel at reduced rates.

Shipments by river or along the coast from a point within the United States to a point also within the United States are under the "Twenty-Eight Hour Law" (see Section 13.3 (d)) instead of the "Export Law." If on these trips the animals are otherwise properly fed and watered, unloading is not required if the animals have sufficient room to enable them to lie down. Travel comfort and safety (Section 13.3) apply to water shipments as well as to road transportation.

Advance consultation with the carrier is necessary if transportation by water is in the planning.

13.7 Air Transportation. Air transport of horses is rapidly gaining popularity, and airport facilities for handling horses are fast improving. The expense of air transportation is offset by the short time involved and the small, minimal amount of stress upon the animals. Air freight is the proven method of travel for race and show horses. They arrive at the destination quickly and are ready for top performance after a short rest.

Tranquilizers are carried and administered by the attendant should an animal become extremely excited during flight. Though the chance of an animal's getting loose and going berserk aboard an airplane is extremely small, the pilot is authorized to destroy an animal that is endangering the lives of the passengers. Aside from this lawful right, the airline will refuse to accept a horse as cargo unless the owner signs a waiver supporting such an emergency measure.

The procedure for shipping horses air-freight is as follows: The freight office of an airline that handles horses will contact a live animal cargo service company, which provides the special crate and preparation needed prior to departure of the flight (Braniff, Northwest, Flying Tiger, TWA, PAM, United, Air France, Alitalia and KLM are among those that handle such shipments). The cargo company will arrange for a specially built plywood crate 80 inches long, 30 inches wide and 5 feet high to be at the flight pickup point (flown in if necessary). The owner of the horse must provide an attendant to accompany the horse and must deliver a recent health certificate and a brand inspection certificate as well as any other credentials required at the destination, all of which must travel with the animal. The shipper must transport the animal to the airport at the designated hour. Depending upon facilities, the horse may be led up a ramp and into the plane, where it is loaded into a shipping crate, or it may be walked into a crate attached to a low pallet outside the plane—after which the crate, horse and attendant, weighed together as a package, are raised by a lift up to the freight door, where the unit is wheeled into the airplane and the pallet is locked into place. Most airlines and shippers like to have a veterinarian on hand to mildly tranquilize the animal before loading.

The airline will supply both feed and water for the animal; however, to prevent digestive disturbances the shipper should provide familiar feed and water.

Most airports are not equipped with facilities for stabling horses; the shipper must know this in advance and, if necessary, arrange for the pickup immediately after landing. Horses being shipped abroad that must first be flown across the United States, are usually held over for a one-day rest at the last stop within the United States. For example, a horse being flown from San Francisco, California, to Frankfurt, Germany, would be rested in New York at the airline's expense.

If a horse is shipped entirely within the United States, usually no quarantine is required; however, horses entering the United States or returning from other countries are met by a veterinarian who must take a blood sample before the horse can leave the plane (see Section 13.1).

The shipping charge is usually by the hundredweight, but a minimum charge is required, since nothing else can be placed on the pallet holding the crate. These charges vary according to destination and handling facilities. Most airlines include the attendant in the weight charge; however, some require the handler to pay passenger fare. In either event, he is furnished a seat and refreshments. Crate rental charge ranges from $40 to $50 per day measured from the time the crate leaves storage until it returns to storage. This may involve one to three days, depending on flight connections. The weight of the unit—horse, crate and man—usually approximates 1,500 pounds for light horses (crates weigh between 300 and 400 pounds).

Costs vary greatly from area to area. As of March 1972, approximate costs, which included the handler, were as follows:

San Francisco, Calif. to Atlanta, Ga.	$1,100
San Francisco, Calif. to New York, N.Y.	1,375
San Francisco, Calif. to Europe	2,500

When a large number of animals are to be flown, it is sometimes cheaper to charter a cargo plane. An average cargo plane can accommodate about 40 horses and costs approximately $12,000 for a flight from New York to England.

The "Twenty-Eight Hour Law" (see Section 13.3 (d)) seems literally to apply to flights within the United States, but as a practical matter the short-flight makes it nearly impossible for the law to be violated in regard to unloading.

If horses are to be exported by air from the United States, the "Export Law" (see Section 13.6) would seem applicable, giving the Secretary of Agriculture jurisdiction over the shipment.

Subject to the airline rules and regulations, the comfort and safety (Section 13.3) of traveling horses should not be dispensed with unless in conflict with sound flight practices.

As with other modes of commercial transportation, the air carrier should be consulted well in advance of the anticipated flight departure.

13.8 Trailer Qualities to be Considered in a Purchase. Most horsemen, regardless of the breed they raise, prefer trailers designed for Thoroughbred horses. These trailers are longer, wider and taller than regular trailers and are much more comfortable for the horses.

The running gear for trailers is basically of two types: tandem (4-wheels), and single axle (2-wheels). The safety and convenience of the tandem commend it over the single-axle type. A blowout or the loss of a wheel is less likely to result in disaster; the one remaining companion wheel will bear the weight and the trailer can usually be hauled on to a garage for repairs. The tandem, however, tends to drag around turns; for that reason, horsemen who haul in mountainous country often prefer a single-axle trailer.

Trailer bodies set on leaf springs usually have more rocking motion from front to rear and corner to corner than do those resting upon torsion springs. In any event, the trailer must be mounted upon springs; solid riding is very tiring and jarring bumps are extremely hard on horse legs and cause more wear and tear on the trailer. Widely spaced springs provide much better vehicle balance and make towing considerably easier than closely set springs. The springs should be heavy enough so that the trailer, when not loaded, has nearly the same firmness as a trailer lacking springs, in order to bear the weight of the trailer when loaded.

Trailers must be equipped with an adequate braking system; laws and good sense require good brakes. Hydraulic and electric braking systems are the two principal types used. Each system has advantages and neither is without disadvantages. Hydraulic brakes attach directly into the brake system of the towing vehicle and function with it. They are smooth operating and are free from locking; however, a leak in most hydraulic systems will leave both vehicles virtually without stopping power. Fatal accidents have been attributed to such failure. The hydraulic line connection between the vehicles should be of a type that will separate and automatically seal the towing vehicle's fluid line in the event the vehicles should break apart while moving. Electric trailer brakes have the advantage of affording a second breaking system. Although they work automatically in conjunction with the brake pedal of the towing vehicle, the driver can, by the use of a hand lever, apply the trailer brakes independently when the need arises. The dual braking system make stopping possible, though somewhat more difficult, when one system fails. The additional advantage of the electric system is that the independent control enables the driver to stabilize the trailer if it should begin to sway. He can do this with the hand lever by lightly applying the trailer brake to force the trailer to drag slightly and follow the forwardpower of the towing vehicle. Swaying usually occurs on down grades or levels where the towing power has the least influence over the trailer. The brakes should be set to operate at the exact time or a fraction before the towing vehicle's brakes become effective. It is far better for the trailer to be pulling back than for its weight and momentum to be pressing against the leading vehicle. Such an adjustment will also give the driver the means of straightening out the swaying trailer by a light touch upon the brake pedal.

Electric brakes, operated through impulses as opposed to pressure, have a tendency toward grabbing, especially in wet weather, and sometimes a surge of electricity may cause locking. Probably the greatest drawback to electric brakes is not in the function or performance but in the hazard created if the electric lead lines are erroneously connected to the wrong circuits of the towing vehicles. An expert automotive electrician familiar with trailer hookups should perform that task. Serious accidents have occurred simply because the current, by some fault in connecting, bypasses the rheostat switch controlled by the brake pedal and hand lever. The most common mistake is in tying into the nightlight circuits, which are dead until the lights are activated. The brakes will be useless until the light switch is engaged, and then, of course, the trailer brakes will unexpectedly lock under the full electric power. If this happens while the vehicle is traveling rapidly, the consequences are often disastrous. This is not the only error possible; as mentioned before, only an expert should perform this electric hookup.

A third type of brake, surge brakes, has been developed. These brakes become engaged as the towing vehicle slows down and are released as it speeds up. The main advantage to this system is that the towing vehicle need not be connected to the trailer braking system. The disadvantages are that such brakes would be useless should the lead vehicle's brakes fail, and they cannot be engaged manually to correct trailer sway. The vehicles, as a traveling unit, should be thoroughly tested at very low speeds under safe circumstances before the brakes are considered in proper working order. This test should include forward movement with the headlights on.

Almost all trailer hitches are designed with a two-inch metal ball and socket; smaller ones are illegal in most states. A safety chain, required by most states, must be strong and form a link between the frames of both vehicles. The hitch should be adequate for the gross load and tongue weight and welded to the frame of the towing vehicle. The popular "gooseneck" trailer, so called because of its appearance, is attached well within the truck bed. Both truck and trailer are greatly stabilized by this combination. The overlapping of the gooseneck into the truck greatly lessens the truck-carrying capacity for hauling feed, saddles and miscellaneous articles, but the sturdy union of the two vehicles and the stability afforded justify both the additional cost and the loss of truck capacity.

State laws generally require trailers to be equipped with tailgates high enough to prevent manure from falling onto the highway. Trailers are built with and without loading ramps as a feature of the tailgate. The choice seems to be merely a matter of personal preference since the horses show only slight preference for a ramp-loading trailer over a "step-up."

Trailers must by law be equipped with tail lights, brake lights, license-plate lights and turn-signal lights.

Escape doors are safety features that all trailers should have. Doors located on the sides of the body should be about six inches above the floor. Such placement will avert the possibility of a foot slipping out of the doorway in the event the door should come open and reduces the temptation for a horse to try using it as an exit.

The panes of the front windows should be of tinted safety glass or Plexiglas to reduce the glare. Some horses become frightened when they see objects flying past them. The windows can be blocked out, if necessary, but should not be dispensed with, for most animals seem to enjoy looking out.

A manger is a convenience that will increase the travel comfort of the horses. It should have smooth edges high enough to hold the feed during eating and traveling. A solid tie-ring should be placed in the manger, well fastened to the body frame and in an easily accessible position for release in case of emergency. A screen or pipe manger partition should be provided so that the horses have companionship but are restricted from playing or fighting. If one is hauling stallions, the manger partition should be solid. A drainage hole in the bottom of the manger will ease the labor of cleaning and reduce the absorption of water into the wood.

The lower two-and-a-half feet of the inside walls should be lined with heavy-gauge sheet metal. If they are not, those portions of the walls will soon be badly battered. The side walls, dividers and front wall should be padded at body height to protect the horses from bumps and bruises.

The interior of the trailer should be painted with a light, nonglaring color that presents an open, airy appearance to the horse, which makes loading an easier chore. Loading a horse into a dark, dingy, tight, box-like area can be a trying and disagreeable task. Trailers with solid metal roofs and walls conduct heat and cold inwardly as well as exaggerate noise, all of which increases travel stress. To help keep the noise level down, preference should be given to wooden-lined walls and a ceiling of other than a simple sheet of metal. The trailer will be in the direct rays of the sun during travel. It should be kept in mind that a dark outside color does absorb heat and a light outside color deflects it. Color properties and the local climate should be considered when selecting an outside paint.

Weak trailer floors have been the cause of some tragic accidents. Even new trailers should be closely examined for solid, very strong flooring. Animals have broken through trailer floors and lost legs before the driver has had any indication of the problem. Floor break-through accidents spell total loss, not just injury.

A trailer should be equipped with two types of interchangeable stall dividers: one solid, the other partially solid. The solid divider is used when hauling a kicking animal as a protection for the companion horse; however, this type restricts foot movement needed for good balance and may result in scrambling. Balance is very difficult for the horse if a solid divider is used in a narrow trailer. The partial divider is for use where kicking is not a problem; it affords more freedom for the feet. Even in a narrow trailer the horses are not likely to step on each other. However, bell boots can be worn on the inside feet as further insurance against injury.

A topless trailer that does not have a high front windbreak is dangerous to horses' eyes; insects and dust are certain to strike the eyes and can cause serious injury. Such a trailer is also an invitation for a horse to try jumping out.

A trailer should not be purchased until it has been road tested. It may be that a secondhand trailer can be load tested, but this is highly unlikely if the trailer is new. Either kind, however, should be tested by two people—one driving, the other in the horse compartment of the trailer. The trailer should be towed at all speeds up to the legal maximum, over normal road conditions and around turns in both directions and at different speeds. In this way rattles, exhaust fumes, ventilation, swaying and general road stability can be noted. Of course the travel will not be smooth; it is more likely to be a jarring ride because the springs must be stiff in an empty vehicle in order for it to be capable of carrying a heavy load smoothly.

13.9 Transporting "Sored" Horses. In December of 1970 the Federal Congress enacted Public Law 91-540,84 Statute 1404 entitled "Horse Protection Act of 1970." The law uses the "commerce clause" to prevent the use of cruel or inhumane methods and devices for affecting the gaits of horses. The federal government has reached into state jurisdiction as effectively as it can to prevent the "soring" of horses.

The statute says that a horse is "sored" if for the purpose of affecting its gait any of the following is used:

Blistering or chemical agents, burns, cuts, lacerations, tacks, nails, chains or boots, or any other cruel or inhumane method and device.

It provides that no "sored" horse shall be "moved in interstate commerce" or into the United States for the purpose of showing or exhibition, and further that no "sored" horse (even though a local horse not moved in interstate commerce) shall be entered in any show or exhibition if any other horse exhibited was moved in interstate commerce, even though the other horse is not "sored."

Anyone who participates by shipping, receiving or carrying, or by entering or permitting entry into a show or exhibition of a "sored" horse is in violation of this statute, which provides for a fine of $2,000 or imprisonment for six months or both for each violation.

It seems quite likely that some commercial carriers will require a sworn statement from the shipper declaring that the horse has not been "sored" since the effective date of the statute, before accepting it for shipment. Even with such a statement a horse is likely to be rejected for shipment if the carrier's inspector has any doubt about the truth of the affidavit.

Some states now have similar laws and many undoubtedly will soon follow the federal precedent.

14
PAPER WORK

(SYNOPSIS)

This chapter discusses business structure, insurance, registration, contracts, marketing and advertising, and suggests farm records and forms indispensable to the successful organization and operation of a breeding enterprise.

The following synopsis by section numbers is provided as an aid for quick reference to specific subjects.

14.1 General Statement. Paper work is an extremely broad term, but as used in this chapter it is not intended to include the general business administration or bookkeeping that is a part of all businesses. The purpose of this chapter is briefly to point out and discuss the records,

349

contracts, registration and advertising peculiar to the business of raising, breeding, registering and marketing horses. The official forms
required by breed associations cannot be altered, but most of the other
forms set out are merely suggested formats that have been found to
be extremely useful in practice although some additions or modifications may be desirable, depending upon procedures and the record system adopted by the farm.

14.2 Business Structure. The horse farm is either a business or a hobby.
Which it is will have a great impact upon tax consequences for almost
all horse-farm owners. Usually tax laws are much more severe on
hobbies than on businesses, but the owner may not be able to make a
choice; the decision usually lies with the government and it is up to the
taxpayer to supply the business information to the government. If
records are incomplete and ill-kept, the taxpayer will find it very difficult to convince the Internal Revenue Service that he is in the horse
business for profit, not for a pastime. Not all horse ventures will be
recognized by the government as business pursuits. Even if the venture
is structured as a corporation, partnership, limited partnership, trust,
joint venture, coownership or other association, it still may be classified
as a hobby for tax purposes. A tax accountant should be consulted and
the venture should be structured so that the owner will have the most
favorable tax advantages and will be in the most flexible position for
financing. Whether the farm is classified as a business or a personal
hobby, it is imperative that the farm costs, expenses and income be
maintained separately from any other business or personal accounting.
The owner, if he expects a favorable ruling from the tax authorities,
must be able to substantiate by proper records and bookkeeping the
complete independence of the business. If the farm bookkeeping is combined with personal matters or other ventures, it is unlikely that any
tax concessions will be granted. Even if the owner is willing to accept
the tax consequences of a hobby, complete, independent and accurate
business records are necessary in order for him to evaluate the success
of the venture and to recognize areas of economic inefficiency.

It is very difficult to estimate the investment required prior to launching a horse-breeding enterprise. Land costs (purchase price or rent),
building costs, equipment costs, financing costs and price of the animals
will vary widely. Except for rent, cost of land, interest upon financing,
insurance costs, attorney's fees and outside stud fees, the costs in the
operation of a small business having ten mares and one stallion can be
expected to closely approximate the following itemized estimates:

Personnel	$ 3800.00
Training	3000.00
Feed	1650.00
Pasture maintenance	750.00
Exhibiting fees and transportation	600.00
Public utilities	600.00

Advertising	450.00
Shoeing and trimming	350.00
Registration and membership fees	300.00
Medical services and supplies	300.00
General Maintenance	200.00
	$12,000.00

The estimate anticipates that the owner will do much of the work and on that basis is an indication of the cost budget required until the farm has been operating long enough to develop its actual costs through adequate accounting.

Income is of course so variable that no estimate can be attempted. On the profit side is the fact that a good stallion will be able to service 20 to 30 outside mares in addition to the ten ranch mares. The stud fees of a well-advertised, high-quality stallion will do much to offset costs. Good colt crops should put the farm on a small profit basis within two or three years and should show a return on investment within five or six years.

The breed and quality of horses have a great bearing upon profit. The maintenance cost is nearly the same for any light horse, but the top-quality animal will bring many times the price of the mediocre horse.

14.3 Insurance. The farm is of course concerned with fire, casualty, theft, vehicle, employee accident and unemployment insurance, but in addition it should consider insurance against the special hazards that arise from the nature of the enterprise. If the owner has any reason, even though slight, to believe his horse is dangerous, liability for damages done by that horse to persons or property—whether the injury occurs on or off the farm—may be imposed upon the owner and, perhaps, the horse's handler. Persons who come to the farm are invitees to whom a special duty of care is owed, and children injured by the horses or other attractions on the farm may be able to recover damages although they themselves are careless. Insurance to cover these special risks is needed protection.

When borrowing to purchase breeding stock, lenders almost always require life insurance pledged as security, up to the amount of the loan upon the horses.

Owner's insurance upon the lives of the ranch horses is controversial. The cost of such insurance is very high; but in spite of the cost most ranches do cover the lives of their most valuable animals.

Annual insurance rates vary from 1% to 6% or more of the animal's value, depending upon the type of policy and risk assumed; death from any cause is more expensive than accidental death. A life policy should provide for loss by death from natural causes, accidental causes and humane destruction deemed necessary by a veterinarian. Age limits for insurance vary, but most companies that will insure set the age limits as 30 days to 12 or 15 years. Premiums for new-born foals are more for the first few months and decrease with the foal's age. Many companies will not insure barren mares over eight years of age against foaling hazards.

Before issuing a policy most companies require a veterinarian's statement of health and the owner's declaration of the animal's value, age and identification. If a group of horses is to be insured under one policy, such a veterinarian's statement and owner's declaration must be submitted for each horse. A stable rate for all of the ranch horses is obtainable at a comparatively lower cost, but the coverage is quite limited. Short-term coverage is available for special events such as shows, surgery, specific foaling and transporting (see Section 13.1). Foaling policies must be secured two weeks or more before the foaling date. Permanent sickness or injury insurance, excluding minor incapacities, can sometimes be obtained; the application for such insurance requires that the attending veterinarian's report of the horse's condition be submitted.

Various horsemen's organizations provide many types of coverage. Residents of California, for instance, can obtain insurance through the California State Horsemen's Association, P. O. Box 1189, Santa Rosa, California 94502, and through the Pacific Coast Quarter Horse Association, P.O. Box 5253, Sacramento, California 95817. Brochures on the available insurance group policies can usually be obtained by request from the various organizations in the area.

It is wise for the horse farm to consult two or three reputable insurance agents for details and price for the insurance program. The farm's attorney should review the policy forms and program before the matter is settled.

14.4 Registration. Breed associations are not governmental in character; they are private associations, although recognized and protected under the laws of most states. The legality and enforcibility of contracts, liens, chattel mortgages and ownerships involving horses must be tested by the law of the government having jurisdiction. This is usually the state laws, but in some cases the governmental body may also be federal. While it is true that the courts give much weight to registration and the documents supporting registration, the primary purpose of registration is to qualify the horse for the advantages afforded by the breed association.

Proper registration of all horses produced by the farm is imperative. Registration is the best guarantee of the correctness of pedigree for sale and breeding purposes and establishes eligibility to compete in many events recognized by the association for points toward association awards.

Almost all associations that maintain a registry follow the same basic procedures, with a few modifications such as requiring photographs or refusing to recognize artificial insemination. An up-to-date copy of the rules and regulations of the association that governs the breed raised by the farm is indispensable if registration mishaps are to be avoided.

Many associations that maintain a registry require photographs from front, side and rear.

A full discussion of all of the numerous associations would not be practical in this work. The American Quarter Horse is the largest breed in the United States and since its registration is typical, the American Quarter Horse Association (AQHA) rules have been selected for discussion, which will acquaint a breeder with the general requirements and procedures for most breeds.

The AQHA registration certificate form formerly issued is shown below, but it has since been revised in a few points, principally by deletion of the reference to "permanent" registration which, under the change of AQHA rules effective January 1, 1962, has no significance, and by showing leg markings in more detail. The form presently in use is also shown immediately following the former one. The back of each form provides space for use by AQHA only, in which each transfer of ownership is endorsed.

Several associations of concern to the Quarter Horse breeder are:
American Quarter Horse Association (the most important).
P. O. Box 200, Amarillo, Texas 79105
(This association has affiliated regional associations not shown here.)

Palomino Horse Breeders of America
P. O. Box 249, Mineral Wells, Texas 76067

The Palomino Horse Association, Inc.
P. O. Box 446, Chatsworth, California 91311

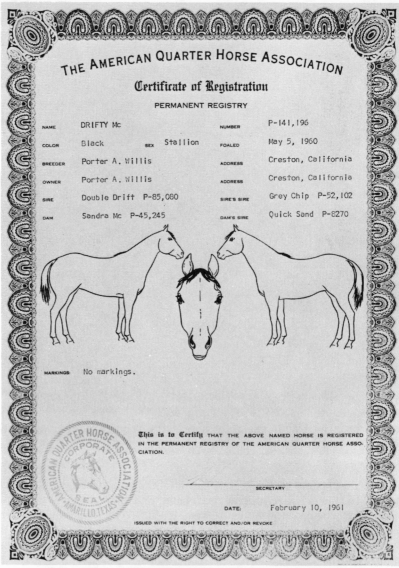

THE AMERICAN QUARTER HORSE ASSOCIATION

Certificate of Registration

PERMANENT REGISTRY

NAME	DRIFTY Mc	NUMBER	P-141,196
COLOR	Black SEX Stallion	FOALED	May 5, 1960
BREEDER	Porter A. Willis	ADDRESS	Creston, California
OWNER	Porter A. Willis	ADDRESS	Creston, California
SIRE	Double Drift P-85,080	SIRE'S SIRE	Grey Chip P-52,102
DAM	Sandra Mc P-45,245	DAM'S SIRE	Quick Sand P-8270

MARKINGS: No markings.

This is to Certify THAT THE ABOVE NAMED HORSE IS REGISTERED IN THE PERMANENT REGISTRY OF THE AMERICAN QUARTER HORSE ASSOCIATION.

SECRETARY

DATE: February 10, 1961

ISSUED WITH THE RIGHT TO CORRECT AND/OR REVOKE

Old AQHA CERTIFICATE OF REGISTRATION

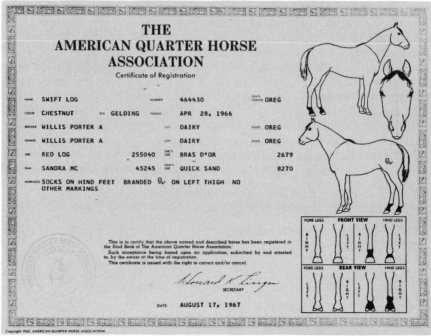

THE AMERICAN QUARTER HORSE ASSOCIATION

Certificate of Registration

NAME	SWIFT LOG	NUMBER	464430	STATE FOALED OREG
COLOR	CHESTNUT SEX GELDING	FOALED	APR 28, 1966	
BREEDER	WILLIS PORTER A	CITY	DAIRY	STATE OREG
OWNER	WILLIS PORTER A	CITY	DAIRY	STATE OREG
SIRE	RED LOG 255040	SIRE'S SIRE	BRAS D'OR	2679
DAM	SANDRA MC 45245	DAM'S SIRE	QUICK SAND	8270

MARKINGS: SOCKS ON HIND FEET BRANDED &c ON LEFT THIGH NO OTHER MARKINGS

This is to certify that the above named and described horse has been registered in the Stud Book of The American Quarter Horse Association.
Such acceptance being based upon an application, submitted by and attested to, by the owner at the time of registration.
This certificate is issued with the right to correct and/or cancel.

SECRETARY

DATE: AUGUST 17, 1967

FORE LEGS	FRONT VIEW	HIND LEGS
RIGHT	LEFT	RIGHT	LEFT

FORE LEGS	REAR VIEW	HIND LEGS
LEFT	RIGHT	LEFT	RIGHT

Copyright 1965, AMERICAN QUARTER HORSE ASSOCIATION

Current AQHA CERTIFICATE OF REGISTRATION

National Palomino Breeders Association, Inc.
East Dixie Street, London, Kentucky 40741

American Buckskin Registry Association, Inc.
P. O. Box 1125, Anderson, California 96007

Pacific Coast Quarter Horse Association
(maintains the California-bred registry and is
affiliated with AQHA)
P. O. Box 5253, Sacramento, California 95817

(The American Horse Council, Inc., 1776 K St., N.W., Washington,
D.C., 20006, can furnish addresses of other breed associations.)

The palomino and buckskin colors are recognized by the AQHA;
hence, a Quarter Horse of one of these colors can be cross-registered
with the appropriate color association, provided it meets the requirements of the other association. A cross-registred animal has the advantage of being eligible for participation in the events of both of the associations in which it is registered, and cross-registered stallions attract
mares in both registries.

Since 1962 the Official Stud Book (Registry) of the AQHA has con-
sisted of two parts—Numbered Horses and new and old Appendix
Horses. The New Appendix is composed of half Thoroughbred and half
Quarter Horse animals foaled January 1, 1962, or afterward, and the
Old Appendix contains those foaled prior to January 1, 1962. All animals that were previously registered with the National Quarter Horse
Breeders Association (NQHBA), now dissolved, or had a tentative or
permanent AQHA number, are now all included in the numbered section of the AQHA Stud Book. (The AQHA now has possession of the
records formerly maintained by the NQHBA.) Any animal lacking a
permanent number is not eligible to be used for breeding; i.e., its get
or produce is not eligible for registration.

A stallion, mare or gelding foaled in 1962 or after, whose sire and
dam are both numbered Quarter Horses, is eligible for registration with
no official inspection; one that has a numbered Quarter Horse parent
and a parent registered in the Jockey Club of New York City is eligible
for registration in the New Appendix Registry. After two years of age,
a New Appendix horse is eligible for advancement to a permanent
number by first qualifying for one of the AQHA Registers of Merit
(ROM) and then passing conformation inspection. Any animal foaled
after January 1, 1962, is not eligible for registration in any part of the
Stud Book if one of its parents is registered in the Appendix Registry,
even though the other parent is numbered; however, such a horse automatically becomes eligible for numbered registration if the Appendix
parent advances to the numbered part of the registry, provided the other
parent is a numbered horse. Animals in the New Appendix Registry
are eligible to compete in all recognized performance contests and in
halter classes within certain age limits.

Animals up to and including two years of age (three years, after

Moolah Bardell

This fine registered Quarter Horse stallion, owned by Mr. Charles Schreiner, demonstrates the results a Quarter Horse breeder can achieve by crossing top Thoroughbred and Quarter Horse lines. Moolah Bardell made his points toward permanent registration at the race track, receiving a Register of Merit (ROM). Now retired from racing, he is standing at stud at the Long Training Stables in Martinez, California and is being shown in both pleasure and halter classes. Moolah Bardell, at seven years of age, is already a proven producer, passing on speed, maneuverability, conformation, cow-sense, disposition and color. With success like this, the time and expense involved in proving an appendix-registered Quarter Horse is well worth while.

January 1 of the foaling year) may be registered in the normal manner by sending to AQHA a completed and properly signed application for registration, together with the breeder's certificate (both reproduced below) and the proper fee. All AQHA forms, as well as others, must completely and accurately represent the true facts. Incomplete or unsigned forms, failure to enclose the fee, or inaccuracies are reasons for rejection or delay of registration, and intentional false statements may lead to suspension from the association, in which case the offender can no longer register Quarter Horses or participate in association-recognized activities. In some instances such false statements are also state crimes.

PEDIGREE

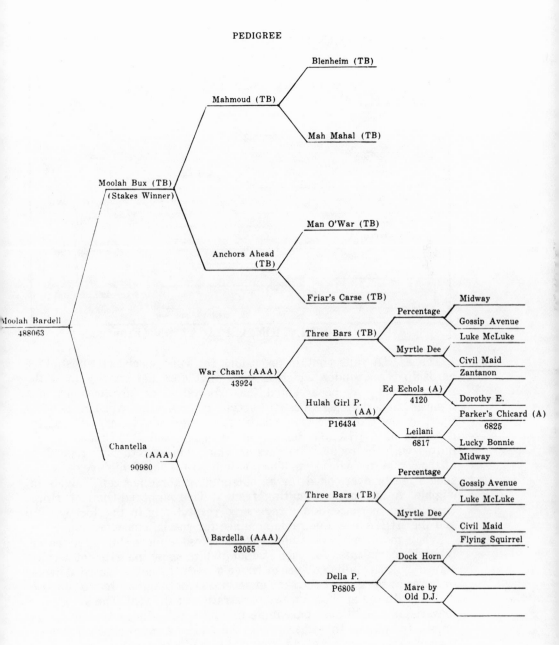

AQHA REGISTRATION APPLICATION (Front side)

The AQHA rules contain provisions for some special circumstances, hardship cases, under which a horse that does not strictly meet the requirements may be accepted for registration or advancement to a number. Generally a request for special consideration will not be acted upon by the Executive Committee unless the horse is inspected at the direction of AQHA. To defray the costs involved, $200 must be submitted with the request, no part of which will be returned, regardless of the committee's decision. The possibility of a "hardship" registration should not be overlooked for an outstanding horse not conclusively ineligible. A fine horse resulting from a Thoroughbred-Quarter Horse cross when the parents are registered respectively in the Jockey Club and the AQHA is a favorable nominee for special consideration.

When more than one stallion has serviced a mare during a season, breeder's certificates for each stallion, which show the dates of service, signed by each stallion owner or lessee of record and the recorded owner or lessee of the dam at the time of service must be submitted to the association with the application for registration of the foal. The association may require additional parentage proof as well—blood typing for example. In spite of this liberal rule, the foal of a mare exposed to more than one stallion by breeding, artificial insemination, pasturing, accident or otherwise without an interval of 30 days is not eligible for registration in any part of the registry. The rule applies whether the mare is exposed individually or in a band of mares. Exposure means opportunity for breeding whether or not breeding is known to have occurred.

INSTRUCTIONS — IMPORTANT: Accurately kept private records are essential. In any case where an application is regarded as questionable, the burden of proof with regard to same shall rest upon the applicant who shall sustain his claims by a preponderance of evidence.

LOCATION OF DAM AT TIME OF FOALING

_____
Name of Ranch and Exact Location

_____
City   State

OWNER OF DAM AT TIME OF FOALING

_____
Name of Registered Owner

_____
Complete Mailing Address   City   State

BREEDER'S CERTIFICATE If the owner of the horse pending registration is not the owner of the sire and owner of the dam at the time of breeding, then a breeder's certificate shall accompany this application.

LOCATION Please give either location of your ranch or farm from nearest post office or advise if your horse is located in same town as shown. In event horse is not located in same town as shown, so state, and give definite directions for convenience of inspector.

_____

_____

_____
Day Telephone   Night Telephone

STALLION BREEDING REPORT FORM MUST BE FILED BY THE OWNER OF THE STALLION

REGISTRATION FEES

(1) APPENDIX	Members	Non-Members	(2) NUMBERED REGISTRY—DIRECT	Members	Non-Members	(3) ADVANCEMENT FEE— Appendix to Numbered Registry	Members	Non-Members
Weanlings	$ 5.00	$10.00	Weanlings	$10.00	$15.00	Mares	$10.00	$15.00
Yearlings	$10.00	$15.00	Yearlings	$15.00	$20.00	Geldings	$10.00	$15.00
2 yr. olds	$15.00	$20.00	2 yr. olds	$20.00	$25.00	Stallions	$25.00	$30.00

SPACE BELOW FOR INSPECTOR'S REPORT ONLY

Month — Year _____

_____
Signature

Month — Year _____

_____
Signature

Month — Year _____

_____
Signature

Reverse side of AQHA REGISTRATION APPLICATION

BREEDING RECORD

NAME OF STALLION

STALLION'S REG. NO.

NAME OF MARE SERVED

MARE'S REG. NO.

DATES SERVED — YEAR

COLOR OF MARE

This breeding certificate was issued to

_____

_____

on _____
MONTH   DAY   YEAR

This form may be obtained from the

AMERICAN QUARTER HORSE ASSOCIATION
P. O. Box 200
AMARILLO, TEXAS 79105

BREEDERS CERTIFICATE

FOR REGISTRY IN A.Q.H.A.

This is to certify that

_____
NAME OF STALLION GIVING SERVICE   REG. NO.

was bred to _____
NAME OF MARE SERVED   REG. NO.

on _____
LIST DIFFERENT DATES (INCLUDING YEAR) MARE WAS BRED   YEAR

Color and markings of mare _____

_____

AT TIME OF BREEDING →

_____
SIGNATURE OF OWNER OR MANAGER OF STALLION AT TIME OF BREEDING

_____
SIGNATURE OF OWNER OR MANAGER OF MARE AT TIME OF BREEDING

BREEDERS CERTIFICATE

The AQHA will register animals conceived through artificial insemination only if the semen is used immediately following its collection and at the place or premises of collection. If parentage is doubted, the association may require the sire, dam and foal to be blood typed as an aid to settling the doubt.

A farm planning to use artificial insemination must notify the AQHA in writing annually prior to beginning insemination. The premises and practice are subject to inspection by the AQHA. Upon acquiring the services of a suitable laboratory for blood typing a large number of horses, the AQHA will require that the blood type of all stallions to be used in the practice of artificial insemination be on file with the association as a further precaution against mistake or fraud.

Before October 1 of each year, the recorded owner or recorded lessee of a stallion must file with AQHA an accurate and complete written Stallion Breeding Report (see below for form furnished and required by the association), showing the names, registration numbers and exposure dates of all mares, regardless of ownership, that have been exposed to that stallion since the last October 1. The same report is required for all Thoroughbred stallions to which Quarter Horse mares have been exposed. The association carefully compares the applications for registration of each of the stallion's offspring with the stallion breeding report for that season.

MUST BE FILED ON OR BEFORE THE FIRST DAY OF OCTOBER	AMERICAN QUARTER HORSE ASSOCIATION P. O. Box 200 Amarillo, Texas 79105 STALLION BREEDING REPORT	IMPORTANT See Other Side For Complete Instructions

The following mares were bred to the stallion whose name is shown below

Stallion's Name		AQHA Reg. No.		Thoroughbred Reg. Number	

IMPORTANT: Give the registered name and number of the stallion.   REPORT ONLY ONE BREEDING SEASON ON THIS FORM. ↓

	Registered Name of Mare Bred	Reg. No.	Recorded owner of mare at time of service	Dates mare was bred (if pasture bred, so state and give dates in and out of pasture)	Year
1.					
2.					
3.					
4.					
5.					
6.					
7.					
8.					
9.					
10.					
11.					
12.					
13.					
14.					
15.					

CERTIFICATION: I do hereby certify that the above named mares were bred to this stallion on the day(s) shown above.

Written Signature of RECORDED OWNER of Stallion or Written Signature of AUTHORIZED Agent or Manager of Stallion _____

Date submitted: _____   Address _____

AQHA STALLION BREEDING REPORT (Front side)

1. Keep a separate record-sheet on each stallion used for breeding purposes.

2. Report shall be on file in the office of the American Quarter Horse Association not later than October 1st of each calendar year.

> "On or before October 1 of each year, the owner or lessee of every stallion (both Quarter Horse and Thoroughbred to which Quarter Horse mares have been exposed) must make a written report showing the names and registration numbers of all mares exposed to said stallion since the previous October 1. The report also shall indicate all breeding dates or dates of exposure.
>
> A. "This report shall include all mares owned by the owner of the stallion as well as mares owned by other parties.
>
> B. "The report shall be made on a form provided free of charge by the AQHA."

3. Demand to see the registration certificate of each mare bred. This will assist you in the preparation of this report.

4. Be sure the last recorded owner of mare is the man who contracts the breeding to this stallion.

5. Give exact date(s) bred—if more than one service is given mare—be sure and show the different dates.

6. If pasture bred—give date the mare was turned into the pasture and date she was taken out of pasture.

7. Keep a copy of this report for your files and future use.

Reverse side of AQHA STALLION BREEDING REPORT

A horse without a name cannot be registered. The length of names acceptable to the AQHA is limited to eighteen characters (including letters, punctuation and spaces). An initial may not precede or follow the name and the only internal punctuation permitted is an apostrophe. The name of another horse registered with AQHA, living or dead, cannot be used.

The registered owner's name may contain initials, but must not exceed thirty characters; this rule applies to firms as well as to individuals.

AQHA computes the age of the horse by calendar years, beginning with January 1 of the year foaled. For example, regardless of the time of year the animal is foaled, it is a weanling until the last day of that year, is a yearling on the January 1 immediately following, a two-year-old the next January 1, and so on.

Since registration fees increase with age and horses not registered until three years old require a $200 inspection fee under the hardship rule, it is advisable to submit applications as early as is reasonably possible.

Incorrect certificates of registration should be brought to the attention of the AQHA and corrected. Errors should be corrected as soon as discovered to avoid accidental reliance upon the incorrect certificate. If markings or other information are incorrect, the animal may be turned away from shows or sales. The association furnishes forms for corrections. If the error is serious, the mistake should immediately be called to the association's attention by telephone or letter.

A lost or destroyed certificate can be replaced with a duplicate. The association will require notarized affidavits from the recorded owner and all persons who have knowledge of the circumstances. Four photographs (each side, front and rear views) are required and a fee must accompany the request. Proper identity of the horse and owner must be established to the satisfaction of the association.

In the event a registered stallion is gelded, the certificate of registration together with the recorded owner's statement of the facts and date of the castration should be submitted to AQHA. The association will note the matter on its records, endorse the fact on the certificate and return it. Certain state racing officials may make such a gelding endorsement on the certificate and must report the fact in writing to the association.

The registered name of a horse not listed in the stud book of the association may be officially changed if it has not started in a recognized race or has earned less than one-half point in an approved show or contest. The fee for name change at present is $25 up to age 18 months from actual date of foaling, thereafter $100.

A recorded owner whose appendix horse is eligible for advancement to a number is required to surrender the appendix certificate before a numbered certificate will be issued.

When a registered horse dies, the recorded owner must surrender the certificate for cancellation; however, the owner is entitled to the return of the canceled certificate if he desires to have it.

It is possible to sell a registered horse without papers, but of course this must be fully understood by the buyer. In such a case the seller is required to notify the association and deliver the registration certificate to the association for cancellation endorsement and notation upon the records. The seller is then entitled to the return of the canceled certificate.

All sales, private or public auction, of registered horses must be reported to AQHA and, unless such a horse is expressly sold without papers, the seller is responsible for sending to the association the certificate of registration, a fully completed AQHA Transfer Report form (see below) signed by the recorded owner and buyer, and the $5 transfer fee. If the papers are not in order, the association may hold them until the requirements for changing the recorded owner are met. Although the manager of an auction sale may supply part of the information required from the seller, it is nevertheless the seller's responsibility.

The owner who sells a foal before applying to AQHA for its registration must provide the purchaser with a proper registration application completed as if no sale had taken place, and a transfer form unless the foal is expressly sold without papers.

Recorded transfers of horses purchased through claiming races are normally initiated by the track official. The signature of the recorded owner of a claimed horse is not required.

TRANSFER RECORD

Registered Name of Horse

Registration Number

Sire

Dam

Date Sold

This transfer report was issued to

on
Month   Day   Year

This form may be obtained from the

AMERICAN QUARTER HORSE
ASSOCIATION

P. O. Box 200
AMARILLO, TEXAS 79105

TRANSFER REPORT

THE AMERICAN QUARTER HORSE ASSOCIATION
P. O. Box 200   Amarillo, Texas 79105

_____ , 19___
DATE SOLD — IMPORTANT

_____ Number_____
REGISTERED NAME AND NUMBER OF HORSE SOLD

I, (Seller) _____ (Print)

STREET   CITY   STATE   ZIP CODE

hereby sell the above named registered horse

To (Buyer) _____ (Print)
PRINT YOUR NAME EXACTLY AS IT SHOULD APPEAR ON CERTIFICATE

STREET   CITY   STATE   ZIP CODE

PHONE_____

Location of Animal _____

We certify that the horse sold is the horse registered with the Association as described in the certificate of registration delivered to the buyer. We authorize the Association to record this transfer of ownership.

Signature of } Signed_____
the Seller   SAME NAME AS LAST APPEARING ON   REG. CERT.

Signature of } Signed_____
the Buyer

NOTICE

It is forbidden by the rules of the Association to transfer animals to any other person than the actual purchaser. Sellers doing so are liable to the penalties imposed by the Constitution and Bylaws of the American Quarter Horse Association.

TRANSFER FEE $5.00	OFFICE USE ONLY

Transfer Fee paid by_____

Certificate Mailed_____ , 19 ___

To_____

Address_____

© Copyright 1966, AMERICAN QUARTER HORSE ASSOCIATION   DP-1

AQHA TRANSFER REPORT

Horse owners sometimes die, become bankrupt, fall into receivership or lose the horse through judicial process. Any such involuntary transfer must be supported by court order, court sale, decree or appointment, of which a certified copy should be sent to AQHA and will be required before any recorded ownership change will be recognized by the association. These documents must contain the same information as do voluntary transfers. The farm legal counsel should review the court proceedings before the farm purchases a horse whose ownership is or has been involved in judicial proceedings.

An agent of the owner, a lessee or contract purchaser of a registered horse will not be recognized, and documents executed by him will not be honored by AQHA unless his power of attorney or his interest is documented in writing and on file. Likewise, if the power of attorney or other right has terminated, written proof of such termination must be filed before the owner will have complete control over dealings involving the horse.

Only a small part of the rules, regulations and information contained in the *Official Handbook of the American Quarter Horse Association* has been mentioned. Every Quarter Horse breeder should have at least one up-to-date copy of the handbook at his disposal. Questions not answered by the handbook can usually be resolved by a letter or telephone call to the association in Amarillo, Texas.

14.5 State-bred Registries. Registries for horses bred within the boundaries of a state are not usually in competition with breed associations. Actually, the state laws that set up such registries are designed to promote the state horse industry and augment the breed associations by requiring eligibility for registration or registration in the recognized breed association before the horse can be registered as a state-bred horse.

The first of these state-bred registries was started by California law in 1933 as an inducement to attract breeders of fine Thoroughbreds to California after horse-race betting was legalized in the state. Later the law was amended to provide the same advantages to breeders of fine Quarter Horses. The principal inducement to the breeders of California-bred horses is an award of 10% of the winning purses due and payable to the breeder regardless of ownership at the time of winning.

A foal cannot be registered in the California-Bred (Cal-Bred) registry unless it was conceived within the state, the foal and its dam have remained within the state from conception until after weaning and it is registered with either The Jockey Club of New York City or the American Quarter Horse Association. The law provides for Quarter Horses as follows: "The Official Stud Book and Registry of the American Quarter Horse Association shall be recognized as the sole official registry for Quarter Horses."

Within the last few years California has extended state-bred registration to Appaloosa and Arabian horses. The law supporting this expansion provides that "The Stud Book of the Appaloosa Horse Club shall be recognized as the official registry of Appaloosa horses. The Stud Book of the Arabian Horse Club Registry of America as approved by the International Arabian Horse Association shall be recognized as the sole official registry for Arabian horses. The [California Racing] Board shall, however, by rule provide for the registration of California-Bred Appaloosa horses and Arabian horses. . . ."

Many large eastern farms, especially Thoroughbred farms, have moved to California in order to take advantage of the continuing monetary returns. The California-Bred Registry idea has been very successful and is showing influence over events other than racing. At present there are California-Bred sales, futurities and halter and performance classes in which a horse cannot compete unless registered in the California-Bred Registry.

The California-Bred Registry for Quarter Horses is maintained by the Pacific Quarter Horse Association, P.O. Box 5253, Sacramento, California 95817. The form presently required for registration of a

California Bred Quarter Horse Application For Registration
(Effective Jan. 1, 1969)

Address all Communications to:

PACIFIC COAST QUARTER HORSE ASSOCIATION
P. O. Box 4822 • Sacramento, California 95825
Phone (916) 922-9857

Cal Bred #	Date Rec'd
Seal on Cert.	Fee Paid $
Cards Made	

Do not write here

NAME .... AQHA REGISTRATION NO. ....
(Do not complete if AQHA Certificate has not been received)

COLOR .... MARE ( ) FOALING DATE ....
CHECK ONE { STALLION ( )   Month   Day   Year
{ GELDING ( ) EXACT LOCATION OF FOALING ....

SIRE .... # ....

DAM .... # .... DAM'S SIRE .... # ....

OWNER OF STALLION .... ADDRESS ....
(When bred to mare)

FARM WHERE SIRE STOOD WHEN BRED TO MARE ....

AFTER DROPPING THIS FOAL,
TO WHAT STALLION WAS MARE BRED BACK .... DATE ....

FARM AT WHICH HE THEN STOOD ....

OWNER OF FOAL ....

ADDRESS ....
(Street)   (City)   (State)   (Zip Code)

CAL BRED BREEDER OF FOAL ....
(Owner of mare at the time of foaling)

ADDRESS ....
(Street)   (City)   (State)   (Zip Code)

☐ Please telephone a CAL-BRED number down to the race track immediately upon receipt of my completed application and fees.

☐ Please stamp the registration certificate on your next trip to the race track.

The AQHA registration certificate for this animal is located at the following race track: ....

☐ I will send the registration certificate to the PCQHA office.

RULES FOR REGISTRATION

ALL QUARTER HORSES BRED IN CALIFORNIA MUST BE REGISTERED WITH THE PACIFIC COAST QUARTER HORSE ASSOCIATION BEFORE THEY WILL BE ALLOWED TO START IN ANY CALIFORNIA-BRED RACE. In order for a breeder to legally claim the 10% breeders' bonus for a California-Bred winning a race within this State, such horse shall, previous to being entered in such race, have been registered and certified as a California-Bred Quarter Horse with the Pacific Coast Quarter Horse Association. The Breeder, owner or his agent is eligible to file application.

CALIFORNIA-BRED DEFINED

"California-Bred" shall be deemed to be a foal dropped by a mare in California after being bred in California and remaining in the state until the foal is weaned.

PCQHA MEMBER FEES FOR REGISTRATION OF FOALS

A charge of $10.00 will be made, to be paid at time of application, for each foal registered by December 31 of the foal's weanling year (year foaled), $20.00 for each foal registered by December 31 of the foal's yearling year; thereafter the registration fee will be $40.00.

PCQHA NON MEMBER FEES

Weanling $20.00; yearlings $30.00; 2 year olds & older $50.00. Application may be made by mail or in person at the office of the Pacific Coast Quarter Horse Association. Application blanks must be completed legibly in ink or on typewriter and should be accompanied by fee and by American Quarter Horse Association certificates. IF DESIRED, APPLICATION CAN BE MADE IN ADVANCE OF RECEIPT OF AMERICAN QUARTER HORSE ASSOCIATION CERTIFICATE TO ESCAPE PENALTY FOR LATE FILING.

WARNING

Neither the mare nor the foal may be removed from California from the time of original service until the foal is weaned. The owner of the dam at the *time of service* is the *A.Q.H.A. BREEDER*. The owner of the dam at the *time of foaling* is the *CAL-BRED BREEDER*. The applicant assumes full responsibility for proper identification of the horse as a California-bred; and agrees that if it should later be proved ineligible, all monies won in California-bred races and all breeders' awards will be forfeited; and agrees to repay such monies on demand by the Pacific Coast Quarter Horse Association, Inc.

I hereby certify that the foal herein described was dropped by a mare in California after being bred in California and remaining in the State until said foal was weaned; that according to the Provisions of Section 19565 of the Business and Professions Code of the State of California this is a California-bred foal.

Signature ....
Check one
( ) Owner   ( ) Agent for Owner
( ) Cal-Bred Breeder   ( ) Agent for Cal-bred Breeder
Witness ....
Street ....
City and State ....

CALIF. BRED APPLICATION FOR STATE REGISTRATION

Quarter Horse in the California-Bred Registry is shown above and must be very carefully and accurately completed and signed.

The state-bred concept has been so well received by breeders that other states are enacting similar laws. Oregon has recently begun the Oregon-Bred Registry.

Breeders who raise horses qualified for state registries should not forgo the opportunity of increased income or sacrifice eligibility for the events restricted to state-bred horses.

14.6 Contracts Generally.

a) The farm cannot operate successfully or establish that it is more than a hobby if its legal affairs and contracts are concluded by handshakes. Good business demands that legal rights and obligations be documented and preserved in writings that clearly embody all of the terms and conditions of each transaction and that meet the legal requirements as to form, substance and execution (signature). A contract that is ambiguous or uncertain in any point will be most strongly construed against the person who drafted or insisted upon the ambiguous portion as written, if a contest over its meaning should arise. A legal advisor familiar with the horse industry should be consulted on all unusual matters that involve rights or obligations.

The sole purpose of this section is to familiarize the breeder with the types of contract usually encountered that pertain to horse breeding (servicing contracts), boarding, leasing, selling, training and exhibiting. The breeder, unless he is an attorney, should not draft a contract or accept a contract without the advice of the farm counsel. The most practical arrangement for establishing the contract forms to be used frequently is to develop them in consultation with the farm counsel. Thereafter any deviation for a specific transaction should be reviewed by counsel before the contract is signed or any part of the payment is made or accepted; in law acceptance of part or total payment may be as binding as if the contract were signed.

Every contract for each horse involved should contain the horse's registered name, its registration number and the name of the association in which the animal is registered. The papers should be signed by the registered owners or lessees and should show the complete addresses and telephone numbers of the contracting parties. The complete terms, which include commencement and termination dates, consideration (price and other benefits), and conditions, provisions, forfeitures, options and guaranties clearly set forth and leaving no room for reasonable minds to differ as to meaning, should be included in every contract. A farm that does an extensive business is almost sure to be required at some time to support its legal position in a court of law. A well-drawn, unambiguous, lawful agreement is the best assurance for friendly relationships and is indispensable when controversy arises.

b) *Stallion Service Contract.* A Stallion Service Contract form is needed by every farm standing a stallion to outside mares. Some forms in use are extremely elaborate, leaving nothing to be implied, while others are quite concise and leave much to be implied from the custom of the industry. Some forms take a position between extremes, thus implying usage and custom of the industry except for those terms expressed in the contract which are contrary to usual procedures or requirements. No opinion about the sufficiency of any form in use is intended; however, the following matters are suggestions that should be considered for inclusion in a form unless the farm and its counsel decide otherwise.

In addition to proper identification of the stallion and mare, the period and place for breeding, the fee, date and method of payment and the execution of the contract by the registered owner or lessee of each horse, the Stallion Service Contract may properly include provisions concerning the following:

1. The time of delivery of the mare to the breeding place and consequences of failure to deliver her;
2. A warranty of the mare's breeding soundness;
3. A veterinarian's certificate of the mare's breeding soundness and payment of the veterinarian's fee;
4. A warranty that the mare is manageable and a statement of dangerous traits or tendencies the animal may have or is suspected of having;
5. The health care of the mare, including treatment, worming and

other preventative measures and the payment of veterinarian fees for such care necessary or desirable for the mare's well-being during the breeding stay;

6. Boarding care for the mare and foal at side, and the charges and time of payment. (Boarding charges for mares alone usually run from $2.50 to $3.50 per day, with charges for other care added. Boarding care for a valuable Thoroughbred mare may be well over $100 per month, due to the special attention demanded);

7. Health care of foal accompanying the mare and the charges and time of payment;

8. The death or unfitness for breeding of either the mare or the stallion. (Many contracts provide that in either event the contract becomes null and void);

9. The sale or termination of a lease of either horse and the consequences of such occurrence;

10. Foal guaranty by stallion owner. (Guaranties are in the nature of return privileges or a live foal that can stand and suck. Return privilege clauses are often written so that the mare not in foal may be returned for the remainder of the season, or she or a substitute may be returned until a live foal is produced. Live foal guaranties often do not require payment until a foal that can stand and suck is produced; however, some clauses call for full payment when the mare is pronounced in foal with the right to return or refund if the foal is born dead or the mare is empty and the fact is certified by a veterinarian. Flat breeding fees are not often encountered, because the entire risk of conception without return privileges or refund is upon the mare's owner);

11. Removal of stallion from the area before the guaranty of a foal is fulfilled;

12. Refund, if any, if a mare is moved out of the area without having conceived;

13. Castrating a colt produced. (In an effort to reduce the number of mediocre stallions, some contracts provide for a return of part of the fee if a colt is castrated within a certain time limit);

14. Disclaimer of stallion owner's responsibility for accident, sickness or death of the mare or the foal accompanying her resulting from any cause except gross neglect or intentional harm or abuse;

15. Authority of the stallion owner to exercise his own judgment in the care and treatment of the mare and her accompanying foal;

16. Obligation of mare owner to pick up the mare and foal at side when the time arrives and to pay charges of transportation for the mare and foal if for any reason transportation by the ranch is necessary;

17. Occasionally the stallion owner is to receive a coownership interest in the foal produced, in place of the stud fee. Such a contract ordinarily requires special preparation. It is not advisable to attempt to modify the common stallion service contract form to do this. Counsel should be consulted in such a specific case;

18. It is wise to include an agreed value of the mare so that it cannot be disputed in a later controversy;

19. Contracts usually do not require the stallion owner to agree to deliver the Breeding Certificate or other papers required for registration of the foal. The custom of the industry requires this cooperation and the breed association rules demand it.

The above suggestions need not be included in every contract, but should be viewed as a check list that will reduce the chances of over-looking an important matter. Many farms cover most of these points orally with the mare owner and use a less-complicated form of contract; however, when very valuable animals are involved, written agreements can prevent many costly arguments.

A simple form of Stallion Breeding Contract in use is shown below as an example, but is not necessarily adaptable for use by all farms. The farm counsel should be the one who develops the contract form for farm use.

c) *Boarding Contract.* From time to time farms will be boarding outside horses. Boarding agreements for visiting broodmares are almost always incorporated in the Stallion Service Contract. Separate boarding contracts should be required for other horses and should be signed by the owner of the horse and a ranch representative.

The contract may provide for:

1. The price of board and due dates;
2. A lien on the horse for unpaid due bills, and method of enforcement;
3. Health care, medical and veterinarian costs;
4. Special care and unusual feeding requirements;
5. Notification to owner in case of death, sickness, injury, straying or theft;
6. Discretion of farm in treating the horse in an emergency or in the event the owner does not respond to the notice;
7. Nonresponsibility of farm for death, sickness, injury or damage to the horse unless it is the result of intentional harm, gross neglect or abuse;
8. Insurance coverage of horse and public liability for damage done by the horse;
9. Agreed value of the horse so that the question of value is settled in advance of any dispute;
10. Any right of the farm to the use of the horse;
11. Written authority from the owner to the farm before the horse may be used by or delivered to anyone else, but protecting the ranch in case of trespass or theft;
12. Statement of dangerous traits or tendencies the animal may have or is suspected of having.

d) *Breeding Leases of Stallions or Mares.* Stallions and mares are often leased for breeding purposes. Such leases are perhaps the most complicated documents of all that pertain to horses. Usually these must be tailor-made by the farm counsel for each case. Since the lessor is generally deprived of all physical control and possession of the animal,

HADAN LIVESTOCK COMPANY

P. O. BOX 93 - LOCKEFORD, CALIF. 95237

Owners Phone (209) 727-5234 • Stud Manager (209) 727-5281

STALLION SERVICE CONTRACT

This certifies that ....
(Mare Owner's Name)

Address ....

has engaged one service to the stallion ....

for the mare .... Wet ☐
                  Mare's Name   Number   Dry ☐
                                      Maiden ☐

....
     Sire of Mare   Number   Dam of Mare   Number

for the 197........ season at $...

1. $........................ booking fee must be paid when mare owner returns this contract.

2.   Balance must be paid upon owner's receipt of veterinarian's certificate that mare is safe in foal (45 day check). Mare must be picked up at ranch immediately after receipt of pregnancy statement, unless prior arrangements have been made.

3.   Mare shall be in healthy and sound breeding condition, as certified by veterinary examination at ranch upon arrival. Owner authorizes and agrees to pay for this examination. Mare owner further agrees to promptly ·pay all veterinary expense (i.e., worming, treating for infection, preventative measures, etc.) necessary to insure well-being of mare. It is also agreed that there will be a charge of $25.00 for foaling mare out at ranch prior to service.

4.   All mares must be halter broke and gentle.

5.   Board, veterinary, and other expenses must be paid monthly.

6.   All board and other bills must be paid in full before mare departs.

7.   Hadan will care for mare and/or foal in a diligent manner, and reserves the right to employ veterinary care at owner's expense in case of illness or injury. Owner agrees that Hadan will not be responsible for accident, sickness or death to owner's mare and/or foal.

8.   Hadan guarantees a live foal from this mating, meaning the foal shall stand and suck. Should it not, owner will be entitled to a return privilege for the same mare the following season only. This guarantee will apply only if Hadan is notified, by registered mail, within 48 hours of foal's death. Statement of details by a licensed veterinarian must follow within 10 days or guarantee becomes null and void. This clause also binding in case of abortion.

9.   Should said stallion die, become unfit for service, be sold, or should the lease on this particular stallion terminate, stud fee or deposit paid will be refunded.

10.   When Hadan Livestock Company signs and returns one copy of this contract to mare owner, it will then be a binding contract on both parties, subject to all above terms and conditions. This contract shall not be assigned or transferred by mare owner.

Owner's selection of care (check one):   Signed:

☐ Mare in pasture - $1.50 @ day.

☐ Mare and foal in pasture - $2.00 @ day.   ....

☐ Mare in private paddock - $2.50 @ day.   Owner of Mare

☐ Mare in private stall and paddock - $3.00 @ day.   Date: ....

                                      Hadan Livestock Company

STALLION SERVICE CONTRACT

the lease should elaborately provide for the rights, duties and responsibilities of both lessor and lessee and the protection of the lessor against damage done by the animal or to the animal for the duration of the lease.

As in any other contract, a lease should carefully identify the parties and the animal leased. It should state the purpose, place where the animal is to be kept, dates of commencement and termination of the lease, date of delivery of the animal, consideration (rental) to be paid to the lessor and the dates of payment. Many leases require that a deposit be left with the lessor as security for performance of the lease until the animal is returned in approximately the same condition as delivered.

Options of renewal or purchase are sometimes included and should provide the procedure for exercising the option together with any credit of past rental to be applied upon the purchase. Any warranties made by the lessor should be spelled out in the lease and the lessee should acknowledge the "as is" condition of the animal when accepted. The obligations for delivery and return as well as the responsibility for payment of transportation should be stated. The lease should commit the care and welfare of the animal to the lessee, who should agree that he will care for and maintain it according to good practice and the usual standards of the industry. The lessor should be allowed reasonable access to the horse and premises where it is to be kept so that he may inspect the health, well-being and conditions in which the horse is maintained. The lease should provide that the lessor is to be free from all costs and expenses whatsoever incurred for the horse during the lease and that the lessee will pay all expenses incurred for the horse, including any property taxes levied on or because of it. A clause stating the responsibility for promoting the animal during the lease term and payment of that expense is often included in the lease. Usually insurance upon the horse is maintained and paid for by the lessor, which should be so stated, but the lessee should agree to do no act that might cause a forfeiture of insurance or establish a defense for the insurance carrier. The lessee should be furnished a copy of the insurance policy so that he will know its conditions and terms. The lease should contain a provision against assignment or subletting unless approved in writing by the lessor, and should provide that the lessee will not mortgage, pledge or sell the animal and that he will keep the animal free from all liens. The lease should very clearly cover the consequences and rights of the parties in the event the horse dies or becomes permanently or temporarily unfit for use for purposes stated in the lease. Such events present questions of whether the lease nevertheless continues in full force, the rents cease or are to be prorated, or the lease is wholly terminated, leaving the parties in the position where they were when the event occurred, and of what happens to the lease security deposit. A provision for immediate notice to the lessor in the event of death, sickness, injury, theft or straying, and in the event that the animal is involved in any damage for which the lessor or the insurance carrier might be liable. The lease should also provide what action may be taken by the lessee in the event that he is unable to reach the lessor for notification. An indemnity (protection) clause in favor of the lessor should be included, whereby the lessee agrees to indemnify the lessor against any matter for which he is responsible under the contract. The lessor should insist upon a provision whereby he warns the lessee of all peculiar traits or dangerous tendencies the animal may have or is suspected of having. The lease should provide for the lessor's obligation to do all things and execute all papers required of him to enable the foals gotten during the breeding lease to be registered, and should grant permission of the lessee to file a copy of the lease with the breed association and present the registration certificate for endorsement of the lease upon it. The lessor may not wish the animal to be bred indiscriminately and to avoid that possibility should include a clause limiting use to certain named animals or a class of animals such as AQHA numbered horses. The lessor of a mare may not want her to be bred on the foal-heat; if so, the contract should so provide. The lessor wishing to maintain a certain level of stud fees for his stallion may provide that the lessee shall not use the stallion for service at a fee below the stated amount, which must be charged in its entirety to the owner of the mare bred. Some stallion leases provide, in addition to the flat rental, for a division of stud fees between the lessor and lessee. It is not uncommon for a lease to provide that all stud fees will be divided on the basis of 50% to each and that the lessor's share is due when earned, regardless of whether or not collected by the lessee. Whatever arrangement is made must be clearly set out. As in other contracts granting possession and control of an animal, the lease should contain an agreed value of the animal so that value is beyond dispute in the event of litigation. Disputes are quite likely to arise over these complicated leases and to avoid litigation an arbitration clause may be advisable. Such a clause usually provides that a dispute is to be submitted to arbitration by three persons—one appointed by each party and the third to be selected by mutual agreement of the two appointed—and that their majority findings shall be binding on the parties. Many leases provide for attorney's fees to the prevailing party in the event of litigation.

The above suggestions demonstrate the intricacies of breeding leases and the need to incorporate special provisions to cover all material matters. Leases may be construed according to the usage and custom of the breeding industry unless a matter is expressly and clearly covered by the lease.

e) *Bill of Sale.* Horses may be sold and transferred on a cash basis by the use of an ordinary Bill of Sale used for other types of personal property. However, it is good planning to develop a printed form that provides blanks for the special identification of a registered animal, so that the inclusion of its registered name, registered number and other desirable information cannot easily be overlooked. Such a form should be approved by the farm counsel before being printed. AQHA provides a suitable form, as shown below.

f) *Installment Sale Contract.* Installment sale contracts require the same identification as Bills of Sale, but are more involved since they provide for future installment payments and usually for interest upon the unpaid balance. If the sale requires four or more future installments (whether interest is charged or not), the transaction comes under the Federal "Truth in Lending" law, which must be complied with in order to avoid severe penalties. Some states have also enacted laws similar to the federal law, which must also be considered. Usury is a charge for the use or forbearance of money or credit in excess of the annual rate allowed by the state law. If the installment contract contemplates interest or charges in excess of what the price for cash would have been, the applicable state usury law must be considered in order to avoid the consequences of a violation. Installment contracts are the lawyer's business and should not be undertaken without his approval of the specific transaction.

g) *Training and Exhibiting Contract.* Training and exhibiting con-

BILL OF SALE

DATE_____

_____
SELLER

_____
BUYER

_____
BUYER'S ADDRESS

CITY _____ STATE

DESCRIPTION:_____

_____

_____

SELLER RETAIN THIS STUB FOR YOUR RECORDS.

This form may be obtained from the

AMERICAN QUARTER HORSE
ASSOCIATION
P. O. Box 200
AMARILLO, TEXAS 79105

BILL OF SALE

I, (Seller) _____ (Print)

STREET   CITY   STATE

sell a _____foaled_____
COLOR   SEX   DATE

To (Buyer) _____ (Print)

STREET   CITY   STATE   PHONE NUMBER

(If Animal Sold Is An Unregistered Foal:)

By _____
SIRE   REG. NO.

Out
Of _____
DAM   REG. NO.

(If Registered:)

Name _____
REG. NO.

_____   _____
DATE   SIGNATURE OF SELLER

© Copyright 1966, AMERICAN QUARTER HORSE ASSOCIATION   DP-3

AQHA BILL OF SALE

tracts tailored to the special needs will be required by most farms. Usually the contract is developed by the professional trainer or exhibitor in consultation with the farm. These contracts require the same care as do leases and should be submitted to the farm counsel before they are signed. The custom and usage of the trade will be implied unless the subject is expressly covered in the contract.

14.7 Marketing. The ultimate goal of horse production for profit lies in marketing. All things that draw the purchasing public to the farm and increase the desirability and value of the animals produced are part of marketing. The successful farm is always conscious of its reputation and uses every opportunity to present the farm and its animals in a favorable light. The progressive farm is very active within the breed and regional horse associations. Such organizations not only sponsor shows, sales and races but also organize tours, lectures and conventions to help keep its members abreast of the latest developments in the horse industry. The ranch should be especially meticulous in maintaining proper registration procedures, since only registered horses have officially recognized pedigrees and only registered horses have the opportunity to establish official records of performance and merit maintained by the breed association in which the animals are registered. Horses having outstanding performance records are the most desired by the buying public. Each breed association has a system of grading for authorized events. These systems are usually based upon point awards ascribed to the various events, but the number of points is often determined by the number of entries in an event and the show classification, which depend upon the number of horses that participate in the show. All of these things are covered by the association's rules and regulations. Points earned are credited toward the total needed for championships and other achievement awards offered by most breed associations. Although horse racing is controlled exclusively by state laws and the state racing boards, some associations do recognize racing performance for point purposes. Some types of AQHA awards require that a certain number of points be earned in racing. Referring further to the AQHA, some of the most coveted titles and awards in order of importance are as follows:

AQHA Supreme Champion

AQHA Champion

AQHA Superior (naming event) Horse

Annual event champions receive the title of

19— Champion (naming event) *Horse*

Honor Roll awards

Registers of Merit for racing or other events

Some awards entitle the horse owner to an appropriate trophy. All awards are noted on the association registry and each is certificated to the animal's owner. The association publication carries the announcement of all major achievements.

Awards are only a few of the many ways in which the farm reputation and the value of the individual horses may be enhanced. The breeding farm must have the official up-to-date breed rules in order to remain aware of the current rules and regulations governing the performance and registration aspects of the industry. Ranches raising running horses should also have the latest official information on racing in the states where participation is anticipated.

Achievement awards should be displayed in the most advantageous location. An attractive trophy case, placed in equally attractive surroundings where visitors are sure to see it, does much to make known the farm successes and quality of the horses. Ribbons and certificates too are very impressive to many segments of the public and should also be displayed.

The individual file kept on each animal should contain a record of its awards so that no achievement is overlooked or misstated at the bargaining table or during the preparation of advertising.

Participation in sales, shows and races brings the horses and the farm name to the attention of the buying public. Favorable notoriety is a big part of marketing. The farm should participate in as many events and as often as it can furnish animals with reasonable expectations of impressive performances. A poor performance can be very detrimental to the farm's reputation. For this reason animals should never be shown at large shows until they are at their best.

Planning for performance events is a painstaking task that must be scheduled well in advance. Good planning cannot be accomplished unless the farm is familiar with the dates, places and types of events most beneficial for its horses. Racing and show dates are usually set aside far enough in advance for the farm to be selective; however, it is imperative that the farm receive early notice. It is well worth while for the farm to be included on the mailing lists of the state racing board secretary and all organizations that sponsor contests, sales and shows in which the farm may participate, so that announcements and entry blanks are forthcoming automatically. Subscriptions to breed magazines are the most reliable sources of advance information and help the farm keep abreast of current breed activities and government legislation affecting the industry. Early nomination is a good precaution to prevent limited events and futurities from slipping past. In fact, early entry for any event will avoid the disappointment of rejection because of late application. Entries for major shows are usually closed well before event day. It is wise for the farm to maintain an up-to-date card catalogue of events that contains all meaningful information, including the name, address and telephone number of all shows, sales and race meets of interest.

Advertising, whenever and wherever it can be done with reasonable economy and good effect, is an indispensable part of marketing. Imaginative advertising is a highly specialized field and will not be discussed in detail; however, it is such an integral part of successful horse production that a few comments seem appropriate.

General, favorable notoriety can be developed rather inexpensively by displaying the farm name, location, telephone number, and the breed raised wherever it will be noticed. The well-kept transporting vehicles with information attractively displayed on the sides and rear identifying them with the farm, advertises wherever it is seen—on the road, at shows, sales and races and in towns. Displays upon vehicles that move must be clear, concise and easily readable if their full usefulness is to be realized.

The farm entrance is extremely important. If the farm fronts upon a heavily traveled route, its sign must be designed so that the main message can be perceived at a glance. This point is not quite so important where the traffic is light and travel is considerably slower. Farm entrances are commonly emphasized with a decorative archway, gazebo or impressive stanchions. Horse figures tell the story more quickly than words and, if handsomely done, are very effective. Life-size horse models are available commercially and are prominent eye-catchers. If these are used, it is suggested that they be mounted as high as possible and securely fastened down to prevent theft.

Highway direction signs inviting visitors and attractive, simple advertisements are advisable along main roads well before the farm will be reached. Some ranches arrange for road signs many miles distant from the farm entrance or turn-off road.

Farms standing noted stallions find it beneficial to incorporate them into the advertisements. They do this with pictures and short statements such as "Home of Great Guy," "Big Boy Standing Here." Even the names of deceased famous horses can add color to the advertising. If a farm has based its breeding program on an outstanding sire, now deceased, such as Man-O-War, Old Sorrel, Poco Bueno, or any other notable stallion, it would be foolish to overlook the color and prestige the name will add to the farm reputation.

Advertising does not stop at the gate; as stated before, the achievement awards should be on display. Stall doors should decoratively display pedigrees, records and exceptional accomplishments of the occupant, whether it is a stallion or a mare.

A prominent sign is a source of advertising as well as a convenience for visitors or customers.

Personnel is representative of the farm. Manners, friendliness and appearance are indicative of farm quality. Whatever task the help is engaged in of course dictates to some extent personal appearance, especially in regard to clothing, but appropriate dress and good grooming in keeping with the job and good taste add much to the personality of the ranch. Bad habits, unnaturally loud voices, profane language and the use of alcohol or cigarettes in the barn area will lower or destroy the desired farm image.

The handling of animals, whether on the farm or off, must always be professional. Unruly horses need training in private, not public places. Abuse, confusion, yelling, profanity, whether at public showings or at the farm, clearly demonstrate lack of adequate and proper training, unqualified personnel or general inefficiency of the farm and its manager. Animal handling, although not planned advertising, can demonstrate the superiority of the farm and, conversely, can give proof to the public of inferiority.

The farm reputation is being enhanced or downgraded whenever the farm is represented. This is especially true during off-ranch public appearances, but is also true for ranch exhibiting and public inspection. Horses and handlers must be well groomed at all times to present the farm and its produce in a favorable light. Poor grooming detracts from the farm in a general way and from the horses in particular. Horses should never be exhibited on or off the farm unless at their best. Animals in poor condition should be kept where they are not likely to be seen by visitors.

In addition to the general farm advertising, specialized advertising through magazines, newspapers, brochures, catalogues or fliers is indispensable. The need for this type of advertising may be seasonal or may depend upon the periods when public sales are being held. Stud service sales are at their height just prior to the breeding season. Heavy emphasis in advertising the stallion should commence well ahead of the breeding season and should not be relaxed until the stud is booked to capacity or the season is ending. Service fees are a source of income necessary to the breeding farm and if the seasonal opportunity to use the stallion to capacity is lost, the farm will have lost income it should have realized. This phase of the business is highly competitive and requires that advertising stress every favorable point of the stallion. Imaginative arrangement and organization of advertising material plays a big part in marketing stud services, but—whatever else is thought advisable—the advertising will fall short unless it incorporates the following information.

> Stallion name, breed, titles, awards, favorable photograph, pedigree, color, stud record and fee. The pedigree should relate back not less than two generations and attention should be called to superior relatives and breeding successes for the stallion as well as his close relations. All worthy achievements should be included. Winnings, racing times, records, competition against outstanding horses, outstanding offspring, all is information that will influence sales. The standing location, manager's name and telephone number are indispensable and a direction map is often helpful, especially if the farm is off of the main highways.

Special advertising preceding a public auction far enough in advance to allow prospective buyers ample opportunity to investigate the farm, its reputation and quality of stock usually has the effect of encouraging bidding activity and consequently brings higher sale prices. Horses of the better-known farms are in more demand and bring prices consistently higher than those of the farms that have not made efforts to catch the buyer's attention prior to the auction.

The ideal time for advertising foals is in the spring of the year when the foal crop is dropped. An eye-catching photograph for foal advertising is that of a young, healthy foal in peaceful country surroundings, or perhaps one of a mare and foal taken to emphasize the foal.

Advertising in general news media or magazines of general interest does not usually prove fruitful. The horse-minded public looks almost exclusively to the one-purpose media specializing in horses. Breed magazines, general horse magazines and livestock journals or newspapers are the most likely to reach the industry's market as well as the horse-minded public.

14.8 Farm Records. The efficiency of the breeding farm depends in large measure upon accurately kept, currently maintained and frequently used records. There are as many different record-keeping systems and forms as there are horse ranches. The information recorded varies greatly, depending on the type of animal raised, size of the farm, event participation, tax structure and manager's preference. No matter

The ranch office reflects the efficiency and progress of the enterprise. A library of information, closed-circuit television, records of ranch animals, awards, photographs and pedigrees can be housed in the office and business agreements may be transacted in professional surroundings.

what kind of information is kept, all successful farms maintain complete, accurate, up-to-date records that meet their particular needs. Many of the more commonly encountered records and forms are discussed and illustrated in the following pages. It should be noted that while the forms presented have proven useful for many farms, other breeders may choose to alter them to better suit their particular operations.

1. A folder (not illustrated) for each ranch horse containing all pertinent information should be maintained in the main office of the ranch. Included in the folder should be purchase papers, registration papers, pedigree, achievement records and awards, health records, advertising copy, breeding records, sale papers and all other information that could have a bearing upon marketing, future health problems, training, performance and campaigning, and a notation of traits peculiar to the animal, such as the tendency to rear, kick, or bite.

2. The Manager's Daily Log (illustrated at the end of this chapter) is a record of the overall daily activities of the farm. It is useful in routine chores and contains information needed for entry into other permanent records.

3. A Location Map (not illustrated) of all animals should be maintained. The most convenient arrangement is a farm map showing buildings, paddocks and pastures, mounted upon a cork bulletin board where the name or other identification (perhaps an assigned farm number) of each horse is pinned at its place of location and moved as the horse is moved. Without such a ready reference, much time is lost and confusion results. A farm having many horses and receiving outside mares as well may have difficulty in identifying and locating them. Many horses look very much alike and even those with distinctive markings are difficult for ranch personnel to keep in mind. An excellent aid to identification of outside mares in particular is to assign a number to each mare, which is placed upon the mare by closely clipping the hair in an obvious place such as the shoulder, hip, or the like. The number should be large enough to be seen from a distance and of course will fade as the hair grows but will usually be evident for the duration of the breeding season (see Section 7.6).

4. A Teasing and Breeding record (illustrated) for each mare is absolutely necessary for breeding operations. Two copies should be kept, one on cards in the barn office for use there, and one for the main farm office for its use and as a permanent record.

5. An Outside Mare—Individual Record (illustrated) should be kept in the main ranch office for each outside mare, as well as a Stall Record (see 6 below for Stall Record).

6. In addition to special records needed, two general records on each resident horse (ranch-owned horse) should be accurately maintained. These are the Ranch Horse—Individual Record (illustrated) kept in the main office, and the Stall Record (illustrated) which originates on the stall door but is turned in to the ranch

main office as it is replaced. While the stall record is not necessarily replaced on a daily basis, it should be replaced often enough so that the main ranch office has up-to-date records of all animals.

7. A Brood Mare Record of Produce (illustrated) should be kept for each ranch brood mare, but of course cannot be maintained for outside mares.

8. Special records kept on stallions are the Stallion Breeding Report and a Stallion Semen Report, both of which are illustrated.

9. The nursery history of each ranch foal should be maintained up-to-date as long as the foal is ranch property. A blank Nursery Record form is illustrated and serves quite well if entries are not allowed to lapse.

10. Each veterinary visit should be supported by a report, as illustrated. Some veterinarians provide their own special forms. Regardless of the form used a report must be obtained at the time of the visit, the items of concern should be entered on the record of each horse involved and the report should be retained for further medical reference and for cross-checking against veterinary bills.

11. Pasture records (not illustrated) showing soil conditions, soil treatment, grass species and all other pertinent information, together with rotation dates and periods of grazing is extremely helpful in planning pasture care and use. Visual pasture inspection coupled with accurately kept pasture records leads to efficient, money-saving pasture maintenance and adequate high-quality green feed.

Almost as important as farm-operation records are marketing records. The office should maintain a card catalogue with name, address and telephone number of each past or prospective patron and organization having an interest in purchasing or a market outlet for horses or stud services.

A calendar record of upcoming performance events and shows will greatly facilitate the selection of events to be attended, timely nominations of entries and scheduling of training and transportation.

MANAGER'S DAILY LOG

Date: _____

Stallions bred _____
(stallion names)

_____

Mares serviced _____
(mare names)

_____

Mares foaled _____
(mare names)

Nighttime attendant to observe these foaling mares _____

_____
(mare names)

Bookings requested _____
(mare owner's names)

_____

Horses arriving _____
(names)

_____

Horses departing _____
(names)

_____

Horses put in training _____
(names)

_____

Horses taken out of training _____
(names)

_____

Horses purchased _____
(names)

_____

Horses sold _____
(names)

_____

Supplies to order _____

_____

Supplies received _____

_____

Visitors _____
(names—total number)

Remarks _____

TEASING AND BREEDING RECORD
(19— Breeding Season)

Name of Mare _____ Reg. No. _____

Sire _____ Dam _____

Owner _____ Age (years) _____

Breeding problems and characteristics: _____

_____

_____

Date	In	Out	Bred to	Date	In	Out	Bred to
March 1		X					
" 2		X					
" 3		X					
" 4		X					
" 5	X						
" 6	X		RED HOG				
" 7	X						
" 8	X		RED HOG				
" 9		X					
" 10		X					

Average ESTRUS period —— days.   Average ESTRUS cycle —— days.
        (number)   (number)

OUTSIDE MARE—INDIVIDUAL RECORD

Name _____ Reg. no. _____

Sire _____ Reg. no. _____

Dam _____ Reg. no. _____

Date foaled _____

Color and markings _____

_____

Owner's Name _____

Address _____Phone no. _____

Breeding history _____

_____

Remarks: _____

_____

PROGRESSIVE REPORT

Date	(Include all incidences of significance: sickness, injury, treatment, medication, hoof care, breeding, etc.)

RANCH HORSE—INDIVIDUAL RECORD

Name _____ Sex _____ Reg. no. _____

Sire _____   Reg. no. _____

Dam _____   Reg. no. _____

Date foaled _____

Color and markings _____

_____

Purchased from _____

Date _____ Price _____

Breeding history _____

_____

Remarks: _____

_____

PROGRESSIVE REPORT

Date	(Include all incidences of significance: sickness, injury, treatment, medication, hoof care, training, breeding, etc.)

STALL RECORD

Name of horse _____

Ranch horse _____ Owner _____
            (check)   (name)

Stall location _____ Sex _____
                                                  (mare, stallion, etc.)

Date	Feed	Breeding	Complications, medication, Hoof Care, etc.
March 1	not fed prior to worming		Tube wormed (Equizol) Breeding stitched
" 2	6 lbs. Omelene 12 lbs. alfalfa supplements		
" 3	"		
" 4	"		hooves trimmed
" 5	"		
" 6	"		
" 7	"		
" 8	"	Red Log	
" 9	"		
" 10	"	Red Log	

BROOD MARE RECORD OF PRODUCE

Name _____ Year foaled _____ Reg. no. _____

Sire _____ Dam _____

Color and markings _____

Breeder's name _____ Address _____

Purchased from _____ Price _____

Date Bred	Sire	Date Foaled	Complications of Gestation	# Days Gest.	Foal Sex	Foal Color	Foal Name	Date Weaned	Sold to	Price	Date

Remarks about offspring produced:

STALLION BREEDING RECORD

Name _____ Reg. no. _____

Year foaled _____

Color and markings _____

Purchased from _____ Price _____

Date Bred	Name of Mare	Owner of Mare	Remarks about Mare

STALLION SEMEN EVALUATION

Name _____ Reg. no. ____ _____ _____

Date examined _____Examiner _____

Motility _____

Live-Dead count _____

Morphology _____

Sperm concentration _____

Bacteriological culture results _____

_____

PH _____
(tissue fluids)

_____
Evaluator's signature

Veterinarian's recommended treatment _____

_____

_____

_____

_____

NURSERY RECORD

Name of dam _____

Name of foal _____ Reg. no. _____

Date foaled _____ Sex _____

Color and markings _____

	Weight	Height	Girth	Bone
1st year				
2nd year				
3rd year				

Treatment

Date	Parasites	Foot Corrections	Injuries
March 10	high ascarid count		
March 11	Equizol		
March 22		Trimmed front feet to correct splay foot condition	
March 24			Kicked on knee. Swelling- Ice pack for one hour.
March 25			Swelling on knee reduced. Ice pack 1 hour morning and night.
March 26			no swelling.

VETERINARIAN REPORT

Date _____ Veterinarian _____

Name of Horse	Work Performed	Name and Address of Owner	Ranch Follow-up Treatment

Medical supplies to order _____

_____

Culture report _____
      (Name of mares)

_____

GLOSSARY

Abort, Abortion   Expulsion or absorption of a dead embryo or fetus prior to the time it could survive outside of the mare's body.

Abscess   A cavity formed by disintegrating tissue that is filled with pus.

Action   Movement of the animal's legs. His way of going. How the horse moves.

Acute   Rapid and severe course of a disease.

Aerobic   A microorganism that grows only in the presence of oxygen.

Afterbirth   Placenta.

Aged horse   Technically, a horse over 8 years old. Often used to refer to a horse over 12, or smooth mouthed.

Amino acid   The main component of protein. In order to absorb protein the digestive tract must break it down into its amino acids.

Anaerobic   A microorganism that lives and grows in the absence of oxygen.

Anemia   Lack of fully matured red blood cells in the blood. Low quality or quantity of blood.

Anestrus   Time during which a mare will refuse a stallion. Opposite of estrus.

Annual plant   A plant that lives only one year or season.

Antibiotic   A substance such as penicillin or streptomycin, produced by some microorganism and capable of destroying or inhibiting growth of bacteria.

Antiseptic   A chemical that inhibits the growth of microorganisms without necessarily killing them.

Antitoxin   A substance made by the body and carried in the blood that acts against specific toxins. An injection made from blood serum, containing antitoxin.

Appendix Registry   Registry maintained by breed associations for an animal that will be given a permanent number if it proves itself a worthy representative of the breed. Until it receives its permanent number, such an animal's offspring is not eligible for registration. Example: AQHA Appendix Registry is maintained for animals with one registered Quarter Horse parent and one registered Thoroughbred parent.

Artificial vagina	A tube with properties similar to those of a mare's vagina, used to collect semen from a stallion.
Asterisk (*)	Placed before an animal's name in the official Stud Book, it means the animal is imported.
Astringent	A drug that causes contraction and drying of tissues. Example: zinc oxide, sulfate.
Atrophy	Shrinking away of the muscles due to lack of use.
Autoclave	Instrument used to sterilize equipment by steam-heat and pressure.
Bacterin	A vaccine that provides protection from a bacterial infection.
Balk	Refuse to go.
Ballooned vagina	The vagina remains distended after use of a vaginal speculum. The vagina will return to its normal configuration in a few hours.
Band (broodmares)	A group of broodmares turned loose together.
Barefoot	Unshod.
Barren mares	Mares that have produced foals previously but did not foal the last season, or are not in foal at the end of the breeding season.
Bedding	Material used on the floor of a stall to absorb moisture and provide padding. Example: straw, shavings.
Biennial	A plant that lives for two growing seasons.
Bleached hay	Hay that has been exposed to the sun too long and has lost its green pigment.
Blemish	An unsightly scar or lump that does not impair the performance of the animal.
Blister	The application of a caustic blistering agent to the skin to induce a greater blood supply to the area.
Bloodlines	The ancestry of an animal.
Bloom	Hair that is very healthy, clean and shiny.
Bluestone	Copper sulfate.
Board	The cost of caring for the daily requirements of a horse—feed, bedding, pasture, etc.
Bolting	Eating very rapidly without chewing adequately. Also refers to an animal that is running away.
Bone	The circumference of the front cannon bone measured half way between the knee and the fetlock. A large measurement or "good bone" indicates large bone size and adequate distance between the flexor tendon and the bone itself.
Bottom	The amount of endurance the animal possesses.
Bottom side	Dam's side of pedigree.
Brand	An identifying mark burned into an animal's skin by use of a hot iron.
Breed	A group of animals with common origin and identifying characteristics that are consistently passed on from one generation to the next.

Breed Association	An organization formed to keep records of animals belonging to a particular breed, to issue registration certificates to such animals and to promote that particular breed.
Breeder	Owner of mare at the time of service. Thus, the person who makes the choice of stallions.
Breeder's Certificate	Document given to the breeder at the time of service certifying that the stallion owner bred a particular stallion to the mare on a specified date or dates. The Breeder's Certificate must be presented to the breed association at the time of application for registration of the resulting offspring.
Breeding	The act of copulation between a stallion and a mare; also refers to the animal's pedigree.
Breed publication	A magazine or newspaper that concerns itself with one particular breed. Example: "The Blood Horse" concerns only the Thoroughbred.
Breeding roll	A tube-shaped pad placed just above the penis during breeding to prevent the stallion from penetrating too deeply and damaging the vaginal lining.
Breeding shed	Covered area in which animals are bred.
Breeding stitch	If a mare has been sutured to prevent windsucking, a "breeding stitch" is usually taken at the lower end of the sutured area of the vulva to prevent tearing during breeding. Once the mare is bred, the stitch is removed.
Breedy	Refined, well-made, particularly around the head, forequarters and legs.
Broad-spectrum drug	A drug effective against a number of different types of disease-causing organisms or parasites.
Broodmare	A mare kept for the sole purpose of producing foals.
Bulk	Indigestible fiber found in feed. A certain amount of bulk is necessary to help move feed through the digestive tract.
Bull pen	Auction sales ring.
Caecum	Enlarged portion of the horse's large intestine, analogous to the appendix of the human, where fermentation takes place.
Calcium-phosphorus ratio	The amount of calcium compared to the amount of phosphorus in the diet. A ratio somewhere between 1:1 or 2:1 is conducive to proper bone development.
Calked shoes	Shoes having projections downward from the toe or heel to provide better traction. Sometimes used for corrective shoeing.
Carotene	A substance found in green and yellow feeds such as green grass or corn, which is converted to Vitamin A in the animal's body.
Caslick's operation	Suturing the upper portion of the vulva of a mare to prevent windsucking.

Cast	A horse that has lain or fallen too close to a fence or wall and cannot get up without assistance.
Castration	Removal of the testicles of a male horse.
Catch pen	A small pen built into a part of a large pasture to aid in catching loose animals.
Cat-hammed	Poorly muscled in the hind quarters, especially on the inside of the hind legs.
Chromosomes	Rod-like structures found in the nucleus of a cell during cell division, which contain the genes that produce the animal's characteristics.
Chronic	An ailment or habit that lasts for a very long time or is continually recurring.
Claiming Race	A race in which the horses are for sale. A person may claim or purchase the horse prior to the start of the race.
Climax grasses	The grasses that will dominate an area if there is no interference from man or grazing by animals.
Closebreeding	The breeding of very closely related animals such as sires to daughters or brothers to sisters.
Coarse	A horse lacking refinement, breeding and quality, having common head, hairy fetlocks, flat feet, puffy legs, etc. A coarse feed has a high fiber content.
Cocked ankles	The fetlock joint is bent forward so that the pastern is at about a 90° angle to the ground. The hoof toe is short and the heels are long.
Cold-blooded	An animal with no Oriental or Eastern blood. Such animals are usually quiet, easygoing and not especially excitable.
Colicky	Animal exhibiting signs of pain or discomfort in the abdominal or chest cavity.
Colostrum	The first milk produced by a lactating mare. Contains antibodies and has a laxative quality, both of which are necessary for the well-being of the newborn foal.
Colt	Technically, a male horse under four (sometimes five) years of age. Horse colt—an uncastrated colt.
Compost pile	A place where droppings and stall litter are deposited to allow decomposition.
Communicable disease	A disease that can be transmitted from one animal to another by direct or indirect contact.
Concentrates	Feed classified as a concentrate is low in fiber and has a TDN of close to 75%.
Conception	The fertilization of the egg by the sperm, which begins embryonic development.
Condition	State of health. To improve the animal's health.
Conditioning	Bringing an animal into the peak of condition or best health so that it can perform well.
Conformation	The build of an animal. The way it is put together.
Congenital	Acquired during development within the uterus.

Not due to heredity.

Conjunctiva — A thin membrane that lines the eyelids and partially covers the forepart of the eyeball.

Consideration — Price paid for a thing or a right, or a right given for a price. Contracts are not valid unless for consideration.

Consignment sale — A horse sale in which many different people have contracted through an agency to sell horses prior to the date of the sale.

Consignor — The person who agrees to sell an animal through an agency.

Constipation — Infrequent and difficult bowel movements.

Contagious — Capable of being transmitted from one animal to another.

Contract — An agreement between two or more persons or firms to do or not to do a certain thing, supported by a consideration and binding upon the parties.

Contracted heels — A condition where the frog is shriveled due to dryness, lack of use or disease and the heels are pulled close together.

Counsel — Farm counsel is an attorney licensed to practice law and who handles the farm's legal matters.

Cracked grain — Grain kernel broken into several pieces for ease of chewing and digestion. Cracked corn is the most popular cracked grain.

Creep — Noun: An enclosed or separated area with the opening too narrow or too low for the mare to enter, where the foal can eat free choice.
Verb: Feeding nursing foals extra high protein supplements free choice.

Cribber — An animal that takes a solid object in its teeth, arches its neck and swallows air.

Crossing — Breeding animals of different pedigrees.

Crossbred — Offspring of a sire and dam of different breeds. Example: Arab and Thoroughbred = Angloarab crossbred. Can also mean the offspring of a purebred sire and a grade mare.

Crude protein (CP) — Calculated amount of protein in the feed based on the amount of nitrogen in the feed.

Cryptorchid — A stallion with one or both testicles retained in the abdominal cavity.

Culture (bacterial) — A sample of fluid from a mare's reproductive tract, incubated under laboratory conditions to detect presence of harmful bacteria.

Cull — An inferior animal that should be eliminated from the breeding herd.

Culling — The process of eliminating undesirable animals from the herd.

Cured hay — Hay that has been dried so that it can be stored safely without molding.

Cyst	A protective sac formed by the body around a fluid or solid substance such as a parasitic organism.
Cystitis	Inflammation of the bladder.
Dam	Mother.
Digestible protein (DP)	The amount of protein contained in the feed that is actually used by the animal.
Digestibility	The amount of the feed that can actually be utilized by the animal's body.
Disposition	The temperament of an animal.
Dominant gene	A gene that always shows itself if it is present. The stronger of two genes.
Drench	Noun: A liquid medication given orally. Verb: Forcing an animal to swallow liquid medication.
Drip specimen	A sample of semen collected from the penis as the stallion dismounts from the mare after copulation.
Dropsy	Distension of the underline by accumulation of body fluids.
Dry matter	Material left after all of the water has been removed.
Ear down	Restrain an animal by biting or twisting the ear.
Edema	The collection of abnormally large amounts of body fluid in the spaces between tissues in the body. Because of gravity, most edema occurs on the legs and abdomen.
Ejaculation	The discharge of semen from the stallion.
Embryo	The earliest stage of development of the foal in the uterus.
Entire horse	Male horse not castrated. A stallion.
Equine	Includes all members of the family *Equidae*— horses, zebras, asses, etc.
Equine practitioner	Veterinarian that specializes in treating horses.
Estate Disbursal Sale	A sale of a herd to settle an estate maintaining a horse farm.
Estrogen	Female hormone. Estrogen is found in large quantities in green grass and tends to increase fertility of mares.
Estrus	Heat period. The time when the mare will receive a stallion.
Estrus cycle	The heat cycle. The period from the beginning of one estrus period to the beginning of the next.
Execution of Contract	The date, signature and delivery of a legal paper creating a right or a binding obligation.
Farrier	Horseshoer.
Far side	Right side.
Favor	Limp slightly.
Feces	Semi-solid waste from intestines discharged through the anus.
Fertile	Capable of producing offspring.

Fetus   The latter stages of the development of an unborn animal in the uterus.

Filly   Female horse four years and younger that has never produced a foal.

Fines   Powder residue found in pelleted, crushed, ground or rolled feeds.

First milk   Colostrum or first milk the mare produces for her foal. Contains antibodies that help protect against disease.

Fitted   Animal in the peak of condition.

Flagging   Movement of the stallion's tail as he ejaculates.

Flushing   Feeding a thin mare heavily prior to breeding to increase fertility. Mares are allowed to lose weight during winter months so that they may be flushed prior to the breeding season. The gaining condition results in higher fertility and higher conception rates.

Foal   A young nursing horse of either sex.

Foal colic   Abdominal pain of the mare that often follows foaling due to the rapid contraction of the uterus.

Foal crop   The foals produced by the farm in a given year.

Foal founder   Laminitis. Inflammation of the sensitive laminae of the hoof, due to retention of portion of the afterbirth.

Foal heat   Heat period that in most mares occurs an average of nine days after foaling.

Foaling   The process of a mare giving birth.

Foaling mares   Pregnant mares.

Foaling stall   Large stall especially designed for a mare giving birth.

Forage   To graze. Any kind of roughage.

Foundation herd   The original breeding animals of a farm upon which the farm's breeding program is based.

Freeze brand   An identifying mark placed on the animal by freezing an area of skin, resulting in loss of hair pigment.

Frog   The wedge-shaped structure on the bottom of the foot that helps pump blood through the foot by expanding as the foot strikes the ground.

Fungicide   Chemical that kills fungus.

Furlong   Racing term denoting 1/8th of a mile.

Futurity   A race or horseshow class in which animals to compete must be nominated while very young.

Geld   To castrate a male horse.

Gelding   A castrated male horse.

Gene   The heredity unit found on a chromosome that determines a specific characteristic.

Genetics   The study of heredity.

Genital infection Infection involving the reproductive organs.

Genitals	The reproductive organs, used primarily in reference to the external structures.
Genotype	The genetic make-up of an animal.
Germinal epithelium	Tissue that produces the sperm or egg cells.
Gestation	The period from the time the egg is fertilized to parturition. Pregnancy.
Get (of sire)	Offspring of the stallion.
Good flesh	A healthy animal with adequate covering of flesh, but not fat.
Grade	An animal that cannot be registered with any breed association.
Grading up	Breeding common grade mares to good purebred stallions or vice versa, and that offspring to another purebred, etc.
Graze	Act of an animal eating grass. To graze a horse is to allow it to eat rooted grass.
Halter-broke	Trained to be haltered and led.
Halter horse	A horse especially bred and trained to be shown in a halter class, where conformation is judged.
Hand	Unit of measure for determining the height of a horse from the top of the withers—four inches to the hand. Additional inches are shown at the right of the decimal point. Example: 15.2 means 15 hands 2 inches.
Handler	The person in charge of taking care of an animal.
Handicap	Chances of winning a race are equalized by giving the faster horses more weight (handicap) to carry.
Haul	To transport a horse by trailer or truck.
Hay	Noun: Cured feed high in fiber made from whole plants that have been cut and allowed to dry. Verb: The process of cutting and curing hay is known as haying.
Hay belly	A distended abdomen due to excessive feeding of roughage such as hay, straw, grass, etc.
Headshy	Horse that avoids being handled around the head.
Heat	Breeding: The time during which a mare will accept the stallion. Estrus. Racing: One of two or three separate runnings of a race, which will be decided by the winner of two or more starts.
Heat cycles	The time from the start of one heat or estrus to the start of the next heat or estrus. Also known as estrus cycle.
Heavy feed	High in concentrates, lacking bulk; thus more difficult to pass through the digestive tract.
Heel calks	Projections downward from the heels of a shoe, which help the horse gain traction. May be used to lift the heels of a horse that is low in the heels.

Hemorrhage	Uncontrolled loss of blood from a blood vessel or artery.
Heredity	The transmission of characteristics from one generation to the next.
Hermaphrodite	An animal possessing both male and female sex organs.
Hernia	The protrusion of an internal organ through the wall of its containing cavity, usually meaning the passage of a portion of the intestine through an opening in the abdominal muscle.
Heterozygous	Having two unlike genes for one characteristic.
Homozygous	Having two identical genes for one characteristic.
Hormone	A substance secreted by a gland into the bloodstream that helps control body functions.
Horsing	Mare showing signs of heat.
Hot-blooded	Of Eastern or Oriental blood. Hot-blooded animals tend to be more sensitive, active and temperamental than animals lacking Oriental blood (cold-blooded horses).
Hot walker	A person or a machine that walks horses that have been worked to "cool them out." A mechanical exerciser.
Hybrid	Offspring resulting from a cross of different species. Example: A mule is the hybrid offspring of a jack and a mare.
Hybrid vigor	Phenomenon exhibited by offspring produced from the union of two unrelated inbred horses. Such offspring tend to be healthier and hardier, and perform better and are more fertile than the average equine.
Immunity	Resistance to a specific disease.
Impaction	Blockage of intestine.
Impregnation	The depositing of semen in the female reproductive tract, resulting in fertilization.
Inbreeding	The crossing of animals more closely related than the average of the population. There are great variations in degree of inbreeding. Linebreeding and close-breeding are both forms of inbreeding. A "very inbred" animal is the product of two very closely related horses such as brother and sister.
Incomplete dominance	One gene fails to completely cover the presence of another gene and a characteristic somewhere between the two genes is produced.
Indemnity	Agreement to repay if loss is suffered due to subject of the agreement.
Infection	Invasion of the body by disease-causing microorganisms.
Infectious disease	A disease that can be transmitted from one animal to another.
Infestation	Parasites are preying on the animal.

Inflammation	The reaction of body tissue to damage, resulting in swelling, heat, pain and discoloration.
Inoculate	Noun: Nitrogen-fixing bacteria planted with a legume such as alfalfa to insure that a symbiotic relationship will be formed that will result in a healthy legume stand. Verb: To give an inoculation.
Inoculation	Introduction of a live microorganism into an animal to produce a mild form of a disease in order to develop immunity. Also the introduction of a live microorganism into a cultural medium.
Intradermal	Injection given in the skin layers.
Intramuscular	Injection given in the muscle tissue.
Intranasal	Vaccine or medication administered through the nose.
Intravenous	Injections made directly into a vein.
Involuntary muscles	Muscles that cannot be regulated by the will of the animal. Example: Heart muscle, intestinal muscles.
Jog	Slow trot.
Jumping	Letting a teaser mount a mare without actually breeding her. Act of jumping a horse over obstacles.
Kickboard	Lining on the lower four feet of a stall wall to protect the hoof and wall from blows of a kicking or pawing horse.
Lactation	Milk production.
Laxative	A substance that helps fecal matter to pass through the bowels more easily.
Lease	A legal relationship whereby the owner (lessor) grants possession and control of an item of property to another party (lessee) for a certain time for a certain purpose at a stated price (called the rent or rental).
Lethals	A genetic factor responsible for causing death during fetal development, at birth or in later life.
Leach	The washing of nutrients out of the soil or feed by the action of water.
Leech	A bloodsucking parasite.
Legume	Plant of the pea family with a true pod enclosing the seed and having nodules on the roots containing nitrogen-fixing bacteria that help the plant utilize atmospheric nitrogen. Example: alfalfa.
Lessee	The person who has been granted use, control and posession of a horse or other property.
Lessor	The owner who has granted the use, control and possession of an animal or other property to someone else, called the lessee.
Let-down	1. Allow an animal to lose some of its condition or fitness. 2. The udder relaxes, allowing the release of milk.

3. A term of conformation, *well-let-down* refers to great distance from hip to hock.

Libido	Sex drive.
Lien	A debt that is secured by property (horse), enforceable by foreclosure and sale of the property if not paid.
Ligament	A tough fibrous tissue connection between bones, or supporting an organ.
Linebreeding	Breeding of animals related to a common ancestor.
Litter	Bedding, manure and urine removed from an area where an animal has been kept.
Live-foal guarantee	The stallion owner guarantees that a foal will stand-and-suck. If a foal is not born, or if it dies before it nurses, the owner of the mare is entitled to a rebreeding or a refund.
Load	Put an animal into a transporting vehicle.
Longe (lunge)	The horse is allowed to travel in a circle while on a rope held by the handler in the center of the circle.
Longevity	A long, useful life.
Maiden mare	A mare that has never been bred.
Make a bag	Produce milk in the udder.
Mammary glands	Milk-producing glands.
Manure	Fecal matter.
Mare	Female horse over four years of age or one that has produced a foal.
Masked gene	Term for a particular characteristic that is not seen due to the presence of a dominant, stronger gene that expresses its characteristic over that of the masked gene.
Masturbation	Self-induced ejaculation.
Maternal	Referring to the mare. Motherly.
Meconium	Fecal material accumulated in the colon of the foal during fetal development, which must be passed within a few hours after birth.
Metritis	Inflammation of the uterus.
Microbes	Microscopic organisms.
Morphology	The configuration of the sperm cells.
Mortgage	A lien for a debt secured by property which, if not paid, may be foreclosed, whereupon the property is sold to satisfy the debt.
Mortgagee	The creditor under a mortgage. See mortgage.
Mortgagor	The person who owes the debt and gives a mortgage. See mortgage.
Mothering ability	The ability of the mare to take care of her foal. Good mothering ability consists of ample milk production as well as watchfulness, protection and general concern of the mare for her foal.
Motility	Amount of movement the sperm exhibits.
Mount	The stallion places his forelegs on either side of the mare's back during breeding.

	To get on a horse.
Mouthing	Determining the age of a horse by its teeth.
Mucus	Slimy secretion of the mucous membranes that line various body cavities such as the vagina.
Mummified fetus	Foal that died but remained in the uterus. The soft tissue having been absorbed by the maternal blood and lymph systems, the remaining bones and skin are later expelled as a mummified fetus.
Muscling	The muscular development of an animal.
Mutation	A sudden change in the genes that is passed on to later generations.
Native grasses	Grasses that are natural to an area, not introduced by man.
Navel cord	Umbilical cord. Cordlike tube that attaches to the foal at the navel, through which food and waste materials travel between the placenta and the fetus.
Near-Disbursal Sale	A sale in which a ranch sells all but a few select individuals. Such sales often amount to a massive herd culling in an attempt to drastically upgrade the quality of stock.
Near side	Left side of a horse.
Net energy (NE)	The amount of energy in a feed that is available to the animal for use.
Neural	Pertaining to the nervous system.
Nick	A cross of two different bloodlines that consistently produces superior animals.
Nine-day heat	The heat or estrus period that occurs approximately nine days after foaling.
Nomination for show, sale or race	While a foal is very young it may be entered in or nominated for a future show, sale or race. A nomination fee is usually charged, with additional entry payments spaced between the nomination and the event.
Nurse mare	A mare that will accept and nurse an orphaned or rejected foal. Cold-blooded mares are usually better nurse mares because they are not as temperamental as hot-blooded mares.
Nutritive ratio	The ratio of digestible protein to the energy and fiber components of a feed. The closer the nutritive ratio is to 1, the less the fiber and the greater the usability of the feed.
Nutritional requirements	The amount and type of feed or nutrients required by a horse to meet its body needs. Nutritional requirements will vary greatly according to the amount of work or production demanded from the animal.
Official Stud Book	A book maintained and published by a breed association listing all of the animals ever registered by that association and the parents of each animal. A registry.
Off side	Right side of a horse.

Option   A right granted by the optionor to another (optionee), which the optionee may accept or reject during the period of its existence.

Optionee   See option.

Optionor   See option.

Outcrossing   The breeding of animals of the same breed but that have no common ancestor for at least four to six generations.

Ovary   The female organ that produces the egg (ovum) and the female hormones, estrogen and progesterone. A mare has two ovaries.

Overgrazing   Allowing animals to eat and trample too much of the vegetation in an area.

Ovulation   The releasing of an egg from the ovary.

Ovum   Egg cell.

Paddock   Small pasture or area in which animals are turned loose.

Palatable   Feed that is favored or well-liked. Pleasing to the taste.

Parasite   Plant or animal that obtains its food at the expense of a living animal.

Parturition   Act of giving birth.

Pasture   Noun: An area in which animals are turned loose and allowed to graze at leisure.
                           Verb: Allowing an animal to remain free in a pasture.

Pasture rotation   The moving of livestock from one pasture to another so that the pasture will be given a rest period in which to recover from grazing pressure.

Pedigree   Record of the ancestry of an animal.

Pelleted ration   A ration made up of a mixture of ground grains and chopped hay tightly pressed together into a pellet.

Perennial plant   A plant that lives more than two years.

Performance horse   A horse that is trained to do a job requiring athletic ability. Example: racing, roping, cutting, driving, jumping.

Peristalsis   Wavelike contraction passing along certain tubular organs such as those of the digestive tract.

Permanent pasture   A pasture that is not replanted yearly, usually irrigated and with a long growing season.

Permanent registration   Permanent registration results when a horse is registered with a breed association, is issued a number and is entered in the Stud Book. The number is issued for life. The horse need not prove itself any further to remain in the registry. All of the offspring of a permanently registered horse, when the other parent is permanently registered, are eligible for registration with the breed association.

Phenotype	The visible characteristics of an animal regardless of the genes it possesses.
Physiological abortion	Abortion due to a malfunction of the body processes and not caused by a disease organism.
Pigeon-toed	Toes turn in.
Placenta	Surrounds the foal in the uterus and is responsible for the transfer of nutrients between mare and foal. Often referred to as the afterbirth when expelled after the foal is born.
Pledge	A lien secured by possession of personal property as security until the debt is paid. A pledge may be foreclosed and the property sold if the debt is not paid.
Pointing	One forefoot is extended forward to take the weight off of it. Pointing is an indication of lameness.
Pole barn	A barn with an upright frame made of sturdy poles such as telephone poles, having a roof and an overhang but lacking four solid walls.
Pony	Noun: Horse under 14.2 hands. Noun: A horse used to lead other horses. Verb: To exercise a horse by leading it from another horse.
Post-foaling	After foaling.
Prepotency	The ability of an animal to pass its characteristics on to its offspring.
Produce of dam	Offspring of a mare.
Production Sale	Sale of animals produced by the breeding farm.
Progeny	Offspring.
Proud cut	A male animal that has had part but not all of the testicle removed. It is thus sterile but maintains stallion characteristics.
Proud flesh	Accumulation of excessive scar tissue in and around a wound, especially on the legs.
Proven producer	A mare or stallion that has consistently produced outstanding offspring.
Prussic acid	Toxic substance produced by some strains of grass, such as Sudangrass and flax, when under stress. Hydrocyanic acid.
Pump the tail	The movement of the tail as the stallion ejaculates.
Purebred	Comes from a long line of animals of the same breed. Not synonymous with Thoroughbred breed.
Pyometra	Accumulation of pus in the uterus.
Quarter crack	A crack in the hoof occurring in the half of the hoof wall nearest the heel.
Rasp	Noun: File used to remove portion of the horn of the hoof. Verb: To file down the horn of the animal's hoof.
Recessive	Characteristics that do not appear unless both members of the gene pair are identical.

Registration	The recording of an animal in the Stud Book of a particular breed association.
Registry	Association that records the ancestry of animals of a certain breed or origin.
Retained afterbirth	The placenta or a part of it not expelled within a few hours after foaling. The retained placenta causes an inflammation and can result in laminitis or even death if not removed within six or eight hours.
Return breeding	The owner is entitled to another breeding if a mare does not conceive or if she aborts or produces a dead foal or a foal not able to stand and nurse.
Rolled grain	A grain that has been flattened, thus exposing more digestible nutrients and often making the feed more palatable. Steam present during the rolling process greatly reduces fines (dust) and increases palatability.
Roughage	Feed classified as roughage is high in fiber content and has a TDN of close to 50%.
Ruminant	Any species of mammal having a rumen where food is allowed to ferment prior to entering the remainder of the digestive tract. Example, cow, sheep, goat, deer.
Saliva test	Test often administered to race or show horses to detect the presence of narcotics.
Sand colic	Colic caused by a collection of sand or gravel in the digestive tract. More common in horses grazing on sandy soil.
Savage	Bite.
Scours	Diarrhea.
Scrotal hernia	Portion of intestines protrudes through the inguinal canal in the abdominal wall into the scrotum.
Scrotum	Skin sac surrounding the testicles.
Scrub	A low-quality animal of unknown breeding.
Secondary infection	Infection that follows another infection. Example: Pneumonia can follow or complicate influenza and is therefore a secondary infection of influenza when following it.
Seedbed	Ground that has been tilled and prepared for seeding.
Selective breeding	Careful choice of animals which are to be bred to each other with prior thought to the type of offspring that will be produced.
Selenium	Element found in some very alkaline soils and in plants grown on them. Can be toxic to horses.
Semen	The sperm and fluid discharged during ejaculation.
Seminal fluid	The fluids that transport and nourish the sperm on its trip from the male reproductive tract to the egg.
Septicemic	Pertains to disease-causing bacteria or their toxins that affect the blood.

Service	Process of breeding a mare.
Settled	The mare has conceived and is in foal.
Sex-linked characteristic	Characteristic that appears much more frequently in one sex than in the other.
Shed out	Lose the winter hair coat.
Silage	Feed cut, chopped and stored while green and allowed to ferment before being fed.
Sire	Father of a horse. Male parent.
Slip	Abort a foal.
Smegma	Dried waxy discharge from the sheath of stallions and geldings. Often referred to as a "bean." Sometimes interferes with urination and causes irritation.
Smooth-mouthed	The cups in the teeth have been worn away, indicating the horse is over 12 years of age.
Soil acidity	Amount of acid or free H^+ ions present in the soil. Most plants require a specific pH or amount of acidity to do well.
"Sored" horse	A horse is "sored" if for the purpose of affecting its gait any of the following is used: blistering or chemical agents, burns, cuts, lacerations, tacks, nails, chains or boots, or any cruel or inhumane method or device.
Species	A distinct group of plants or animals having specific common characteristics, which do not naturally interbreed with plants or animals of other species.
Speck	To examine the vagina and uterus through a vaginal speculum.
Sperm	The male sex cell that fertilizes the ovum (egg).
Spermatozoa	Male sex cell that fertilizes the egg (sperm).
Splay-footed	Toes turned out.
Stag	A stallion castrated late in life, retaining stallion characteristics.
Stallion	A male horse not castrated.
Stallion breeding report	A record of all registered mares bred to a specific stallion, submitted to the breed association at the end of the breeding season.
Stand at stud	The stallion is available for breeding.
Sterile	1. Stallion or mare incapable of reproduction. 2. All microorganisms have been killed.
Stillborn	Born dead.
Stud	1. A stallion kept for breeding purposes. 2. A farm that stands a stallion.
Stud colt	Uncastrated colt.
Stud fee	The charge for breeding a mare to the stallion.
Studman	Person in charge of handling the stallion.
Subcutaneous	An injection given in the area between the skin and the underlying muscle tissue.
Subclinical	Illness or infestation of a minor nature, not severe

enough to be obvious but impairing the usefulness and productivity of the animal.

Supplement	A nutrient fed along with the normal ration to increase the health and vigor of the animal. Example: Protein blocks.
	Verb: The act of feeding supplements.
Sutured mare	A mare that has had the upper portion of her vulva stitched closed to prevent windsucking.
Systemic	In the bloodstream.
Tack	Equipment used on a horse. Example: brushes, blankets, saddle.
Tattoo	An identifying mark, usually a number, tattooed on the upper lip of a horse. Especially used for race horses.
Teaser	An animal used to detect mares in heat. May be a gelding, stallion, cryptorchid or proud-cut horse or pony.
Teasing	The process of allowing a teaser to make advances to a mare to detect estrus.
Teasing chute	A chute in which a mare is placed during teasing to prevent injury to the teaser in the event the mare is not in heat.
Teasing record	Record kept on all mares to be bred, indicating when they are in heat and when they go out of heat.
Tendon	A tough, fibrous cord that connects a muscle to bone.
Tetany	A muscle remaining in a contracted state.
Therapeutic	Curative.
Third phalanges	Coffin bone or ospedis, most distal (lowest) bone in the foot.
TDN	Total Digestible Nutrients. Indicates the usable amount of nutrients found in various feeds.
Topline	Contour of the top of an animal.
Top side	Sire's side of pedigree.
Toxemia	Condition caused by presence of bacterial toxins in the blood.
Toxic	Poisonous.
Toxin	Poisonous waste product secreted by certain disease-causing organisms.
Trainer	Person who specializes in training horses.
Transfer of Ownership form	This form must be filled out and sent in to the breed association when a registered animal is sold.
Tremor	Involuntary shaking or quivering of the body.
Truth in Lending	A Federal law requiring a lender or seller for credit to furnish the debtor certain information about the transaction before it is closed. Provides severe penalties for violation.
Twitch	Instrument of restraint that pinches the upper lip

of the horse so that the animal can be held still while undergoing treatment.

Type Characteristics displayed by an animal that identify it with a specific breed. A fine representative of a certain breed would be considered "Typy." Example: Typy Thoroughbred.

Udder The organ containing the mammary glands, which secrete milk.

Ulceration Fluid, blood and other body substances are lost through the skin or mucous membranes, causing a slow degeneration of the tissue involved.

Umbilical cord Cord-like tube attached to the foal at the navel, through which nutrients and waste material travel between the mare's placenta and the fetus. Navel cord.

Umbilical hernia Protrusion of a loop of intestine through the abdominal wall at the navel.

Underline Contour of the underside of the horse from elbow to flank.

Unmasticated Unchewed.

Unsoundness A defect, injury or disease that impairs the usefulness of the animal.

Urea A man-made feed additive that is high in nitrogen.

Urine Liquid discharge of waste material accumulated from the blood by the kidneys and discharged through the urethra.

Usury An annual charge of a greater amount than allowed by law for the use or forbearance of money or credit. Penalties for violations are severe.

Vaccination Injection of a vaccine to develop resistance to specific disease organisms.

Vaccine Killed or attenuated microorganisms injected into the body to prevent or treat an infectious disease.

Vaginal speculum Instrument inserted into the vagina of a mare so that the vagina and cervix may be examined or treated, samples taken or the mare inseminated.

Vaginitis Inflammation of the vagina.

Vector A carrier. Usually a bloodsucking insect that carries disease from one animal to another. Example: The mosquito is a vector of sleeping sickness.

Venereal Carried or transferred by copulation.

Vermifuge Medicine used to remove internal parasites.

Vermin Any small destructive animals such as rats, mice, lice, etc.

Virus A submicroscopic entity capable of reproduction only within living cells, many of which produce diseases.

Viscosity The thickness of a fluid.

Vigor   Vitality. A healthy, active individual demonstrates vigor.

Vitamins   Substance found in various foods required by the body in small amounts. Example: Vitamin A, B, C, D, etc.

Voluntary muscles   Muscles controlled by the brain that can be activated or deactivated by the will of the animal.

Wash racks   Area set aside for the bathing of animals.

Waxing   Prior to foaling the teats of the mare's udder become caked with a yellowish wax-like substance, which later falls or strings off.

Wean   Remove the foal from the mare so that it can no longer nurse.

Weanling   Animal less than a year old that has been weaned from the mare.

Weaving   A horse's rhythmical moving back and forth, bobbing its head and perhaps lifting its front feet alternately. A nervous habit, usually the result of boredom due to confinement.

Webbing   A cloth or rope material woven together in a fishnet pattern with large holes, often placed in front of an open stall door to keep the animal in while allowing the circulation of air. Often referred to as a stall-guard.

Windsucking   1. A vice where the animal arches its neck and swallows air.
                       2. A condition sometimes found in mares where the vulva does not close tightly and is at an incline allowing air and fecal material to enter the vagina. The movement of air in and out of the vagina can often be heard as the mare trots. Such mares are usually sutured to close the upper portion of the vulva to prevent windsucking, which can lead to infection.

Worming   Treating an animal with a vermifuge to remove internal parasites.

Yearling   An animal that has passed its first birthday but has not reached its second.
                       Long yearling: Yearling over 1½ years of age.
                       Short yearling: Yearling less than 1½ years of age.

INDEX

Asterisked subentries appear in the index as main entries.